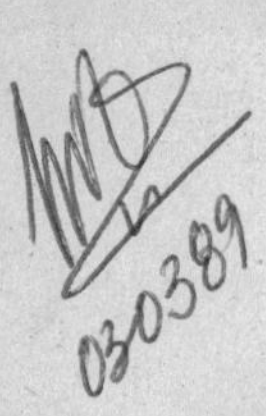

BARRON'S

HOW TO PREPARE FOR THE

COLLEGE BOARD ACHIEVEMENT TESTS PHYSICS

FOURTH EDITION

By
Herman Gewirtz
Chairman, Physical Science Department (Retired)
Bronx High School of Science
New York City

Barron's Educational Series, Inc.
New York • London • Toronto • Sydney

Printed in 1987

All inquiries should be addressed to:
Barron's Educational Series, Inc.
250 Wireless Boulevard
Hauppauge, New York 11788

Library of Congress Catalog Card No. 84-16912

International Standard Book No. 0-8210-2768-X

Library of Congress Cataloging in Publication Data

Gewirtz, Herman.
How to prepare for College Board achievement tests, physics.

Includes index.
1. Physics—Examinations, questions, etc.
2. Universities and colleges—United States—Entrance requirements. 3. Universities and colleges—United States—Examinations. I. Title. II. Title: College Board achievement tests, physics.
QC32.G45 1984 530'.076 84-16912
ISBN 0-8120-2768-X

PRINTED IN THE UNITED STATES OF AMERICA

789 100 9876

Contents

Introduction

Although the chief purpose of this book is to aid in the preparation for the achievement test in physics of the College Board, it should also be very useful in the preparation for any examination in physics.

The College Board does not publish copies of former examinations each year and the scope of the examinations is not restricted by any given syllabus. The examinations are made up each year by a group of experts who are guided by a knowledge of what is commonly taught throughout the country. You will, therefore, be well prepared for the test if you know and understand what is taught in a good secondary school course in physics and if you get a little practice in the types of questions used. It is the aim of this book to help you in both areas.

The questions on the examination are based on the big subject matter areas of mechanics, optics and waves, electricity and magnetism, heat, and modern physics. Very few, if any, of the questions ask for mere recall of knowledge. The questions are designed to see if you really understand concepts and principles, if you can reason quantitatively, and if you can apply scientific concepts and principles to familiar and unfamiliar situations. Most questions are usually of the common multiple-choice variety, but other types have been used. The examinations at the end of this book were constructed to achieve this variety of form and content.

HOW TO PREPARE FOR THE EXAMINATION

The best preparation results from regular study, reading, and review spread over a long period of time. This is done in connection with work in school and at home, assigned work, and voluntary study. This prolonged effort not only helps you remember the facts but also helps you get a better understanding of the concepts and principles. Study this book along with any other your teacher may want you to use for regular classwork and try to answer the questions at the end of each chapter before looking at the printed answers.

As mentioned above, the Examination is not restricted to any specific syllabus. However, an article published several years ago pointed out that the major programs have many topics in common. A good physics student from any course should be able to do well on the examination. Therefore, in order to be helpful to students with different course backgrounds, there is some material in this book which probably has not been covered in your course, and, possibly, will be of little importance on the examination you will take. This may be true of much of the information in Chapters 14 and 15, and of some of the discussion of Simple Machines in Chapter 3. For example, if you prepared for the New York State Regents Examination in Physics or for PSSC, you had nothing on the efficiency of simple machines. Don't waste time on it. Your first objective in reviewing should be to know well what you were expected to know for your course, and then see how well you can answer questions related to that subject matter.

When the time for the examination approaches review all of this material and try one of the examinations at the end of this book. Review further any weak subject areas thus discovered and try additional examinations. To aid in the final review, some items are italicized or in bold type. This should help locate some of the more important facts, formulas, and definitions. In memorizing such facts, it is most helpful to study them and then try to write them down from memory. For example, after you have studied the Numerical Relationships on page 4, you should cover the page and write down the eight facts that are listed. Next, compare your version with the book. Repeat, if necessary, until you know all the facts correctly and completely.

You should be able to obtain a copy of the College Board pamphlet *Taking the Achievement Tests* from your high school advisor. Read the section on Physics, which describes the test and includes sample questions and answers. Because the instructions for the test may have changed, be sure to read them carefully. This will save you time on the actual test.

HOW TO TAKE THE TEST

The examinations at the end of this book are not copies of former tests, but practice with them should be valuable to you. However, be sure to read the instructions given with your test. Note the amount of time allotted and recognize that you may not be able to answer all questions in the available time. The questions are not of equal difficulty, but each question usually gets the same credit. Do not waste time on questions that seem difficult or time consuming. Go back to them at the end of the test, if you have time. (As was pointed out above, some of the questions may be intended primarily for a different type of course.) Time can often be saved in numerical examples by making approximate calculations. Don't worry if you can't answer all the questions. Probably no one taking the test can. Should you guess? Read the instructions. Usually there is a penalty for incorrect answers, but if you know anything about the subject matter of the question and can eliminate some of the choices, it is advisable to guess.

When taking the practice tests in this book, do not stop at the end of 1 hour. Follow the above advice and eventually try to answer all the questions. Practice keeping track of time without wasting it.

Each of the practice tests includes 75 questions. Some of the questions were deliberately made rather difficult to challenge you. On the actual examinations the questions are carefully evaluated, and individual performance is compared with group performance before a score is given.

Also *note* whether the **directions** on your test are the same as those used on the Practice Tests in this book. *When you take the test, be sure to read the instructions given with your test.*

GOOD LUCK!

1

Measurement and Forces

Do you know what is meant by significant figures? Can you combine and resolve forces? Do you understand what is meant by the moment of a force? Test your understanding by trying the following example:

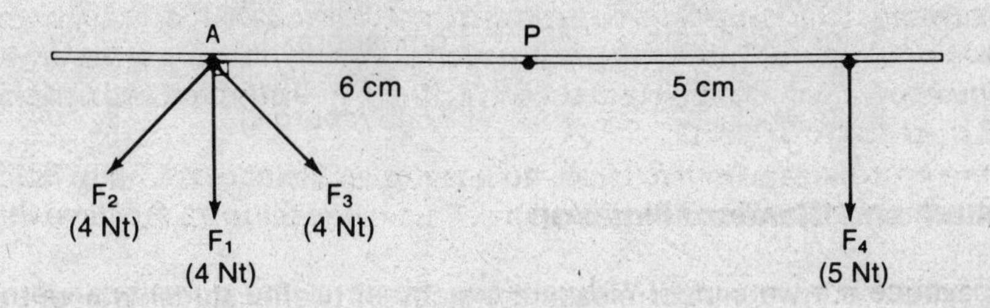

A bar is pivoted at *P* and forces are applied as shown. Which force produces the greatest torque?

In this chapter we shall keep these and other questions in mind as we review forces, vectors, and basic ideas of measurement.

In physics, as in other sciences, a theory must agree with observation. In addition, a theory must lead to predictions which can be verified by experiment. Experimentation in physics usually involves careful measurement.

UNITS AND SYSTEMS OF MEASUREMENT

Powers of 10

$10^0 = 1$	$10^1 = 10$	$10^{-1} = 0.1$
	$10^2 = 100$	$10^{-2} = 0.01$
	$10^3 = 1000$	$10^{-3} = 0.001$
	$10^6 = 1{,}000{,}000$	$10^{-6} = 0.000001$
	$3 \times 10^4 = 30{,}000$	$3 \times 10^{-4} = 0.0003$

Prefixes

mega- (M)—10^6; one megawatt = one million watts
kilo- (k)—10^3; one kilometer = one thousand meters

centi- (c)—10^{-2}; one centimeter = 0.01 meter
milli- (m)—10^{-3}; one millimeter = 0.001 meter
micro-(μ)—10^{-6}; one microampere = 0.000001 ampere
nano- (n)—10^{-9}; one nanosecond = 0.000000001 second

Systems of Measurement

There are three systems commonly used. In the *CGS* system the basic units of length, mass, and time are, respectively, the *c*entimeter (cm), *g*ram (gm), and *s*econd (sec). In the *MKS* system the corresponding units are *m*eter (m), *k*ilogram (kg), and *s*econd. In the English system the three basic units are the *f*oot (ft), *p*ound (lb), and *s*econd (sec)—the *FPS* system. Unfortunately, other units are used as well—for example, the *slug* as a unit of mass equal to approximately 32 lb. Also, the same word that is used for the unit of mass is also often used as a unit of force or weight. Sometimes the difference is clearly indicated—for example, by speaking about a five-pound mass (5 lbm) or a five-pound force (5 lbf) or a five-pound weight (5 lbw). However, usually we speak of a five-pound object, and we shall show when we discuss Newton's second law that this need cause no difficulty.

Some Numerical Relationships

1 meter ≐ 39.37 in.*	1 liter = 1000 cm^3
1 meter = 100 cm	1 cm^3 of water weighs 1 gm (approx.)
1 in. = 2.54 cm	1 ft^3 of water weighs 62.5 lb (approx.)
1 lb ≐ 454 gm*	
1 kg ≐ 2.2 lb*	

Significant Figures and Standard Notation

We can count exactly but we cannot measure exactly. If twenty students count carefully the number of baseballs in a box, we have the right to expect the same answer from each of the students. If they say 45 baseballs, we don't think, maybe it's 45¼ baseballs. However, if we ask the same twenty students to measure very carefully the height of the box, we can expect different answers such as 62.3 cm, 62.4 cm, and 62.5 cm. Each of these answers has three significant figures, but the last digit is doubtful. If, as far as we can tell, the last digit should be zero, we write the zero down and thus indicate the accuracy with which we tried to measure. In "62.0 cm" the zero is significant and we have three significant figures. In "0.0230 cm" we still have only three significant figures, the first two zeros being used only to indicate the location of the decimal point. In the *standard form* the same number would be written as 2.30×10^{-2} cm. This notation consists of two factors, one of which is an integral power of ten and the other has one digit (which is not a zero) in front of the decimal point. It is especially useful when dealing with large or small numbers. The *order of magnitude* of a quantity is its approximation to the nearest power of ten. The order of magnitude of 2.30×10^{-2} cm is 10^{-2} cm; one order of magnitude larger is 10^{-1} cm.

VECTORS AND FORCES

We shall now consider two important vector quantities, displacement and force. A *vector quantity* is completely specified by giving its magnitude and direction. It can be added the way displacements are added. How do we add displacements?

Displacement is a change of position. Suppose we walk 30 meters to the east. We have now given the magnitude, 30 meters, and a direction, to the east. This is conveniently represented graphically by a *vector*, an arrow drawn to a suitable scale in the correct direction as shown in the diagram:

*The equal sign with a dot over it (≐) stands for: is approximately equal to. Another symbol used is ≈.

The vector *AB* represents the displacement of 30 meters to the east. If this displacement is now followed by a displacement of 40 meters to the north, the new displacement is represented by using the same scale for *BD:*

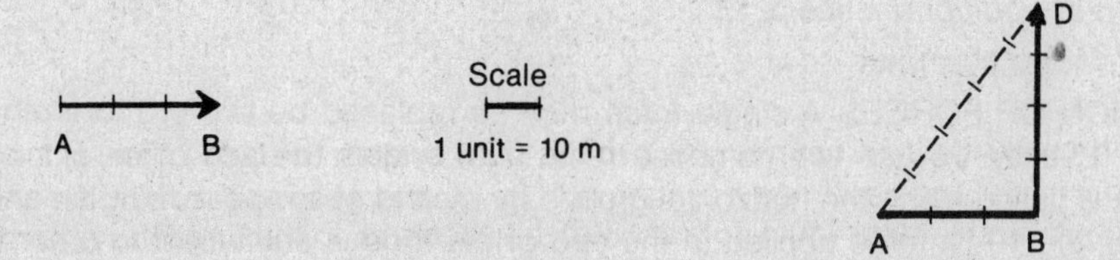

The net result of these two changes is that our change of position is from *A* to *D,* and the vector *AD* gives the resultant displacement. We can get the magnitude of the resultant, 50 meters, by applying the scale to vector *AD* or by using our knowledge of the right triangle (to be reviewed later).

A *force* is a push or pull. It is a *vector quantity*-i.e., to specify it we must give its *magnitude* (number and unit) and its direction. Example: 5 pounds acting westward. Another vector quantity is velocity. A *scalar* quantity is one that is fully specified by giving its magnitude, e.g., mass or speed. The *weight* of an object is a force: it is the pull of the earth on the object; its direction is downward. Some *units* of force are: pound, dynes, and newtons. The weight of one kilogram is approximately 9.8 newtons (Nt or N). In the diagram a three-pound force is represented to the right. A scale is arbitrarily selected and indicated. A certain length is used to represent one lb, and three of these lengths are laid off on a horizontal line to represent 3 lb. An arrowhead indicates the direction.

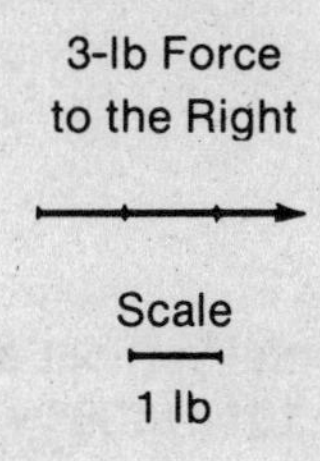

Concurrent Forces

Concurrent forces are two or more forces acting simultaneously on the same point of an object. We are often concerned with getting the combined effect of these forces. The *resultant* of two or more forces is a single force which produces the same effect as these forces (and can therefore be used to replace them). For example, if a 4-lb force and a 3-lb force act in the same direction their resultant is a 7-lb force in the same direction; if they act in opposite directions (at an angle of 180°), their resultant is the difference between the two forces, a one-pound force in the direction of the 4-lb. force.

PARALLELOGRAM OF FORCES When the angle between two forces is other than 0° or 180°, we usually get the resultant by a graphical method. Each force is represented by a vector drawn in the given direction from the same point and using the same scale for both as shown in the diagram. Using these two vectors as sides, the parallelogram is completed. The diagonal drawn from the common point of the two given forces is the resultant vector and will indicate both magnitude and direction of the resultant force. To get the magnitude of the resultant (*R*), we see how many of the scale lengths fit into it. In the diagram this is indicated as about 6½, and since each of these lengths represents one pound, the magnitude of the resultant is about 6½ lb. If the two given forces are perpendicular to each other, the Pythagorean theorem may be used: The square of the hypotenuse is equal to the sum of the squares of the other two sides (see diagram). To find the resultant of three

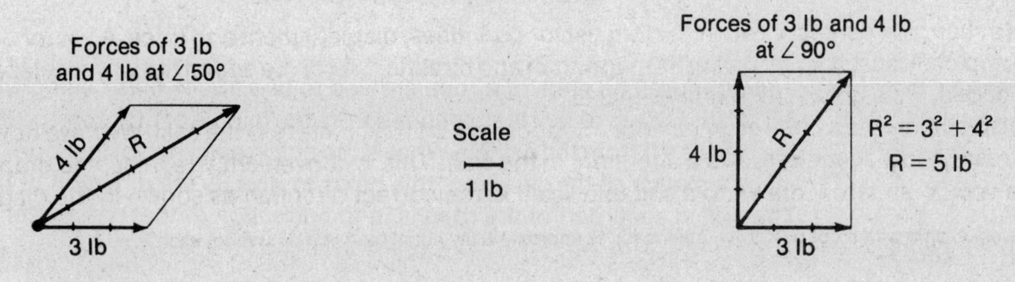

concurrent forces we apply a similar procedure: first replace any two forces by their resultant and then combine this resultant with the third force to get the final resultant.

The *equilibrant* force is the force opposite and equal to the resultant force. If an object is in equilibrium under the action of several forces, any force may be considered the equilibrant of the other forces. (For *equilibrium* see p. 15.)

RESOLUTION OF FORCES A single force may be replaced by two or more other forces—its *components.* Usually we use two components at right angles to each other. For example, if, as shown, a box is pulled along the horizontal ground by means of a rope making an angle of 30° with the horizontal, the 10-Nt force applied to the end of the rope is equivalent to a horizontal force of 8.7 Nt and a vertical force of 5.0 Nt. We get the value of these component forces by constructing a rectangle as shown in the figure.

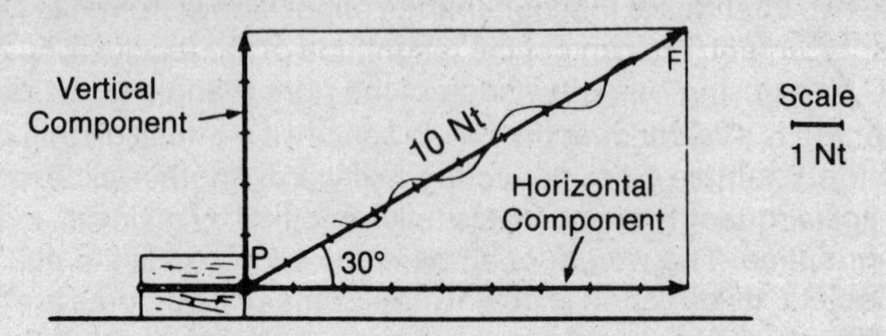

1. First we draw the three direction lines starting from the common point *P:* horizontal, vertical, and at 30° to the horizontal. 2. Using an arbitrary scale, we measure off the known force: 10 Nt at 30° with the horizontal. 3. From the end of this known vector, *F*, we draw horizontal and vertical lines to complete the parallelogram. 4. The intersections of these two lines with the original vertical and horizontal lines mark the ends of the two desired components; we put arrowheads at these ends. 5. We measure the length of each component vector and use the scale to translate the lengths to units of force. The 10 Nt force is shown crossed out because we are replacing it by its components.

You must know and understand the graphical solution. If you already know a little trigonometry, it may sometimes save you time to apply it instead of using the graphical solution. You may recall that, in a right triangle, the sine of the acute angle is equal to the ratio of the side opposite the angle to the hypotenuse.

$$\sin\theta = a/c \qquad \cos\theta = b/c$$

For example, in the previous diagram, the rope pulls on the box at an angle of 30°.

$$\sin 30^\circ = \frac{\text{vertical component of force } F}{\text{force } F}.$$

$$\begin{aligned} \text{vertical component} &= F \sin 30^\circ \\ &= (10\text{ Nt})\ (0.5) = 5\text{ Nt} \\ \text{horizontal component} &= F \cos 30^\circ \\ &= (10\text{ Nt})\ (0.866) = 8.7\text{ Nt} \end{aligned}$$

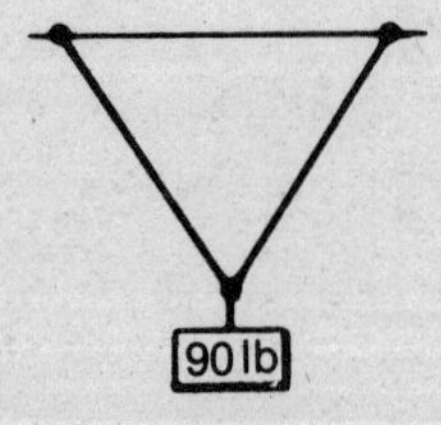

Illustration: The ends of a rope are tied to two points on a ceiling which are 6 ft apart. A 90-lb object is suspended from the middle of the rope; the middle of the rope is 4 ft from the ceiling. What is the pull in each half of the rope? First we make a sketch of the situation; this helps to give us the direction of the rope. The pull in each part of the rope is in the direction of that part of the rope.

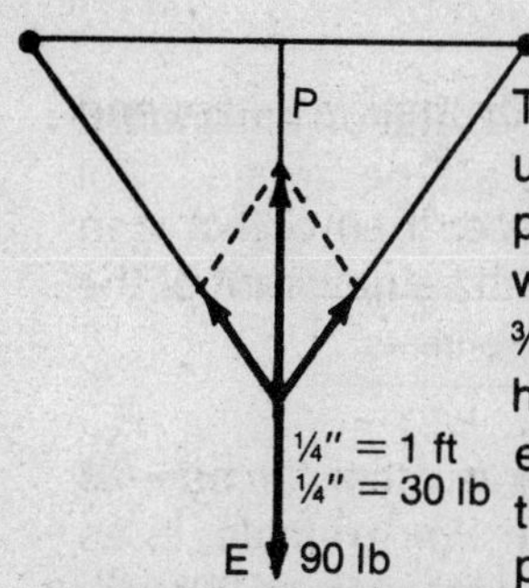

Then we are ready to draw the vector diagram. To get the proper directions we use a scale drawing, such as: Let ¼ in. equal 1 ft. To represent the *forces* properly we need another scale. Let ¼ in. equal 30 lb. The only force about which we know everything is the weight of the 90-lb object. We represent it by a ¾ in. line downward. The 90-lb weight is the equilibrant of the pull in the two halves of the rope. We then get the resultant (*R*) of these pulls by extending the equilibrant vector backwards for the same length, 3/4 in. Next, we think of R as the diagonal of a parallelogram. From the head of *R* we draw two lines, each parallel to one of the lines representing the rope. The intersections of these lines mark the end of the vectors representing the pull in the ropes. To determine the magnitude of this pull, we measure the vectors. Each vector is slightly more than ½ in. long. Since ¼ in. represents 30 lb., the pull in each half of the rope is slightly more than 60 lb in the direction shown.

The successive steps in the method are summarized by the series of diagrams:

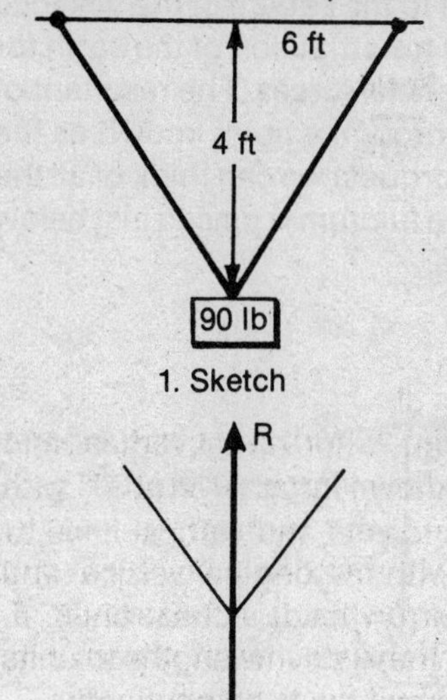

1. Sketch

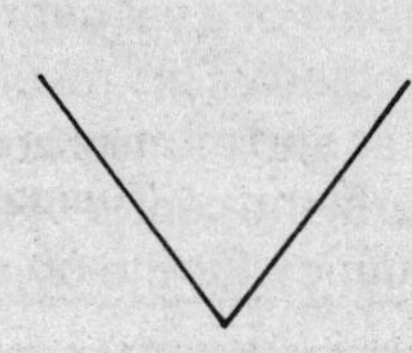

2. Apply scale for lengths

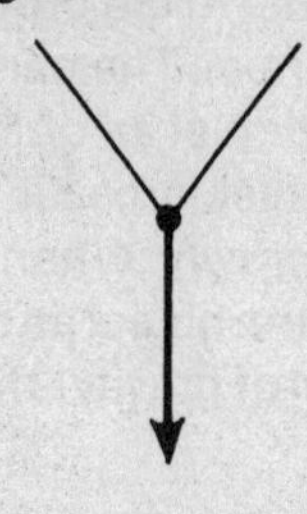

3. Apply scale for known force(s)

4. From known equilibrant get resultant

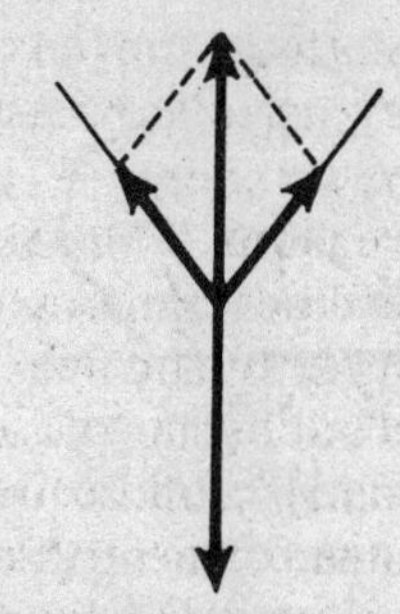

5. Complete parallelogram

6. Measure vector: calculate F from scale for forces.

Parallel Forces

If two equal and opposite parallel forces (nonconcurrent) are applied to an object their resultant force is zero, but the object will rotate unless some other forces are properly applied. The *torque* or *moment of a force* is the effectiveness of a force in producing or tending to produce rotation. The magnitude of this

moment of a force = force × length of moment arm

The length of the moment arm is the perpendicular distance from the *fulcrum* or *pivot* (*P*) to the direction of the force. In the following diagram, the moment arm of the 2-lb force is 8 in. The moment arm of the 4-lb force can be found graphically and is approximately 3 in.

In general, the moment arm, and therefore the torque, is greatest when the force is perpendicular to the bar.

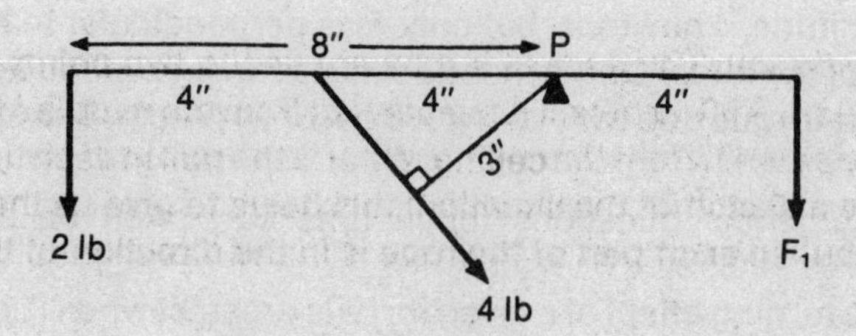

Here, we see that the 2-lb and 4-lb forces tend to produce counterclockwise rotation around the fulcrum or pivot at *P*. To balance these torques, clockwise moments must be applied. For equilibrium:

the sum of the clockwise moments = the sum of the counterclockwise moments

Therefore, in the diagram, $F_1 \times 4'' = 2\text{ lb} \times 8'' + 4\text{ lb} \times 3'' = 28\text{ lb-in.}$ $F_1 = 7\text{ lb}$

Center of Gravity

Newton's *law of universal gravitation* states that two objects attract each other with a force that is proportional to the product of their masses and inversely proportional to the square of the distance between them. The earth's attraction for objects is known as *gravity.* This attraction of the earth for every portion of an object, therefore, consists of a set of practically parallel forces. The resultant of these parallel forces is the *weight* of the object. This resultant goes through a point known as the *center of gravity:* When talking about forces and moments of forces (torques) we can think of all the weight of the object as being concentrated at the center of gravity. If the fulcrum is placed just below the center of gravity, there will be no tendency for the object to rotate.

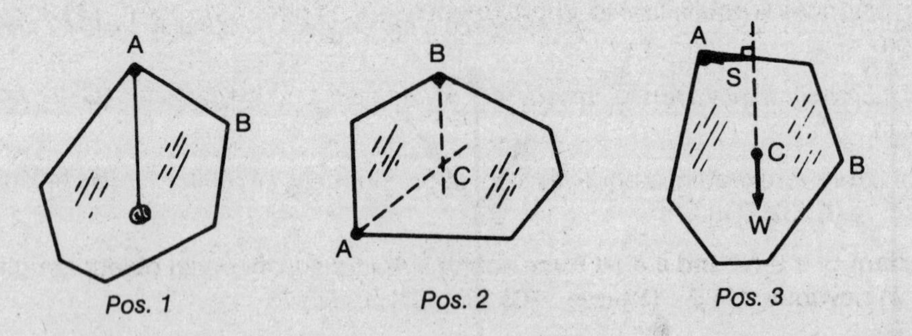

Pos. 1 *Pos. 2* *Pos. 3*

How can the center of gravity of an irregular object be determined? Consider the object depicted above. As in Pos. 1, first suspend the object so that it swings freely about point *A*. Repeat this process at least once by suspending the object from another point such as *B* (Pos. 2). The point where the vertical lines intersect is the center of gravity, *C*. Note that, if the object is in some position such as Pos. 3, with the pivot at *A*, the weight of the object acting at *C* produces a clockwise moment around *A* (equal to $W \times S$), which causes the object to swing to Pos. 1, where the length of the moment arm equals zero, i.e., the object comes to rest with its center of gravity directly below the point of suspension.

The above indicates that the stability of objects can be increased by building them with centers of gravity as low as possible. The stability of objects is also increased by providing them with as big a base as possible.

ANSWERING THE QUESTIONS Now let us consider briefly the questions asked concerning the first diagram at the beginning of the chapter F_1 produces a moment greater than F_2 or F_3 because all three have the same magnitude, 4 newtons, but only F_1 is perpendicular to the bar. It therefore has the greatest moment arm of the three. Only a mental calculation is required to show that F_4 produces a larger torque than F_1: The torque due to F_1 is 4 Nt $\times$ 6 cm or 24 Nt-cm. The torque due to F_4 is 5 Nt $\times$ 5 cm or 25 Nt-cm.

Note that on these numerical questions much time can often be saved by making only approximate calculations.

Observe, in general, the turning effect of a given force is greatest when it is applied at right angles

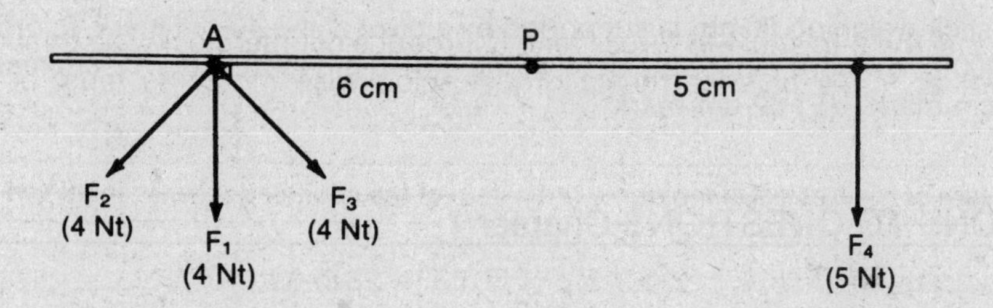

to the bar. (When the 4-Nt force is in positions such as F_2 or F_3, only a component of the 4-Nt force acts at right angles to the bar to produce a torque. The other component produces zero torque.)

QUESTIONS Chapter 1

Select the choice which will answer the question best.

1. A length of 45 mm is equivalent to, approximately, **(A)** 1.8 in. **(B)** 6.2 in. **(C)** 11 in. **(D)** 18 in. **(E)** 23 in.

2. A mass of 5 ounces is equivalent to, approximately, **(A)** 11 gm **(B)** 50 gm **(C)** 90 gm **(D)** 120 gm **(E)** 140 gm

3. A length of 3.5 in. is equivalent to, approximately, **(A)** 2.5 cm **(B)** 8.9 cm **(C)** 11.4 cm **(D)** 13.9 cm **(E)** 16.5 cm

4. An area of 2.5 square meters is equivalent to, approximately, **(A)** 100 in.^{2} **(B)** 140 in.^{2} **(C)** 180 in.^{2} **(D)** 220 in.^{2} **(E)** 3900 in.^{2}

5. The resultant of a 3-Nt and a 4-Nt force acting simultaneously on an object at right angles to each other is, in newtons, **(A)** 0 **(B)** one **(C)** 3.5 **(D)** 5 **(E)** 7

6. Two forces act together on an object. The magnitude of their resultant is least when the angle between the forces is **(A)** 0° **(B)** 45° **(C)** 60° **(D)** 90° **(E)** 180°

7. The resultant of a 5-lb and a 12-lb force acting simultaneously on an object in the same direction is, in pounds, **(A)** zero **(B)** 5 **(C)** 7 **(D)** 13 **(E)** 17

8. A uniform rod 30 in. long is pivoted at its center. A 40-lb weight is hung 5 in. from the left end. Where must a 50-lb weight be hung to maintain equilibrium? **(A)** 5 in. from right end **(B)** 6 in. from right end **(C)** 7 in. from right end **(D)** 8 in. from right end **(E)** 9 in. from right end.

9. A uniform rod 30 in. long is balanced at its center. A 40-lb weight is hung on the rod 5 in. from the left end, and a 20-lb weight 7 in. from the left end. How far to the right of the fulcrum must a 50-lb weight be placed to maintain equilibrium? **(A)** 5 in. **(B)** 7 in. **(C)** 9 in. **(D)** 11 in. **(E)** 13 in.

10. A sled is pulled by means of a rope 6 ft long; one end of the rope is attached to the sled and the other end is held 4 ft above the level ground. If the force pulling the rope is 8 Nt, the horizontal component of the force (neglecting the height of the sled) is **(A)** 3 Nt **(B)** 4 Nt **(C)** 6 Nt **(D)** 8 Nt **(E)** 10 Nt

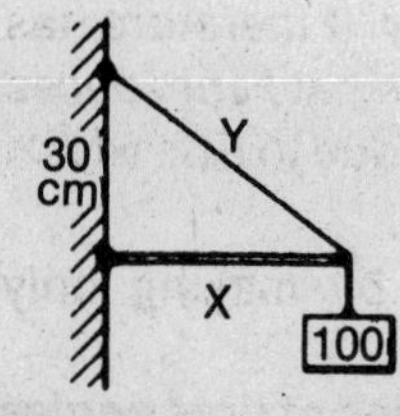

11. An object weighing 100 Nt is suspended from one end of horizontal rod *X*. The other end of the rod rests against a vertical wall. The weight of the rod is negligible. The system is kept in equilibrium by rope *Y*, one end of which is attached to the same end of the rod as the weight. The other end of the rope is attached to a point on the wall 30 cm above the other end of the rod. The rod is 40 cm long. The tension in rope *Y* is **(A)** 70 Nt **(B)** 90 Nt **(C)** 100 Nt **(D)** 130 Nt **(E)** 170 Nt

12. A meter stick weighing 80 gm is supported by a pivot at the 40-cm mark. In order to maintain equilibrium a 100-gm weight must be placed at the **(A)** 10-cm mark **(B)** 16-cm mark **(C)** 32-cm mark **(D)** 50-cm mark **(E)** 62-cm mark

EXPLANATIONS TO QUESTIONS Chapter 1

Answers

1. **(A)**	4. **(E)**	7. **(E)**	10. **(C)**
2. **(E)**	5. **(D)**	8. **(C)**	11. **(E)**
3. **(B)**	6. **(E)**	9. **(D)**	12. **(C)**

Explanations

1. **(A)** We notice that the choices are not close in value. Therefore our calculation need not be very precise; we can thus save time. We recall that 1 meter = 39.37 inches. Therefore, 1000 mm $\doteq$ 40 inches, 10 mm $\doteq$ 0.40 inch; 45 mm $\doteq$ 1.8 inches.

2. **(E)** We recall that 1 pound = 16 ounces = 454 gm. Therefore 5 ounces $= \left(\frac{454}{16} \times 5\right)$ gm = 140 gm.

3. **(B)** We recall that one inch = 2.54 cm. Therefore, 3.5 inches = (3.5×2.54) cm = 8.9 cm. Again, the choices are not close in value; it would have been enough to note that the answer is between 3×2.54 and 4×2.54; i.e., between 7.6 and 10. This would give (B) as the answer.

4. **(E)** One meter = 39.37 in. $\doteq$ 40 in. One square meter $\doteq$ 40 in. $\times$ 40 in. = 1600 in.^{2} Therefore 2.5 square meters $\doteq (2.5 \times 1600)$ in^2 = 4000 in.^{2} The only answer close to this is (E).

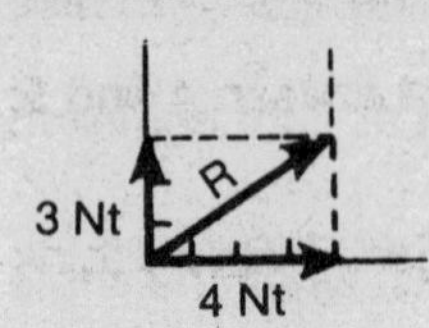

5. **(D)** We can draw the two forces to *scale*, letting ⅛ in. = 1 Nt. If we complete the parallelogram and draw the diagonal, we notice that the resultant is 5 Nt. You should also recognize the *3-4-5-right triangle;* the resultant forms the hypotenuse of a right triangle whose arms are, respectively, 3 and 4. Therefore the hypotenuse is 5.

6. **(E)** When two vectors act in opposite directions (an angle of 180°), their resultant is the difference between the two vectors. When two vectors act in the same direction (an angle of 0°), their resultant is the sum of the two vectors. As the angle between the two vectors increases from 0° to 180°, their resultant decreases from the sum to the difference of the two vectors; to get the actual value for any of these angles, we may use a parallelogram.

7. **(E)** When two forces act in the same direction, the resultant of the two forces is their sum. If you said 13, you were too hasty. You should recognize the *5-12-13 right triangle,* but the two given forces were not at right angles to each other. Always read the question carefully.

8. **(C)** Make a simple, neat diagram putting in the given information. For equilibrium, clockwise moments = counterclockwise moments

$$\text{moment} = \text{force} \times \text{moment arm}$$
$$50x = 40 \times 10$$
$$x = 8'' \text{ from fulcrum}$$

Ans: (15 − 8)″ or 7″ from rt. end.

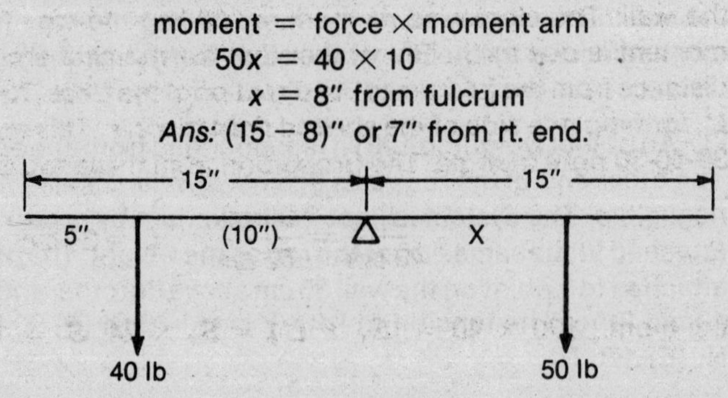

9. **(D)** Make a simple, neat diagram; insert given information. Be sure to obtain distance of direction of force from the *fulcrum.* (Length of moment arm is measured from the fulcrum.)

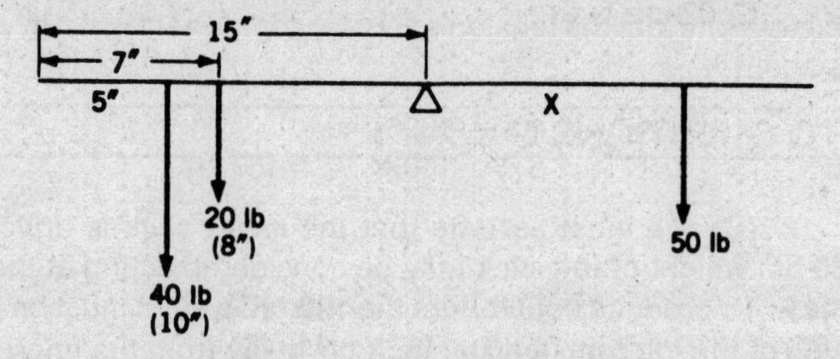

$$\text{moment of force} = \text{force} \times \text{moment arm}$$
$$\text{sum of clockwise moments} = \text{sum of counterclockwise moments}$$
$$50\,X = (40 \times 10) + (20 \times 8)$$
$$50\,X = 400 + 160 = 560$$
$$X = 560/50 = 11.2 \text{ in.}$$

Choice (D) is the closest.

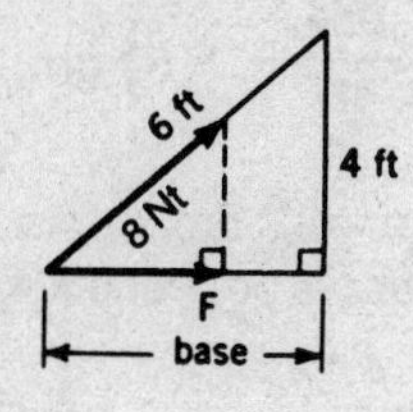

10. **(C)** You may make a scale drawing and obtain a graphical solution. You will probably save time by making a rough drawing, inserting the given values, and using a little algebra to obtain the solution. From the end of the vector representing the 8-Nt force we drop a perpendicular to obtain F, the horizontal component of the 8-Nt force. The two right triangles are similar. Therefore, $\frac{F}{8} = \frac{\text{base}}{6 \text{ ft}}$. The length of the base $= \sqrt{(\text{hypotenuse})^2 - (\text{height})^2} = \sqrt{36 - 16}$; base $= \sqrt{20'}$. $F = \frac{8}{6}\sqrt{20} = 5.95$ Nt. Choice (C) is closest. It is not necessary to take exact square roots. You note that the square root of 20 is between 4 and 5; using 4.5 is adequate.

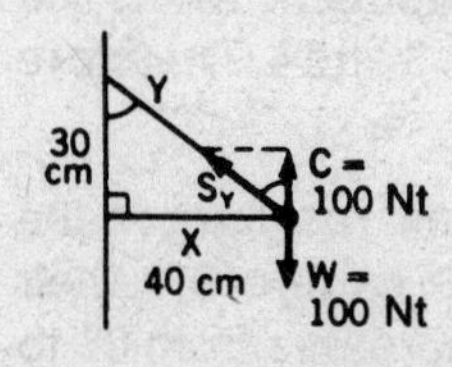

11. **(E)** First we insert into the diagram the given information. Then we think of a method for solving the problem.

Method 1—Components

For equilibrium, the upward component of the rope's tension, S_Y, must equal the downward pull of 100 Nt. The diagram shows two right triangles—the force triangle and the dimension triangle, a 30-40-50-right triangle. These two triangles are similar; a pair of equal acute angles is marked (alternate interior angles). We then write the proportion for the two similar triangles:

$$\frac{S_Y}{100 \text{ Nt}} = \frac{50 \text{ cm}}{30 \text{ cm}}; \quad S_Y = \frac{5}{3} \times 100 \text{ Nt} = 167 \text{ Nt}$$

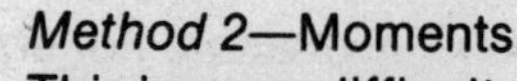

Method 2—Moments

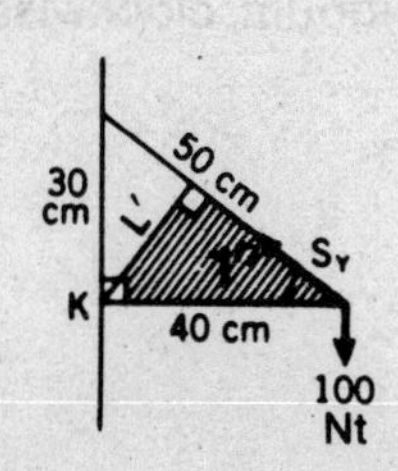

This is more difficult. Take moments about K, the point where the rod touches the wall. The clockwise moment = 100 Nt × 40 cm. The counterclockwise moment is due to S_Y. Remember that the moment arm is the perpendicular distance from the fulcrum to the direction of the force. This distance is given by L', forming one side of the shaded right triangle. This triangle is similar to the 30-40-50 right triangle. The proportion from these two similar triangles is:

$$\frac{L'}{40 \text{ cm}} = \frac{30 \text{ cm}}{50 \text{ cm}}; \quad L' = 24 \text{ cm}$$

clockwise mom. = ccl mom.; $100 \times 40 = (S_Y \times L') = S_Y \times 24$; $S_Y = 167$ Nt.

NOTE: Method 2 is longer than Method 1, but it is the more general method. If the rope were not tied to the same point on which the other force(s) act, Method 1 could not be used.

If you know trigonometry, note that the tension's vertical component is $S_Y \sin \angle 1$, which equals $S_Y \times {}^3/_5$. Taking moments about K,

$$S_Y \times {}^3/_5 \times 40 = 100 \times 40$$
$$S_Y = 100 \times {}^5/_3 = 167 \text{ Nt}$$

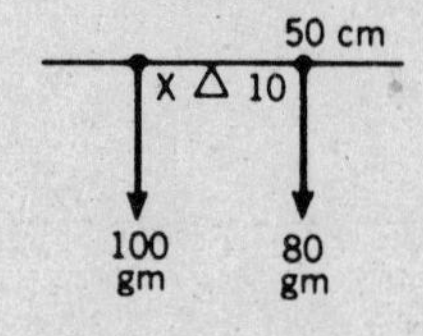

12. **(C)** We must assume that the meter stick is uniform. Therefore, the 80-gm weight of the stick may be considered acting at the center, the 50-cm mark. To produce equilibrium the 100-gm weight must be placed on the other side of the fulcrum, a distance X cm away from the fulcrum. Ccl moments = clockwise mom. $100X = 80 \times 10$; $X = 8$ cm from the 40-cm mark. This is at the 32-cm mark.

2

Motion and Forces

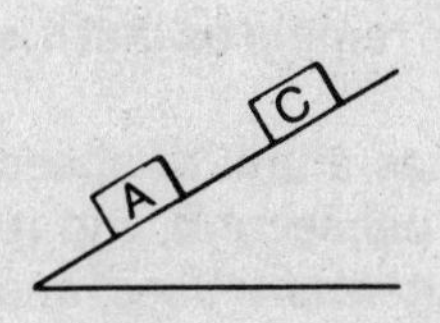

Object *A* moves with constant speed down the inclined plane. Object *B* on the table (not shown) moves with constant speed. Do you realize that the unbalanced force acting on *A* is zero but that an unbalanced force may be acting on *B?* If object *C* moves with decreasing velocity down the plane, what is the force acting on *C?* In this chapter we shall consider some of the important types of motion and the forces producing them.

SPEED AND VELOCITY

When we speak about motion or rest, the earth is our *reference point* unless some other frame of reference is specifically indicated. For example, in the above paragraph, the plane and table are assumed to be stationary with respect to the earth, and the speed of *A* is constant with respect to the plane.

Speed is distance covered per unit time. *Velocity* (v) of an object is its speed in a given direction. Speed is a scalar quantity. Velocity is a vector quantity—it has magnitude and direction. The magnitude of the velocity is the object's speed. The velocity changes if either the speed or the direction of motion, or both, change. You can think of direction as being given by the points of a compass. You can then see readily that if an object is moving at constant speed around a horizontal circle its direction, and therefore its velocity, is constantly changing; at one instant the object may be moving towards the north and a little later it will be moving towards the south.

The motion of an object can sometimes be thought of as the result of a *combination of velocities.* For example, if you walk at 4 miles per hour from the last car of a train towards the front while the train is moving at 60 miles per hour, your velocity with respect to the earth is 64 mi/hr in the direction of motion. (Note that it was implied that the 4 mi/hr was with respect to the train.) If the two given velocities are not in the same direction, we can get the combined velocity by using combination of vectors as we did for forces on page 5. *Example:* A plane is moving north in still air at 100 mi/hr. A steady west wind starts blowing at 200 mi/hr. What will be the plane's velocity? The solution is shown in the following diagram. (Recall that a west wind blows from the west towards the east.)

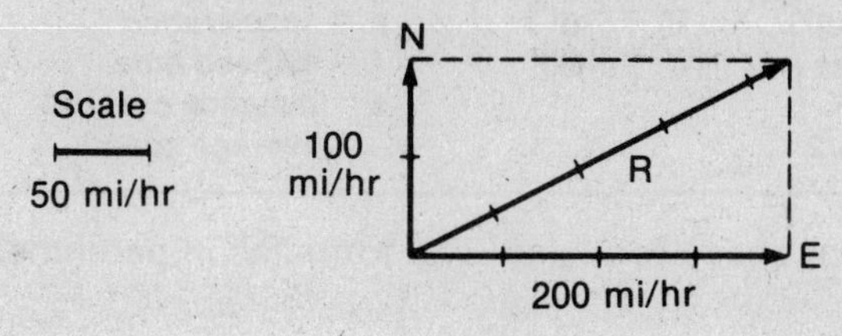

The plane's velocity will be approx. 220 mi/hr in the direction shown by *R*. (It would be wasteful to work for a more accurate solution.)

distance covered = speed × time

$$s = vt$$

ACCELERATED MOTION

Uniform motion is motion in which the velocity is constant. If the velocity changes the motion is said to be *accelerated. Acceleration* (*a*) is the rate of change of velocity. Because of the definition of velocity, acceleration may result from a change in direction as well as from a change in speed. Acceleration is a vector quantity.

$$\text{acceleration} = \frac{\text{change in velocity}}{\text{time required for change}}$$

$$a = \frac{v_f - v_i}{t} = \frac{\Delta v}{t}$$

Δv (delta *v*) = change in velocity. Recall that we may use an arrow over the letter to remind us that it is a vector quantity.

The formula indicates that the magnitude of the acceleration is expressed as a unit of speed divided by a unit of time. Examples: 6 miles per hour per second, 4 ft per second per second, and 10 meters per second per second. To save space the last two examples would be written as 4 ft/sec^2 and 10 m/sec^2. To emphasize the meaning of acceleration it is a good idea to read such expressions as "four feet per second every second" and "10 meters per second every second."

If an object slows down we sometimes say that the object is *decelerated.* Sometimes we also say that such an acceleration is negative. This usage is not recommended but should be interpreted to mean that the direction of the acceleration is opposite to the direction of the initial velocity.

Uniformly accelerated motion is motion with constant acceleration. If the object starts moving from rest, the initial velocity is zero, and the expression for its acceleration becomes $v_f = at$. If an object is allowed to *fall freely* near the surface of the earth (its initial velocity is zero and no forces other than gravity act on it during its fall), the acceleration of the object remains constant and is independent of the mass of the object. The letter *g* is used universally for this acceleration. The value of *g* varies slightly from point to point on the surface of the earth. Unless told to do otherwise, in the MKS system use $g = 9.8$ m/sec^2; in the English system use $g = 32$ ft/sec^2.

If an object is projected into the air in some direction other than the vertical, air resistance being negligible, the path is a parabola. The motion can be thought of as a combination of two separate motions: a horizontal motion in which the velocity remains constant and has the value of the horizontal component of the velocity with which it is projected, and a vertical motion due to gravity. If it is projected horizontally, the vertical motion is the same as in free fall. You should memorize the following formulas, but be sure you memorize at the same time the situations for which each formula applies.

Motion with constant acceleration (starting from rest)

$v_f = at$	$(v_f = gt)$	v_f = final velocity
$s = \frac{1}{2}at^2$	$(s = \frac{1}{2}gt^2)$	a = acceleration
$v_f^2 = 2as$	$(v_f^2 = 2gs)$	t = elapsed time
		s = distance covered
$v_{av} = v_f/2$		v_{av} = average speed

Additional notes on use of above formulas: The formulas in parentheses are for freely falling objects near the surface of the earth; in this case the acceleration has a known value and, as mentioned before, we use the letter g to indicate this. (Air resistance is neglected.) The same formulas may be used *for objects not starting from rest* (initial velocity not zero) if the final velocity is zero. In that case let v_f stand for the initial velocity. *Example:* A ball is thrown vertically upward and rises to a height of 4.9 m. With what speed did the ball leave the hand? *Solution:* Let v_f represent this speed.

$$v_f^2 = 2gs$$
$$v_f^2 = 2 \times 9.8\ \text{m/sec}^2 \times 4.9\ \text{m}$$
$$v_f = 9.8\ \text{m/sec}$$

For objects thrown vertically upward, the time required to reach the maximum height is equal to the time required for the object to fall back to the starting level.

For linear motion with constant acceleration not starting from rest, the original four formulas become:

$$v_f = v_i + at$$
$$s = v_it + \tfrac{1}{2}at^2$$
$$v_f^2 = v_i^2 + 2as$$
$$v_{av} = \frac{v_i + v_f}{2}$$

FOR ALL TYPES OF MOTION

average speed = distance covered / time required

If a speed is given in miles per hour it can be converted quickly to feet per second: multiply by 22/15. *Example:* 30 mi/hr = (30 × 22/15) ft/sec = 44 ft/sec.

Air resistance or friction of the air tends to slow an object down. An object dropped from a balloon may actually stop accelerating, reaching a constant downward speed known as its *terminal velocity.*

NEWTON'S FIRST LAW OF MOTION

Statement: If the net force acting on an object is zero, the velocity of the object does not change (that is, speed and direction of motion remain constant). The term *net force* means the same thing as resultant force or unbalanced force. Newton's first law is often stated in other ways: When a body is at rest or moving with constant speed in a straight line, the resultant of all the forces acting on the body is zero.

Of course, an object at rest and remaining at rest is in equilibrium, but note that an object may be moving and still be in equilibrium. An object at rest tends to remain at rest; an object in motion tends to keep moving in a straight line unless acted on by an unbalanced force.

Illustration: Passengers in a car lurch forward when the driver applies the brakes. *Application*–In the diagram, a 200-lb object is suspended from a ceiling by means of a rope. The object is at rest; what are the forces acting on it? Since the object is at rest it is in equilibrium. Therefore Newton's first law applies and the resultant of all forces acting on it is zero.

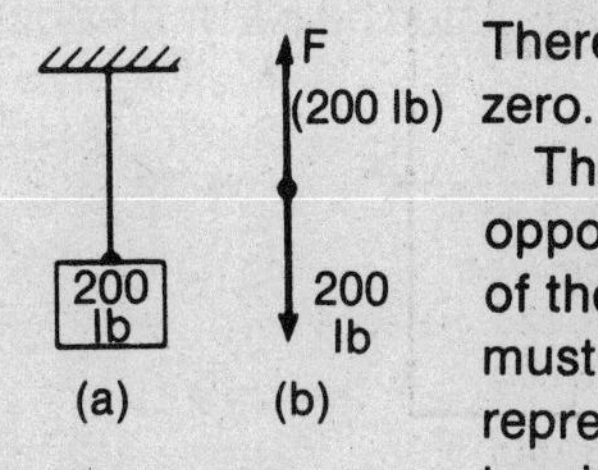

The earth supplies a 200-lb force acting downward (gravity). An equal and opposite force must be applied to the object. This is done by the rope. This pull of the rope is referred to as the *tension* in the rope. The tension in this rope must be 200 lb. This is shown by the vector diagram, where the object is represented by the dot and the rope's upward pull by the vector F. Note that the tension of the rope also acts downward on the *ceiling,* but we are making no use of this because we are concerned with the forces acting on the *object.*

NOTE: It also follows from Newton's first law that, if an object is to remain at rest, any force acting on it in one direction must be balanced by an equal force on it in the opposite direction. For example, refer to the illustrative problem on page 6. Each half of the rope exerts a pull on the object. This pull can be replaced by a horizontal component and a vertical component. The sum of the two vertical components must be 90 lb in order to balance the pull of the 90-lb object.

NEWTON'S SECOND LAW

STATEMENT 1 If the net force acting on an object is not zero, the object will be accelerated in the direction of the force; the acceleration will be proportional to the net force and inversely proportional to the mass of the object. *Example:* A push that is enough to give a 3000-lb car an acceleration of 4 ft/sec^2 will be able to give a 6000-lb truck an acceleration of 2 ft/sec^2. Twice as great a push is needed to give the car an acceleration of 8 ft/sec^2.

We can think of the net force as being used to overcome the inertia of the object. The greater the mass of the object, the greater is the force required to produce a given acceleration, or to "budge the object." This leads to definitions: *INERTIA*—the property by which an object resists being accelerated; *MASS*—the measure of an object's inertia. The proportion between resultant force and acceleration can be represented as:

$$\frac{F_1}{F_2} = \frac{a_1}{a_2}$$

When the unbalanced force acting on the object is equal to the weight of the object, as in free fall, the acceleration is equal to *g*, the acceleration due to gravity. This leads to the useful proportion:

$$\boxed{\frac{F}{w} = \frac{a}{g}}$$

Note that the weight *w* and the resultant force *F* must be in the same units and the *g* and the acceleration *a* must be in the same units. *Example:* An automobile weighs 3200 lb. What unbalanced force will produce an acceleration of 4 ft/sec^2?

$$\frac{F}{3200 \text{ lb}} = \frac{4 \text{ ft/sec}^2}{32 \text{ ft/sec}^2} \qquad \therefore F = 400 \text{ lb}$$

Newton's second law also leads to the equation $F = ma$, where m is the mass of the object. Also, the weight of an object is equal to the product of its mass and the value of the acceleration due to gravity:

$$w = mg$$

There are some situations you have to watch. If you use Newton's second law, $F = ma$, different units must be used for force and mass, as indicated in the table below. If you use the relationship $F/w = a/g$, the same units are used for force and weight. Of course, if an object goes sufficiently far away from the earth, the gravitational pull of the earth on the object will become significantly less although the mass of the object remains constant. An orbiting astronaut is still attracted by the earth; his weightlessness is only apparent. (The definition of weight used in this book is the one used in most high school and college textbooks.) Also see the discussion on Newton's law of gravitation later in the chapter.

At this point some students may wish to review the material on proportional relationships and graphs presented in Chapter 17. For reference, the appropriate units are tabulated below; memorize the first row.

F & w	m	a	g	v
newtons	**kilograms**	**m/sec^2**	**9.8 m/sec^2**	**m/sec**
dynes	**grams**	**cm/sec^2**	**980 cm/sec^2**	**cm/sec**
pounds	**slugs**	**ft/sec^2**	**32 ft/sec^2**	**ft/sec**
poundals	**pounds**	**ft/sec^2**	**32 ft/sec^2**	**ft/sec**

Note: The abbreviation for meter is *m;* the abbreviation for mile is *mi.*

STATEMENT 2 Newton's second law is sometimes stated as: The rate of change of momentum is proportional to the net force acting on the object.

MOMENTUM is defined as the product of mass and velocity. It is a vector quantity.

Example 1. A 500-gm object is allowed to drop. After 2 seconds of free fall its speed will be (980×2) cm/sec ($v = gt$). Its momentum will be (500 gm) (980×2) cm/sec = 980,000 gm · cm/sec downward. If we use MKS units: momentum = (0.5 kg) (9.8×2) m/sec = 9.8 kg-m/sec downward.

Example 2. An object moves along a straight line with a momentum of 5 units. If it slows down without change of direction so that its momentum is 3 units, its change of momentum is 2 units. However, if it had hit something and bounced back with a momentum in the opposite direction of 3 units, its change in momentum is 5 units − (−3 units) = 8 units in the new direction.

NOTE: At this point, you, like many others, may raise a difficult question: When you speak of a 50-lb object, do you mean that the object's mass is 50 lb or that the object's weight is 50 lb. Where it is used it makes no difference! This is true because one pound of weight is defined as the earth's pull on a mass of one pound at a certain place on the earth (sea level at 45° N. latitude). If this object is moved around on the surface of the earth, or near it, its mass will remain exactly one pound; its weight will change slightly, but for most situations this is negligible. A 50-lb object has a mass of 50 lb and also a weight of 50 lb. Unfortunately, the pound is a unit for both mass and weight.

Example 1. How large a force is needed to give an acceleration of 9.8 m/sec^2 to an object having a mass of one kilogram? $F = ma$; $F = (1\text{kg})\ (9.8\ \text{m/sec}^2) = 9.8$ newtons. You can now see why we said earlier that the weight of a one-kilogram mass is 9.8 newtons. The weight of an object on earth is the earth's gravitational pull on it. When that pull is the only force acting on the object, the object is falling freely. The acceleration of a freely falling object, we know, is 9.8 m/sec^2.

Example 2. How large a force is needed to give an acceleration of 3 m/sec^2 to a 5-lb object?

We know that a 5-lb object falling freely has an acceleration of $g = 9.8$ m/sec^2 and that during its fall the net force acting on it is its weight. The 5-lb object has a weight of 5 lb.

$$\frac{F}{w} = \frac{a}{g}$$

$$\frac{F}{5\ \text{lb}} = \frac{3\ \text{m/sec}^2}{9.8\ \text{m/sec}^2}$$

$$F \doteq 1.5\ \text{lb}$$

Note that both accelerations must have the same units. Since we are given an acceleration of 3 m/sec^2, we use $g = 9.8$ m/sec^2. Likewise both units of force must have the same units. Since we are given that the object has a weight of 5 lb, the unit of the answer is the pound. Note also that we did not have to be concerned about the mass of the object.

IMPULSE AND CHANGE OF MOMENTUM

The *impulse* acting on an object is defined as the product of the net force acting on the object and the length of time that this force acts. The impulse acting on an object is equal to the change of momentum which it produces:

$$Ft = \text{change in momentum}$$

Example: If a 10-newton force acts on a 3-kilogram object for 2 seconds, the impulse acting on the object = (10 Nt) (2 sec) = 20 Nt-sec in the direction of the force. This is also the change in momentum of the object. (We can use the equivalent unit, a newton-second = a kilogram-meter/second.) What is the resulting change in the object's velocity?

$$\text{Change in momentum} = \text{mass} \times \text{change in velocity}$$
$$20\ \text{kg-m/sec} = (3\ \text{kg})\ (\text{change in velocity})$$
$$\Delta v = 6.7\ \text{m/sec (in the direction of the force)}$$

NEWTON'S THIRD LAW

When one object exerts a force on a second object, the second object exerts an equal and opposite force on the first object. (This is sometimes stated as: Action equals reaction.)

Note that two different objects are involved and that the two forces mentioned do not act on the same object. Therefore, they do not cancel each other.

Illustration: A rowboat stops at a pier and the boy in the boat steps toward the pier. The boat moves away from the pier and the boy falls into the water or barely makes the edge of the pier. Explanation: In order to move forward from rest the boy must have a force acting on him (Newton's second law); this force is supplied by the boat: When the boy pushes against the boat with his foot, the boat pushes against the boy's foot with an equal force (Newton's third law). Both the boat and the boy will move with respect to the spot where the boat had stopped, but the boy will probably have misjudged the needed push.

Measurement of the speed with which the boy and the boat start to move would show that the momentum of the boy equals the momentum of the boat, except for direction. The resultant of the two momenta is zero. This is an illustration of CONSERVATION OF MOMENTUM. When no external forces act on a system of objects, the total momentum of the system is unchanged. (This is true for collisions, explosions, etc. The objects in the system exert forces on each other, but these are internal forces. Also remember, momentum is a vector quantity.)

Newton's third law is also the principle underlying the operation of rockets and jet aircraft. As the hot gases are pushed out from the rear, they exert a forward push on the object from which they escape.

Example: In the diagram, block *B* is at rest on a horizontal, frictionless surface. It has a mass of 20 kg. Block *A* slides towards it with a velocity of 6.0 m/sec. It has a mass of 10 kg. After block *A* hits block *B* they remain together.

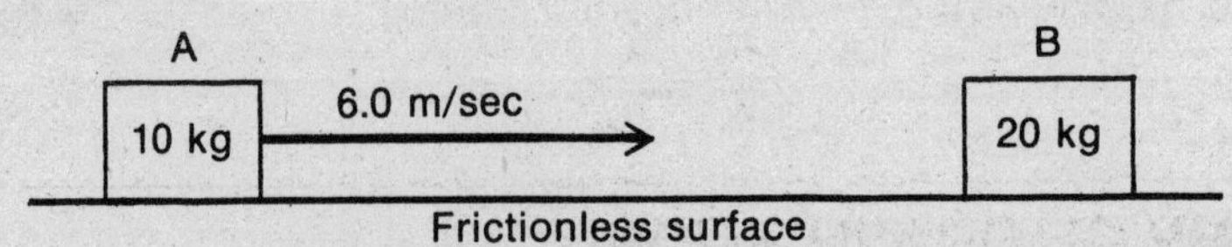

Before the collision the momentum of block A = (10 kg) (6.0 m/sec) = 60 kg-m/sec. The momentum of block B is zero, since it is stationary. After the collision the momentum of the 2 blocks together still has to be the vector sum of the two blocks before the collision, or 60 kg-m/sec. The combined mass is 30 kg. They move off with a velocity of 2.0 m/sec, since

$$\text{momentum} = \text{mass} \times \text{velocity}$$
$$60 \text{ kg-m/sec} = (30 \text{ kg})\ (v)$$
$$v = 2.0 \text{ m/sec}$$

CENTRIPETAL FORCE

When an object moves with constant speed around a circle, the object's velocity is constantly changing because its direction is constantly changing. Because the velocity is changing the object is accelerated. According to Newton's second law, this acceleration is produced by an unbalanced force. This is the *centripetal force:* the force acting on an object moving in a circle which keeps it moving in the circular path. The force is directed towards the center of the circle. The diagram represents a stone whirling at the end of a string. At the instant the object is at *A*, it is moving to the left; at *B* it is moving to the right. The centripetal force is always towards the center as shown by F_c. If the string should tear, the object would fly off tangent to the circle (because of inertia: Newton's First Law); if this happens when the object is at *C*, the object flies off in the direction of *CD*.

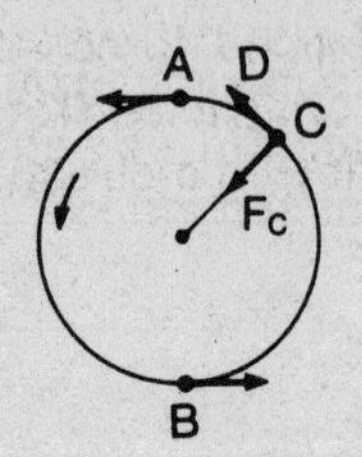

The centripetal force is proportional to the mass of the object and to the square of its speed and inversely proportional to the radius of its circular path. If the speed of an object moving around a circle is increased from 4 ft/sec to 8 ft/sec, the centripetal force is multiplied by 4.

There is no centrifugal force acting on the object moving in a circle. If the string pulls on the whirling stone, the stone reacts and pulls on the string. This equal and opposite force pulling on the *string* may be called a centrifugal force.

The centripetal force on satellites is the gravitational force.

The *centripetal acceleration* produced by the centripetal force is proportional to the square of the object's speed around the circle and inversely proportional to its radius.

FRAME OF REFERENCE

To someone riding in a car which is moving at high speed around a curve, things seem different from the way they seem to someone standing at the side of the road. The rider will feel a push or force toward the outside of the curve, towards the right if he is facing forward.* We, who are watching from the outside, see it merely as an illustration of Newton's first law of motion. His inertia tends to keep him moving along a straight line and his unrestrained upper body will do this somewhat. (Seat belts and friction between him and the seat of the car provide the centripetal force to make him go around the curve.) For us, watching in the street or working in the laboratory, the reference for motion or rest is called an *inertial frame of reference*, a point or set of coordinates which is at rest or moves with uniform velocity, constant speed in a straight line. The laws of physics are the same in all inertial frames of reference. As stated on page 13, the earth is usually a satisfactory reference frame in spite of the rotation around its axis and revolution around the sun.

NEWTON'S LAW OF GRAVITATION

Any two particles attract each other with a force which is proportional to the product of their masses and inversely proportional to the square of the distance between them. This accounts for the weight of an object on earth: the earth's pull on the object.

The more general definition of weight is: The *weight* of an object is the net gravitational pull on it. For objects near the surface of the earth, the weight of the object is practically the earth's gravitational pull on it. The distance is measured from the center of gravity of the earth to the center of gravity of the object. If the earth's radius is taken as 4000 miles, an object weighing 20 lb on the surface of the earth will weigh only 5 lb at a distance of 4000 mi from the earth, because the distance from the center of the earth has been doubled. Of course, the mass of the object is not changed.

ANSWERING THE QUESTIONS Now let us return to the problem at the beginning of the chapter. Object *A* is moving down the plane with constant speed. Since *A's* speed and direction are not changing, its acceleration is zero. Therefore, according to Newton's first and second laws no unbalanced force can be acting on *A*. On the other hand, nothing was said about *B's* direction of motion. For example, *B* might be moving in a circular path. Then an unbalanced force (the centripetal force) would have to act on it. If *B* were moving in a straight line, then no unbalanced force could act on it. Since *C* is decelerating, an unbalanced force must be acting on it with a direction opposite to the direction of motion. This could be the force of friction.

*in the diagram on the bottom of page 18.

QUESTIONS Chapter 2

Select the choice which will answer the question best.

1. The speed of an object at the end of 4 successive seconds is 20, 25, 30, and 35 mi/hr, respectively. The acceleration of this object is **(A)** 5 ft per sec^2 **(B)** 5 mi per hr per sec **(C)** 5 mi per hr^2 **(D)** 5 mi per sec^2 **(E)** 20 mi per hr per sec

2. A bomb is dropped from an airplane moving horizontally with a speed of 200 mi/hr. If the air resistance is negligible, the bomb will reach the ground in 5 sec when the altitude of the plane is approximately **(A)** 80 ft **(B)** 100 ft **(C)** 400 ft **(D)** 800 ft **(E)** 1000 ft

3. While an arrow is being shot from a bow it is accelerated over a distance of 2 ft. At the end of this acceleration it leaves the bow with a speed of 200 ft/sec. The average acceleration imparted to the arrow is **(A)** 200 ft/sec^2 **(B)** 400 ft/sec^2 **(C)** 500 ft/sec^2 **(D)** 1000 ft/sec^2 **(E)** 10,000 ft/sec^2

4. An object whose mass is 100 gm starts from rest and moves with constant acceleration of 20 cm/sec^2. At the end of 8 sec its momentum is, in gm-cm/sec, **(A)** 500 **(B)** 8000 **(C)** 16,000 **(D)** 33,000 **(E)** 64,000

5. A rock is dropped from a high bridge. At the end of 3 seconds of free fall the speed of the rock is, in cm/sec, **(A)** 30 **(B)** 100 **(C)** 500 **(D)** 1000 **(E)** 3000

6. A person standing in an elevator which goes up with constant upward acceleration exerts a push on the floor of the elevator whose value **(A)** is always equal to his weight **(B)** is always greater than his weight **(C)** is always less than his weight **(D)** is greater than his weight only when his acceleration is greater than g (approx. 9.8 m/sec^2) **(E)** is greater than his weight only when his acceleration is less than g

7. Two small masses, when 10 cm apart, attract each other with a force of F newtons. When 5 cm apart, these masses will attract each other with a force, in newtons, of **(A)** $F/2$ **(B)** $F/4$ **(C)** $2F$ **(D)** $4F$ **(E)** $5F$

8. When a certain object is 100 ft above the surface of the earth, the earth's attraction for it is 10 lb. In order for the earth's attraction for the same object to be only 5 lb, the object must be taken to a distance from the surface of the earth of **(A)** 140 mi **(B)** 200 mi **(C)** 1600 mi **(D)** 5600 mi **(E)** 16,000 mi

9. A car going around a certain curve at a speed of 25 mi/hr has a centripetal force acting on it of 100 lb. If the speed of the car is doubled, the centripetal force **(A)** is quadrupled **(B)** is doubled **(C)** is multiplied by the $\sqrt{2}$ **(D)** is reduced to ½ of the original value **(E)** is reduced to ¼ of the original value.

10. Two freely hanging weights, each having a mass of 60 gm, are connected by a light thread which passes over a fixed pulley. The mass of the pulley and frictional losses are negligible. If a 10-gm weight is now added to one of the weights, its downward acceleration, in cm/sec^2, will be, approximately **(A)** 32 **(B)** 80 **(C)** 160 **(D)** 320 **(E)** 980

11. A plane travels from point A for an hour towards the west at 400 mi/hr and then travels towards the north for an hour at 300 mi/hr, arriving at point B. A second plane, starting at A at the same time as the first plane and traveling towards B along a straight line, will arrive at B at the same time as the first plane if its speed is **(A)** 100 mi/hr **(B)** 250 mi/hr **(C)** 500 mi/hr **(D)** 700 mi/hr **(E)** 1200 mi/hr

12. A handball is tossed vertically upward with a velocity of 19.6 m/sec. Approximately how high will it rise? **(A)** 15 m **(B)** 20 m **(C)** 25 m **(D)** 30 m **(E)** 60 m

EXPLANATIONS TO QUESTIONS Chapter 2

Answers

1. **(B)**	4. **(C)**	7. **(D)**	10. **(B)**
2. **(C)**	5. **(E)**	8. **(C)**	11. **(B)**
3. **(E)**	6. **(B)**	9. **(A)**	12. **(B)**

Explanations

1. **(B)** acceleration $= \dfrac{\text{change in velocity}}{\text{time in which change takes place}}$

$= 5$ mi per hr/1 sec

$= 5$ mi per hour per sec.

This means that each second the speed is changing 5 mi/hr.

2. **(C)** The horizontal velocity does not affect the vertical velocity. A suitable expression for free fall, since we know the time of falling and want to calculate the distance covered, is $s = \frac{1}{2}gt^2$; $s = \frac{1}{2}\ (32\ \text{ft/sec}^2)\ (5\ \text{sec})^2 = 400$ ft.

3. **(E)** The arrow is accelerated from rest. Its final speed is 200 ft/sec; it is accelerated over a known distance, 2 ft; $v^2 = 2\,as$; $(200\ \text{ft/sec})^2 = 2\,a\ (2\ \text{ft})$; $a = 40{,}000/4 = 10{,}000\ \text{ft/sec}^2$. The term *average accel.* is used to take care of fluctuations; don't divide by 2.

4. **(C)** The momentum of an object = its mass × its velocity.
At the end of 8 sec, its velocity:

$v = at = 20\ \text{cm/sec}^2 \times 8\ \text{sec} = 160\ \text{cm/sec}.$

momentum $= mv = 100\ \text{gm} \times 160\ \text{cm/sec} = 16{,}000$ gm-cm/sec

5. **(E)** A simple problem! For a freely falling object, starting from rest, final velocity = acceleration × time; $v = gt = 980\ \text{cm/sec}^2 \times 3\ \text{sec} \doteq 3000$ cm/sec. Notice that time can be saved by getting an approximate answer. Be careful about the value of g that you use.

6. **(B)** First apply Newton's third law: the man's push on the elevator is equal to the push of the elevator on the man. The push of the floor of the elevator must be enough to support his weight and in addition must accelerate him upward. Therefore, the push of the elevator on the man is greater than his weight by an amount necessary to give him acceleration $\left(F = W + W\frac{a}{g}\right)$.

7. **(D)** The gravitational force varies inversely as the square of the distance between the objects. When the distance is reduced to one-half of its previous value, the force is quadrupled.

8. **(C)** A little tricky. Let us get an approximate answer first. This should usually be sufficient for the test. First notice that the distance must be measured to the center of the earth; we'll take that distance as 4000 mi or 6400 km. If that distance were doubled, the earth's attraction would be one-fourth, or 2½ lb. Therefore, 8000 mi from the center of the earth, or 4000 mi from the surface of the earth, is too great. You may have a feeling that the first two choices are too low, leaving choice (C). The inverse square law tell us that

$$\frac{F_1}{F_2} = \frac{{d_2}^2}{{d_1}^2} = \frac{2}{1}$$

Therefore $d_2 = d_1\sqrt{2} = 4000\ \text{mi} \times 1.4 = 5600$ mi from the center of the earth. The distance from the surface of the earth is $(5600 - 4000)\ \text{mi} = 1600$ mi.

9. **(A)** The centripetal force acting on an object is proportional to the square of its speed. If the speed of the car is doubled, the force acting on it is multiplied by four.

10. **(B)** This is an application of Newton's second law. The unbalanced force is provided by the 10-gm weight. The masses that are accelerated are the 10-gm weight and the two 60-gm weights.

$$\frac{F}{W} = \frac{a}{g}; \quad \frac{10 \text{ gm}}{130 \text{ gm}} = \frac{a}{980 \text{ cm/sec}^2}$$

$$a = 980 \times \frac{10}{130} = 77.7 \text{ cm/sec}^2$$

11. **(B)** A little tricky. The total time required for the first plane is two hours. The distance to be covered by the second plane is 500 miles (along the hypotenuse; arms of right triangle are 300 and 400 miles, resp.).

$$\text{speed of second plane} = \frac{\text{distance covered}}{\text{time}} = \frac{500 \text{ miles}}{2 \text{ hr}} = 250 \text{ mi/hr}$$

Don't confuse this problem with a situation in which two velocities may be combined vectorially. For example, if the two velocities had acted on the plane at the same time, the resultant velocity would have been 500 mi/hr; but in this problem the two velocities acted consecutively and the total time was two hours.

12. **(B)** This is a problem in motion with constant acceleration equal to 9.8 m/sec^2 downward, in which the object has an initial velocity (19.6 m/sec), but its final velocity at the moment we are considering is zero. Since an object released from that height will hit the ground with a velocity of 19.6 m/sec, we can use the appropriate formula for motion starting from rest and having a final velocity of 19.6 m/sec:

$$v_f^2 = 2gs$$

$$(19.6 \text{ m/sec})^2 = 2\,(9.8 \text{ m/sec}^2)\,(s)$$

$$s = 19.6 \text{ m}$$

3

Work, Energy, Simple Machines

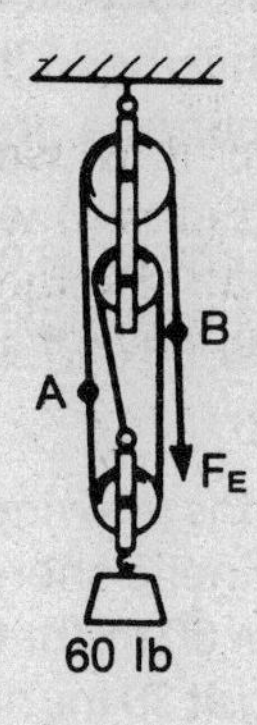

A system of pulleys, arranged as shown, is used under ideal conditions to lift a weight of 60 lb at constant speed with an effort of F_E. What is the pull exerted on the ceiling? What is the pull in the rope at point *A*? If the 60-lb object is raised 4 ft in 2 sec, how far will point *B* move? How much work did Atlas do each day that he held up the earth? How much energy did he use?

WORK AND ENERGY

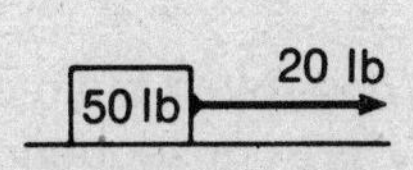

In physics we talk about work done on an object or work done by an object or by a force. When a force *moves* an object, the force does work on the object. If the 20-lb force in the diagram moves the 50-lb object 3 ft, 60 ft-lb of work was done on the object.

Work

Work is equal to the product of the force and the distance the object moves in the direction of the force. If the object moves in a direction other than the direction of the force we must take the component of the force in the direction of motion of the object. For example, if a rope is used to pull a sled, the rope is flexible, and any force, F_1, that we exert on the upper end is to be thought of as a pull which is transmitted to the sled and acts in the direction of the rope. This usually results in moving the sled along the road, and to calculate the work done by F_1 on the sled we find the component *F* in the direction of motion and multiply it by the distance the sled moves while the force is applied.

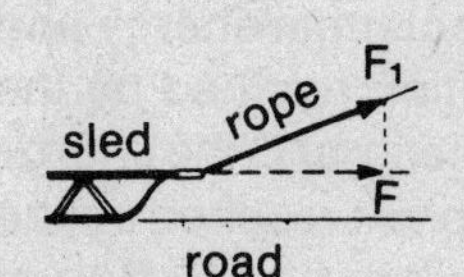

work = force × distance

The *Units* of work are obtained by multiplying a unit of force by a unit of distance. In elementary courses the most common are the *foot-pound* (ft-lb), the *joule* (newton-meter), and the *erg* (dyne-cm).

Energy

In elementary physics *energy* is defined as the ability to do work. If an object does work, it has less energy left. In mechanics, work is done on an object 1. to give it potential energy, 2. to give it kinetic energy, 3. to overcome friction, 4. to accomplish a combination of the above three. Energy used to overcome friction is converted to heat. *Units* of energy are the same as units of work.

POTENTIAL ENERGY is the energy possessed by an object because of its position or its condition. If we lift an object we change its position with respect to the center of the earth; we do work against the force of gravity. As a result the lifted object has a greater ability to do work. This increased ability to do work is the object's potential energy with respect to its original position. This can be illustrated by referring to the diagram. Boys *A* and *B* have lifted the object of weight *w* a vertical distance *h*. The object now can do work: if *A* lets go, the object may now be able to lift *B* thus doing work on *B*. If $w = 200$ lb and $h = 15$ ft, the object gained a potential energy of 3000 ft-lb.

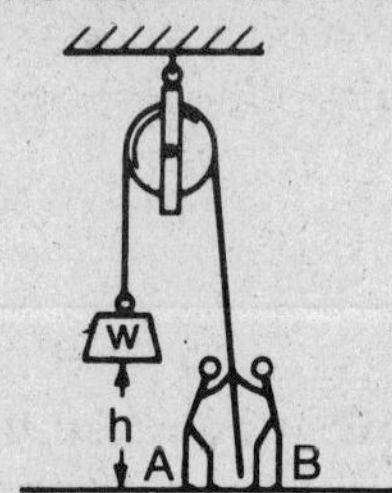

$$\textbf{potential energy} = \textbf{wh}$$

where *w* is the weight of the object and *h* is the vertical height through which it has been lifted. Remember, the potential energy is with respect to the original position. The object has a greater potential energy with respect to a level lower than the one shown in the diagram.

If a spring is compressed or stretched it also gains potential energy. This potential energy is equal to the work which was done in compressing or stretching the spring if frictional losses are neglected and is equal to the work which the spring can now do.

KINETIC ENERGY is the energy possessed by an object because of its motion. Winds (moving air) can do more work than stationary air, as in turning windmills, lifting roofs, etc. The kinetic energy of an object is proportional to the square of its speed. If a car's speed is changed from 30 mi/hr to 60 mi/hr, the car's kinetic energy has become four times as high. A 6000-lb truck moving at 30 mi/hr has twice as much kinetic energy as a 3000-lb car moving with the same speed. Kinetic energy is proportional to the mass of the object:

$$\textbf{kinetic energy} = \tfrac{1}{2}\textbf{mv}^2$$

PRINCIPLE OF CONSERVATION OF ENERGY Energy cannot be created or destroyed but may be changed from one form into another. Mass can be considered a form of energy as a consequence of Einstein's theory of relativity. When mass is converted to forms of energy such as heat, the

$$\textbf{energy produced} = \textbf{mc}^2$$

where *m* is the mass converted and *c* is the speed of light. When *m* is expressed in grams and *c* in centimeters per second (3×10^{10} cm/sec), the energy will be expressed in ergs. When *m* is expressed in kilograms and *c* in meters per second (3×10^8 m/sec), the energy is expressed in joules.

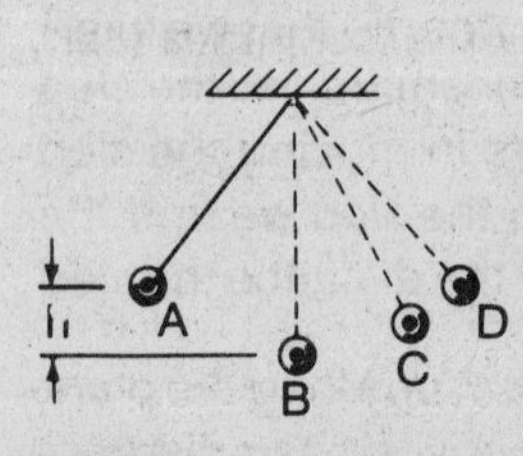

In a swinging pendulum kinetic energy is changed to potential energy and potential energy back to kinetic. If *A* is the highest position reached by the pendulum, the energy is all potential (=*wh*). At *B* it is all kinetic; at *C* the energy is partly potential and partly kinetic, and at *D*, the highest point reached on the other side, the energy is all potential again. If the pendulum has 10 ft-lb of potential energy at *A* (with respect to its lowest position), then the kinetic energy at *B* is 10 ft-lb, provided energy losses due to friction are negligible. The height reached at *D* is the same as it was at *A*. (We'll just note in passing that the motion of the bob is *not* motion with *constant* acceleration.) Energy losses due to friction result in the production of heat.

FRICTION AND WORK *Friction* is a force which always opposes motion or a tendency for motion. For example, in the figure, imagine a block resting on a horizontal surface. If a one-pound pull is applied toward the left and the object doesn't move, a one-pound force must be acting toward the right on the block. If the pull is increased to 2 lb and the block doesn't move, the force to the right must have increased to 2 lb. In each case the force to the right is friction, acting between the block and the surface on which it rests. Once sliding has been produced the force of friction is practically independent of the speed with which the block moves. Friction is also practically independent of the area of contact (unless it is pointlike). Friction does depend on the nature of the surfaces in contact and on the force pushing the surfaces together (the so-called *normal force*). The coefficient of friction is used to describe the surfaces:

Friction

$$\textbf{coefficient of sliding friction} = \frac{\textbf{friction during motion}}{\textbf{normal force}}$$

Example: A 30-Nt object is dragged along a horizontal surface for a distance of 10 m. If the coefficient of friction is 0.2, what is the minimum amount of work that must be done? In this case, the normal force is the weight of 30 Nt. Therefore, during motion, friction is 0.2×30 Nt $= 6$ Nt. Therefore, a pull of 6 Nt will maintain motion at constant speed and the work against friction is $6 \text{ Nt} \times 10 \text{ m} = 60$ *joules.* A larger force could be used (requiring more work); this would result in accelerated motion.

work against friction = friction × distance object moves

Friction during rolling motion is usually less than during sliding motion, if the normal force is the same. In reference tables, therefore, we may find coefficients of rolling friction as well as coefficients of sliding friction.

When an object moves through a fluid (gas or liquid), the force of friction acting on the object does depend on the speed with which it moves through the fluid. For example, as rain drops fall through the atmosphere the frictional force acting on the rain drop increases as its speed increases, until the friction equals the weight of the raindrop. Maximum velocity is then reached and is known as the *limiting* or *terminal velocity.*

INCLINED PLANE Sometimes an inclined plane is used to slide heavy objects up to a platform or other elevated area. This is represented by the diagram in which the height to which the object must be raised is indicated by h and the length of the plane by L. The weight of the object is w, but a smaller

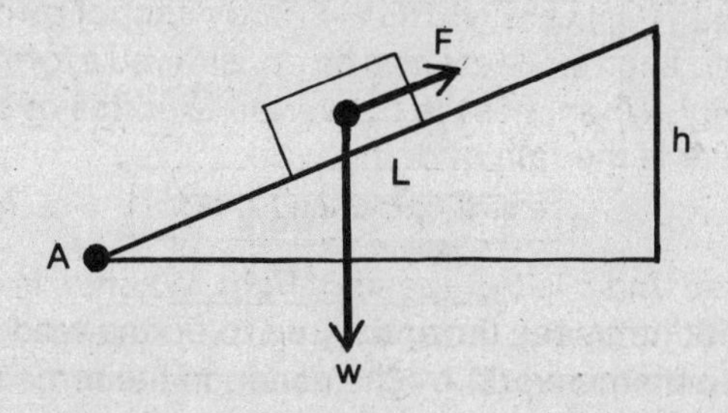

effort, force F, is required if we apply this effort parallel to the plane. It is shown a few pages later, under Simple Machines, that the following proportion is true when friction is negligible:

$$\frac{F}{w} = \frac{h}{L}$$

Since h is less than L, the length of the plane, the *force* F needed to push the object up along the plane at constant velocity is less than the weight of the object. However, we do not save work. The distance L along which the object must be pushed is proportionately larger than h, the vertical distance we want to raise the object.

Suppose the object weighs 1000 Nt. What is the minimum effort needed to slide the object up along the plane?

We could make a scale drawing and get a graphical solution. We could also substitute in the above proportion since we have all the information needed. A third method is useful if we know the angle of the plane with the horizontal, but some of the other information is missing. We notice that the ratio h/L is equal to the sine of angle A:

$$\frac{\text{force parallel to plane}}{\text{weight}} = \sin A$$

$$\frac{F}{w} = \sin A$$

$$F = w \sin A$$

In this example angle A is 30° and the weight w is 1000 Nt. Therefore

$$F = (1000 \text{ Nt})\ (0.5) = 500 \text{ Nt}$$

If we apply an effort of 500 newtons parallel to the plane we can get the 1000-newton object to the top of the plane.

Suppose we get the object near the top of the incline and let go. We know that the object is going to slide down the plane with increasing speed. Why? Gravity, the weight of the object, pulls vertically. We can think of a component of the weight acting parallel to the plane (and another one acting perpendicular to the plane), F_1.

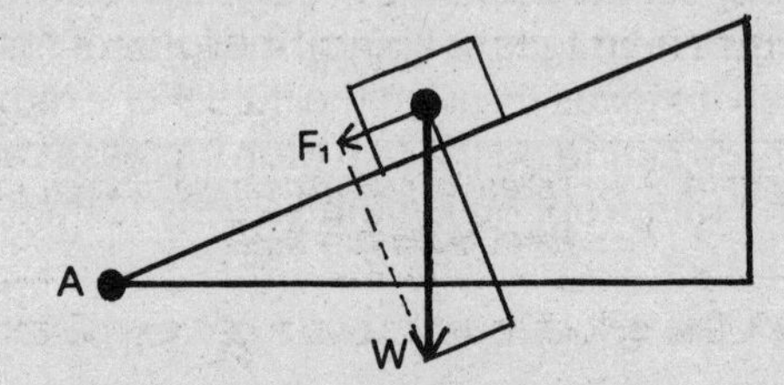

This component must be equal and opposite to F, which we just showed is equal to $w \sin A$. The reason for this is Newton's first law of motion. Then F was the force we needed to slide the object up with constant velocity. The object therefore was in equilibrium. Thus F was opposed by the equal and opposite component of the weight.

POWER AND WORK *Power* is the rate of doing work.

$$\textbf{power} = \frac{\textbf{work}}{\textbf{time}}$$

in which we divide the work done by the time required to do the work. Since work is calculated by multiplying the force used (to do the work) by the distance the force moves,

$$\textbf{power} = \frac{\textbf{force} \times \textbf{distance}}{\textbf{time}}$$

Units of power are: foot-pounds per second, horsepower, watt.

1 hp = 550 ft-lb/sec
1 hp = 746 watts
1 watt = 1 joule/sec

SIMPLE MACHINES*

Probably the most direct way of doing useful work on an object is to take hold of it and lift it. This is often difficult or inconvenient, and we turn to machines to help us. A *machine* is a device which will transfer a force from one point of application to another for some practical advantage. For example, we may want to lift a flag to the top of a pole. It is inconvenient to climb to the top of the pole each day and pull the flag up. Instead, we keep a pulley at the top of the pole and with the aid of a rope apply a downward force at *A* and produce an upward force on the flag at *B*. We have six *simple machines:* pulley, lever, wheel and axle, inclined plane, screw, and wedge. The force which we apply to the machine in order to do the work (such as at *A* in the above case) is known as the *effort*, F_E. The force which we have to overcome (such as the weight at *B*) is known as the *resistance*, F_R.

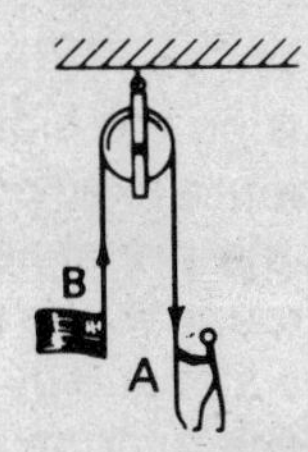

Mechanical Advantage

The *actual mechanical advantage* (AMA) of a machine is the ratio of the resistance to the effort. If we want to lift a weight of 400 lb and need to exert an effort of only 100 lb, the AMA is 4.

$$\textbf{actual mechanical advantage} = \frac{\textbf{resistance}}{\textbf{actual effort}}$$

$$\textbf{AMA} = \frac{F_R}{F_E}$$

If we are interested in lifting an object, the *useful work* done with the machine, or the *work output*, or work accomplished, is equal to the weight of the object multiplied by the distance the object is lifted. In general,

$$\textbf{work output} = \textbf{resistance} \times \textbf{distance resistance moves}$$

$$\textbf{work output} = F_R s_R$$

In order to accomplish this work the effort must move a certain distance, s_E. The *work input* is the work done by the effort.

$$\textbf{work input} = \textbf{effort} \times \textbf{distance effort moves}$$

$$\textbf{work input} = F_E s_E$$

Under ideal conditions there is no useless work, such as work done in overcoming friction. Then,

$$\textbf{work output} = \textbf{work input}$$

$$\frac{F_R}{F_E} = \frac{s_E}{s_R} = \textbf{IMA}$$

where IMA is ideal mechanical advantage.

Note that for a given machine the ratio of the two distances is not affected by anything we do to change friction in the machine; it give us the ideal mechanical advantage even if there is friction. The ratio of the two forces is affected by friction, since for a given situation the required effort is decreased by decreasing friction. The ratio of the two forces gives us the actual mechanical advantage, but if friction is negligible, the actual mechanical advantage equals the ideal mechanical advantage.

Efficiency

With an actual machine some useless work must be done, such as work to overcome friction. The input work equals work output plus useless work. For a machine

$$\textbf{efficiency} = \frac{\textbf{work output}}{\textbf{work input}}$$

*Not taught in some courses. See Introduction.

Sometimes these derived formulas are useful in problems with machines:

$$\textbf{efficiency} = \frac{\textbf{AMA}}{\textbf{IMA}}; \quad \textbf{efficiency} = \frac{\textbf{ideal effort}}{\textbf{actual effort}}$$

The Lever

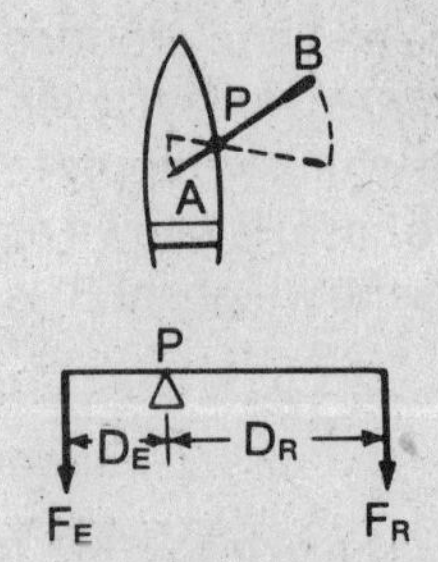

When we use crowbars, bottle openers, or oars, we are using the simple machine known as the lever. A *lever* is a rigid bar free to turn about a fixed point known as the *fulcrum* or pivot. The things we said about the principle of moments (Chapter 1) apply to the lever. The oar can rotate around *P*. We pull on the oar handle at *A* and as a result a push is exerted on the water at *B* by the oar blade. We represent this schematically in the diagram. The fulcrum *P* is represented by a triangle; the pull we apply is the effort, F_E; the push *by the water* on the oar is our resistance, F_R. The moment arms are D_E and D_R, respectively. They are also called *lever arms*. From the principle of moments, $F_E D_E = F_R D_R$. From this we see that $\frac{F_R}{F_E} = \frac{D_E}{D_R}$, which gives us the mechanical advantage of the lever. If the effort lever arm is greater than the resistance lever arm, the mechanical advantage is greater than one. In the case of the lever we often want a mechanical advantage of less than one. This, for example, is true in the above illustration of rowing a boat. The effort is greater than the resistance of the water, but as a result the handle at *A* moves a shorter distance than the blade does during each stroke, as indicated by the dotted lines in the diagram. This gives us a speed advantage when the ideal mechanical advantage is less than one. (The discussion of the boat is somewhat simplified, because in the actual boat the pivot moves with respect to the water.)

Notice, therefore, that the ideal mechanical advantage of a lever depends on where the fulcrum is located with respect to the points where the effort and resistance, respectively, are applied. The IMA may be less than one, greater than one, or equal to one. In the table below are shown three different arrangements of fulcrum (*P*), effort (F_E), and resistance (F_R). Look at the diagrams and, thinking in terms of the possible lengths of the lever arms, see if you can figure out the possible values of the IMA. Then compare your result with the table. It is not necessary to remember what class of lever each case represents.

LEVERS

Class	Diagram	Characteristic	IMA
1st	F_E ↓ — D_E — △ P — D_R — ↓ F_R	P between F_E & F_R	$MA \gtreqless 1$
2nd	P △ ← - - - D_E - - - → ↑ F_E; D_R ↓ F_R	F_R between P & F_E	$MA > 1$
3rd	P △ - - - - - D_R - - - - - ↓ F_R; D_E ↑ F_E	F_E between P & F_R	$MA < 1$

The Inclined Plane

When heavy objects have to be raised to a platform or put into a truck, it is often found to be convenient to slide these objects up along a board. An *inclined plane* is a flat surface one end of which is kept at a higher level than the other. F_E, the effort to pull or push the object up, is usually applied parallel to the plane. The input work, or work done by the effort, equals F_E times the length of the plane, L. The work output equals the weight of the object w times the height of the plane, h. Under ideal conditions the input work equals the output work, and, therefore,

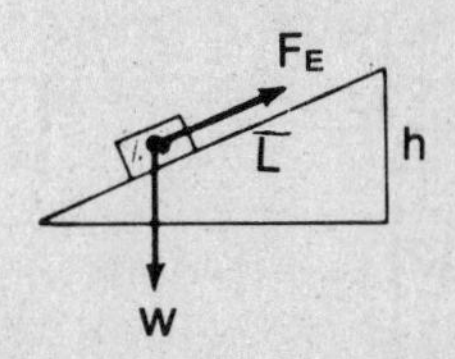

$$\frac{\textbf{weight of object}}{\textbf{ideal effort}} = \frac{\textbf{length of plane}}{\textbf{height of plane}} = \textbf{IMA}$$

Under actual conditions friction opposes the motion. When the object is *moved up* the plane, the actual effort equals the ideal effort plus the force of friction. On the other hand, less force is required to keep the object from sliding down, and, when the object is allowed to *slide down* with constant speed, the effort applied up along the plane equals the ideal effort *minus* the force of friction. (Occasionally the effort might be applied parallel to the base of the plane. Then the IMA equals the base of the plane divided by the height of the plane.)

Example: An inclined plane 13 ft long and 5 ft high is used to lift a 390-lb object to a platform. The ideal MA is 13/5 or 2.6; the ideal effort parallel to the plane is 390 lb $\times$ 5/13 or 150 lb. Actually 200 lb may be required; then the force of friction is 50 lb. The AMA = weight/effort, or 390 lb/200 lb. Therefore the AMA = 1.95. The useful work $= w \times h =$ 390 lb $\times$ 5 ft = 1950 ft lb. The input work $= F_E \times L =$ 200 lb $\times$ 13 ft = 2600 ft lb. The efficiency equals

$$\frac{\textbf{output work}}{\textbf{input work}} = \frac{\textbf{1950 ft-lb}}{\textbf{2600 ft-lb}} = \textbf{0.75 or 75\%}$$

Note that efficiency also equals

$$\frac{\textbf{ideal effort}}{\textbf{actual effort}} = \frac{\textbf{150 lb}}{\textbf{200 lb}} = \textbf{0.75}$$

The term *grade* is sometimes used in reference to an inclined plane, especially a road; it is the ratio of the height of the incline to its length, and is often expressed as a percent.

The Pulley

Especially when heavy objects have to be lifted through a considerable distance, the pulley is a convenient simple machine for the job. The *pulley* is a wheel so mounted in a frame that it may turn readily around the axis through the center of the wheel. If the frame and wheel move through space as the pulley is used, we have a *movable pulley.* Otherwise we say we have a fixed pulley. The wheel is usually grooved to guide the string or rope which is used with it. The term *block and tackle* is used for the commercial assemblage of pulleys and rope.

The ideal mechanical advantage of pulleys can be figured out by determining the distance the effort has to move for every foot the resistance is moved. The IMA $= s_E/s_R$. For many common pulley arrangements this ideal mechanical advantage can be determined visually by counting the number of rope segments supporting the movable pulley(s). Some typical pulley arrangements are shown on the next page. You should be able to reproduce the diagrams quickly from memory and to determine their IMA.

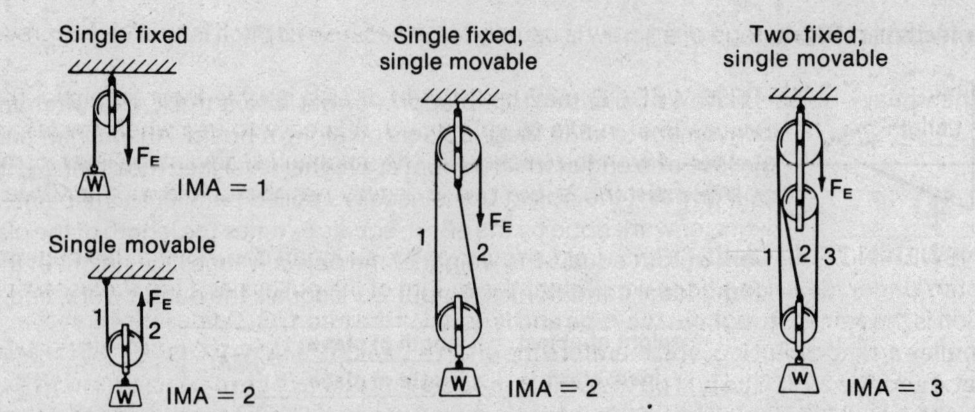

Note that in these diagrams the rope segments are numbered to show how the ideal mechanical advantage can be determined visually from the diagram; the effort is applied to one end of the string while the other end of the string is attached to the weight to be lifted, to the pulley frame, or to a rigid support such as the ceiling. The weight to be lifted (w) represents the resistance if we neglect the weight of the movable pulleys and friction.

Example: A system of pulleys having an IMA of three is used to lift an object weighing 600 lb. The effort required is 300 lb. What is the efficiency? The ideal effort = (F_R/IMA) = (weight/IMA) = (600 lb/3) = 200 lb. The efficiency = (ideal effort/actual effort) = (200 lb)/(300 lb) = 2/3 or 66.7%.

Other Simple Machines

THE WHEEL AND AXLE consists of a wheel or crank rigidly attached to an axle which turns with it. Applications of this are found in the steering wheel of an automobile (the shaft is the axle), the screwdriver, and a doorknob. In the diagram a wheel and axle is used to lift a weight, w. This represents the resistance and is attached to the rim of the axle. The effort, F_E, is applied to a rope connected to the rim of the wheel. The ideal mechanical advantage can again be figured out by thinking in terms of the relative distances moved. During one revolution around the axis represented by P, the effort moves a distance equal to the circumference of the wheel while the resistance moves a distance equal to the circumference of the axle. Therefore,

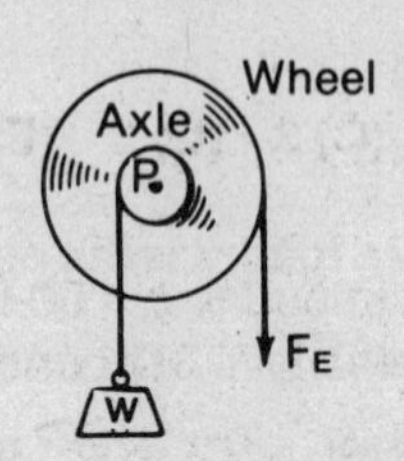

$$\text{IMA} = \frac{\text{circumference of wheel}}{\text{circumference of axle}}$$

Also, since the circumference of a circle is proportional to its radius and to its diameter,

$$\text{IMA} = \frac{\text{diameter of wheel}}{\text{diameter of axle}}; \qquad \text{IMA} = \frac{\text{radius of wheel}}{\text{radius of axle}}$$

THE SCREW The diagram shows a jackscrew, which illustrates the principle of the screw. The effort F_E may be applied at the end of a rod of length L. As the effort is applied through a circumference, $2\pi L$, the screw advances the distance between two adjacent threads, and moves the weight through the same distance. The distance is known as the *pitch* of the screw. The weight represents the resistance. Since IMA equals the distance effort moves divided by distance resistance moves

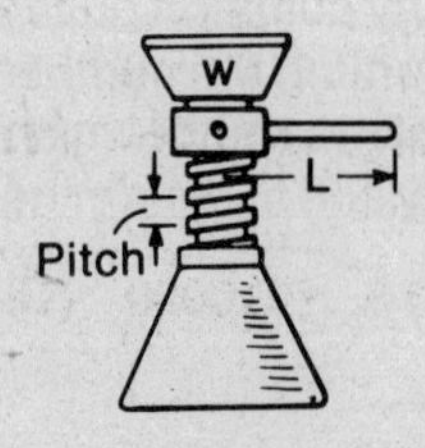

$$\text{IMA} = \frac{2\pi L}{\text{pitch}}$$

The mechanical advantage of a screw is usually large because its pitch is small compared with $2\pi L$.

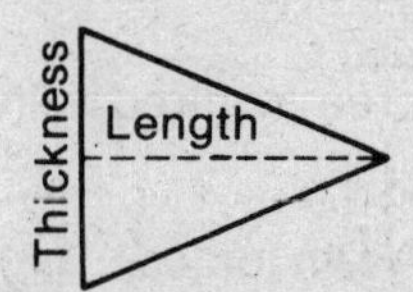

THE WEDGE may be thought of as a double inclined plane. It is used in devices like an axe to split wood. It is easy to use when the length is large compared with the thickness. The mechanical advantage, therefore, depends on these distances, but one is usually not concerned with its value.

ANSWERING THE QUESTIONS Let us return to the situation described at the beginning of this chapter. Under ideal conditions we neglect the weight of the pulleys and frictional losses. Then the tension is the same throughout the rope and is equal to the effort, F_E. As described above, the IMA of this pulley arrangement is 3, and therefore the effort = (weight/IMA) = (60 lb/3) = 20 lb. Therefore the pull at *A* and *B* is 20 lb. The pull on the ceiling is equal to the sum of the downward forces exerted on it = weight + F_E = (60 lb + 20 lb) = 80 lb. (Another way to look at this is to recognize that there are four rope segments pulling down on the fixed pulleys. The pull of each of these segments is 20 lb, and therefore the total downward pull on the fixed pulleys is 80 lb. This has to be balanced by an upward pull of 80 lb exerted *by* the ceiling. Therefore the downward pull on the ceiling is 80 lb.) The distance through which the effort moves, $s_E = s_R \times \text{IMA} = 4\text{ ft} \times 3 = 12\text{ ft}$. Therefore point *B* will move 12 ft. (Time has nothing to do with it.)

Of course, Atlas did no work on the earth as long as he did not move it. (Work = force × distance, and the force is exerted through zero distance.) A giant like Atlas would use a great deal of energy, but none of it would go towards giving the earth more potential or kinetic energy.

QUESTIONS Chapter 3

Select the choice which will fit the question best.

1. If the speed of an object is tripled, its kinetic energy is **(A)** 1/9 **(B)** 1/3 **(C)** 3 **(D)** 6 **(E)** 9 times that its original speed.

For 2-4. A person having a mass of 60 kg exerts a horizontal force of 200 Nt in pushing a 90-kg object a distance of 6 meters along a horizontal floor. He does this at constant velocity in 3 seconds.

2. The weight of this person is approximately, in newtons, **(A)** 40 **(B)** 90 **(C)** 200 **(D)** 400 **(E)** 600

3. The work done by this person is, in joules, **(A)** 540 **(B)** 1080 **(C)** 1200 **(D)** 3600 **(E)** 5400

4. The force of friction is **(A)** exactly 60 Nt **(B)** between 60 and 90 Nt **(C)** exactly 90 Nt **(D)** exactly 200 Nt **(E)** greater than 200 Nt

For 5-7. The 1000-lb weight of a pile driver falls freely from rest; it drops 25 ft and strikes a steel beam and drives it 3 inches into the ground.

5. The kinetic energy of the weight just before it hits the beam is, in ft-lb, **(A)** 3000 **(B)** 25,000 **(C)** 75,000 **(D)** 200,000 **(E)** 800,000

6. The speed of the weight just before hitting the beam in ft/sec is approximately **(A)** 0.25 **(B)** 3 **(C)** 25 **(D)** 40 **(E)** 64

7. The momentum of the weight just before hitting the beam is, in lb-ft/sec, **(A)** 0 **(B)** 3000 **(C)** 25,000 **(D)** 40,000 **(E)** 1,600,000

For 8-10. An inclined plane 5 m long has one end on the ground and the other end on a platform 3 m high. A man weighing 650 Nt wishes to push a 900-Nt object up this plane. The force of friction is 100 Nt.

8. The minimum force he must exert is, in Nt, approx. **(A)** 100 **(B)** 540 **(C)** 640 **(D)** 900 **(E)** 1000

9. In order to hold the object on the plane without letting it slide, the minimum force required is, in Nt, approx. **(A)** 0 **(B)** 100 **(C)** 440 **(D)** 540 **(E)** 640

10. The potential energy gained by the object when it is at the top of the plane is, in joules, **(A)** 100 **(B)** 324 **(C)** 2700 **(D)** 5400 **(E)** 9000

11. A uniform plank 10 ft long weighs 50 lb. The force needed to lift one end of the plank is, in lb, **(A)** 10 **(B)** 12.5 **(C)** 25 **(D)** 50 **(E)** 100

For 12-15. The pulley arrangement shown is attached to a ceiling. A weight of 240 lb is to be lifted. The weight of the pulleys is negligible.

12. If frictional losses are negligible, the force F required to lift the weight at constant speed is **(A)** 48 lb **(B)** 60 lb **(C)** 80 lb **(D)** 120 lb **(E)** 240 lb

13. For the conditions in problem 12, the pull on the ceiling will be **(A)** 60 lb **(B)** 120 lb **(C)** 240 lb **(D)** 288 lb **(E)** 300 lb

14. For the conditions in problem 12 the tension in the rope at X will be **(A)** 48 lb **(B)** 60 lb **(C)** 80 lb **(D)** 120 lb **(E)** 240 lb

15. If the actual force F is 180 lb, the efficiency of the system is **(A)** 25 **(B)** 35 **(C)** 50 **(D)** 75 **(E)** 90

16. The power expended by the man in questions 8-10 in pushing the object to the top of the plane in 3 seconds is **(A)** 400 watts **(B)** 800 watts **(C)** 1070 watts **(D)** 4500 watts **(E)** 4500 joules

EXPLANATIONS TO QUESTIONS Chapter 3

Answers

1. **(E)**	4. **(D)**	7. **(D)**	10. **(C)**	13. **(E)**
2. **(E)**	5. **(B)**	8. **(C)**	11. **(C)**	14. **(B)**
3. **(C)**	6. **(D)**	9. **(C)**	12. **(B)**	15. **(B)**
				16. **(C)**

Explanations

1. **(E)** Kinetic energy $= \frac{1}{2} mv^2$. For a given object the kinetic energy is proportional to the square of its speed. When the speed is multiplied by 3, the kinetic energy is multiplied by 3^2, or 9.

2. **(E)** The weight of an object on earth is the gravitational pull of the earth on it and is equal to the product of the object's mass and the value of the acceleration of a freely falling object:

$$\begin{aligned} \text{weight} &= mg \\ &= (60\text{ kg})\ (9.8\text{ m/sec}^2) \\ &= 600\text{ Nt} \end{aligned}$$

3. **(C)** Work done on an object is equal to the product of the force exerted on the object and the distance moved in the direction of the force:

$$\begin{aligned} \text{work} &= \text{force} \times \text{distance} \\ &= (200\text{ Nt})\ (6\text{ m}) \\ &= 1200\text{ joules} \end{aligned}$$

4. **(D)** Since the force exerted is horizontal, none of it is used to overcome gravity. Also, since the velocity remains constant, none of the force (and work done) is used to give the object kinetic energy. Therefore all of the 200-Nt force used is required to overcome friction.

5. **(B)** The kinetic energy which the weight has just before it hits is equal to the potential energy it had at the top just before falling. The potential energy at the top = weight × height;

$$\text{pot. en.} = 1000 \text{ lb} \times 25 \text{ ft} = 25{,}000 \text{ ft-lb}.$$

6. **(D)** To calculate the speed of an object falling freely from rest if we know the distance of fall:

$$v^2 = 2gs = 2 \times 32 \text{ ft/sec}^2 \times 25 \text{ ft} = 64 \times 25 \text{ ft}^2/\text{sec}^2$$
$$v = (8 \times 5) \text{ ft/sec} = 40 \text{ ft/sec}$$

7. **(D)** momentum = mass × velocity = 1000 lb × 40 ft/sec = 40,000 lb-ft/sec.

8. **(C)** The man has to overcome two forces: the force of friction and the component of the weight parallel to the plane.

$$\frac{\text{parallel component}}{\text{weight}} = \frac{\text{height of plane}}{\text{length of plane}}$$
$$\frac{\text{parallel component}}{900 \text{ Nt}} = \frac{3 \text{ m}}{5 \text{ m}}$$

and the parallel component = 540 Nt. Force required = 100 Nt + 540 Nt = 640 Nt.

9. **(C)** Friction will help in keeping the object from sliding down. The required force, therefore, is equal to the parallel component minus the force of friction = 540 Nt − 100 Nt = 440 Nt.

10. **(C)** Pot. energy = weight × height = 900 Nt × 3 m = 2700 joules.

F
10 ft
5 ft
50 lb

11. **(C)** The fulcrum is at one end, at the left in the diagram. The center of gravity of a uniform plank is in the middle; the weight of 50 lb acts downward 5 ft from either end. Counterclockwise moment = clockwise moment; $10F = 50$ lb × 5 ft; $F = 25$ lb.

12. **(B)** There are 4 strands supporting the movable pulleys; therefore the ideal mechanical advantage is 4.

$$\text{Then IMA} = \frac{\text{weight of object}}{\text{ideal effort}} \text{ and ideal effort} = \frac{\text{weight of object}}{\text{IMA}}$$
$$= 240 \text{ lb}/4 = 60 \text{ lb}$$

13. **(E)** If the weight of the pulleys is negligible, there are two basic forces pulling on the ceiling: force F and the 240-lb object. Pull on ceiling = 60 lb + 240 lb = 300 lb.

14. **(B)** Each of the 4 strands supporting the movable pulleys supplies ¼ of the upward pull on the weight or ¼ of 240 lb. Another way to look at it is to think of the force F as transmitted throughout the rope and therefore being equal to the tension.

15. **(B)** $\text{Efficiency} = \dfrac{\text{ideal effort}}{\text{actual effort}} = \dfrac{60 \text{ lb}}{180 \text{ lb}} = 33\%$

16. **(C)** Power is the rate of doing work: Power = work/time. The work done by the man is equal to the product of the force required and the distance the object is moved in the direction of the force (which in this case is parallel to the plane).

$$\begin{aligned} \text{work} &= \text{force} \times \text{distance} \\ &= (640 \text{ Nt})\ (5 \text{ m}) \\ &= 3200 \text{ joules} \end{aligned}$$
$$\text{power} = \frac{\text{work}}{\text{time}} = \frac{3200 \text{ joules}}{3 \text{ seconds}} = 1070 \text{ watts}$$

(Note that the work required is greater than the potential energy gained because work had to be done against friction as well as to raise the object against gravity.)

Fluid Pressure and The Atmosphere

A piece of iron is put into a beaker partly filled with mercury. To what depth will the iron sink? Oil is poured on top of the mercury. How does this affect the pressure on the iron? How does it change the depth of the iron in the mercury? To answer these questions we need to consider the behavior of fluids.

The term *fluid* refers to both gases and liquids. A *solid* has definite shape and volume. A *liquid* has definite volume but it takes the shape of its container; its top surface tends to be horizontal; (also see *meniscus* in Chapter 5). A *gas* has neither definite shape nor volume and expands to fill any container into which it is put. A cubic foot of concrete weighs more than a cubic foot of water. We say the concrete is denser than the water. The *density* of a substance is the mass of a unit volume.

$$\text{density} = \frac{\text{mass}}{\text{volume}}$$

A solid tends to sink in a liquid of lesser density and to float in one of greater density. The units for density are a unit of mass divided by a unit of volume, such as: lb/ft^3, gm/cm^3, kg/m^3. Instead of speaking about the density of substances it has been found more convenient to speak about their relative density, or density compared to a standard. This is their *specific gravity* (sp. gr.). The standard for solids and liquids is water. The density of water is usually taken as 1 gm/cm^3 or 62.5 lb/ft^3. For solids and liquids

$$\text{sp. gr.} = \frac{\text{density of substance}}{\text{density of water}}$$

$$\text{sp. gr.} = \frac{\text{weight of substance}}{\text{weight of equal volume of water}}$$

$$\text{sp. gr.} = \frac{\text{mass of substance}}{\text{mass of equal volume of water}}$$

The specific gravity of mercury is 13.6; this means that 1 cubic centimeter of mercury weighs 13.6 gm and that 1 cubic foot of mercury weighs (13.6 × 62.5) lb.

LIQUID PRESSURE

Fluids push against the container in which they are placed. *Pressure* is the force per unit area.

$$p = \frac{F}{A}$$

A liquid inside a beaker exerts a force on the bottom of the beaker because of its weight. Therefore, the liquid pressure in this case is equal to the weight of the liquid divided by the area of the bottom of the beaker. A little algebra leads to another relationship: the pressure due to a liquid is equal to the height of the liquid (*h*) times the density of the liquid (*d*). This is true not only on the bottom of the container but also on the sides at any depth within the liquid.

$$p = hd$$

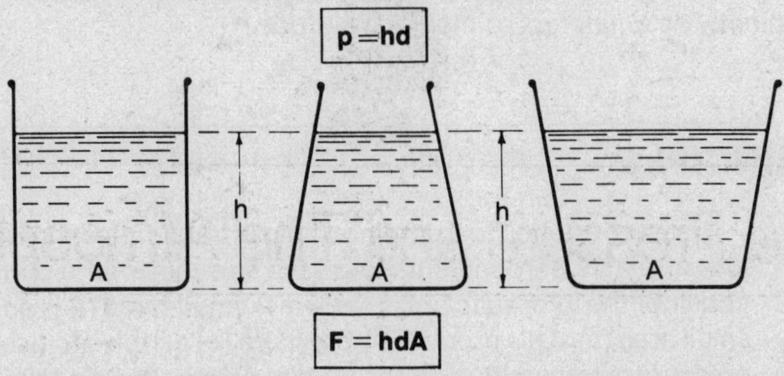

$$F = hdA$$

By combining the above two expressions for liquid pressure we find that on a horizontal area in a liquid the force due to the liquid (*F*) is equal to the height of the liquid times its density times the area on which the liquid pushes. Notice that if the three vessels shown are filled with the same liquid to the same depth, the liquid pressure on the bottom will be the same; if the bottom area is the same, the force on the bottom will be the same although the weights of liquid in the three vessels will be different. Liquid pressure is independent of the size or shape of the container; it depends only on the depth and density of the liquid. Example: A tank is filled with water to a depth of 6 ft. The bottom of the tank has an area of 100 sq. ft. What is the water pressure on the bottom of the tank?

$$p = hd = 6 \text{ ft} \times 62.5 \text{ lb/ft}^3 = 375 \text{ lb/ft}^2$$

What is the force exerted by the water on the bottom of the tank?

$$F = pA = 375 \text{ lb/ft}^2 \times 100 \text{ ft}^2 = 37{,}500 \text{ lb} = 3.75 \times 10^4 \text{ lb}$$

The MKS unit for pressure is the *pascal* (Pa), which is equal to 1.00 Nt/m^2 (exactly).

THE ATMOSPHERE AND GASEOUS PRESSURE

Gases have weight but their density is less than that of liquids and solids. Gases exert pressure. The atmosphere is a mixture of gases. At sea level the pressure due to the atmosphere is about 14.7 lb/in^2. The air therefore exerts a force of about 100 lb on the palm of our hand.

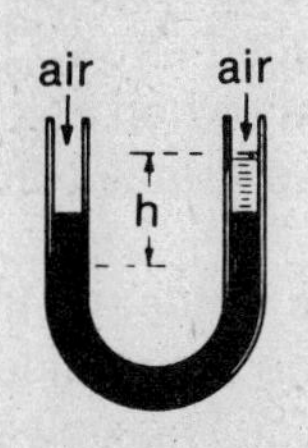

Imagine a U-tube partly filled with mercury. The tube is open at the top on both sides. The mercury is at the same level in both sides since a given liquid will reach the same level in connected containers. The air exerts the same pressure on the surface of the mercury in each arm. However, if air is pumped out of the right arm the greater pressure in the left arm pushes mercury around into the right arm until the pressure due to the excess mercury (= *hd)* is equal to the difference in the air pressure in the two arms. If all the air were pumped out of the arm on the right, the difference in level of the mercury in the two arms *(h)* would become about 30 inches or 76 cm. This is the principle of the mercury barometer. A mercury barometer is made by filling with mercury a glass tube which is about one meter long. The air must be removed from the

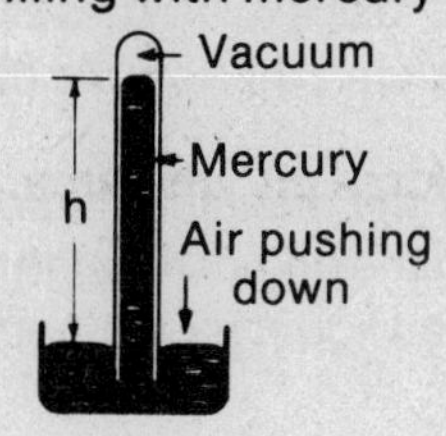

mercury. The tube is then inverted so that the open end will be below the level of some more mercury in a reservoir. Some mercury will flow out of the tube until the height of the mercury column is about 76 cm—a height whose pressure is equal to the pressure of the atmosphere which supports it. The atmospheric pressure is not constant. Its variation is used to aid in weather forecasting. A rising pressure usually indicates the approach of fair weather.

Standard pressure is the pressure of a column of mercury 760 mm high. This is enough to support a column of water 34 ft high. The atmospheric pressure decreases with altitude. Near the surface of the earth the barometer reading drops about 0.1 inch for every rise in altitude of 90 ft. *Aneroid* barometers do not use any liquid. They are sealed, partially evacuated cans. The variation in atmospheric pressure causes the deflection of a needle across a calibrated scale. Such barometers can be used conveniently as *altimeters* to measure altitude.

PASCAL'S PRINCIPLE

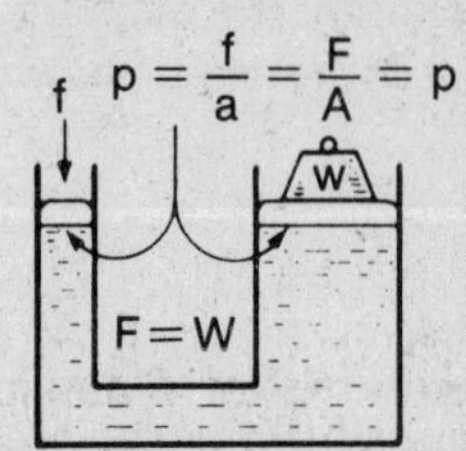

Pressure applied to a confined fluid is transmitted through the liquid without loss and acts perpendicularly on the surfaces of the container. For example, consider the hydraulic press shown. A small force (*f*) is applied to the piston of small area (*a*). This produces a pressure (*p*) on the enclosed fluid and this same pressure is transmitted by the fluid to act on the underside of the large piston, whose area is *A*. The resulting force ($F = pA$) is larger than *f* and is able to support the large weight *W*.

The hydraulic press can be thought of as a machine with *f* representing the effort and *F* the resistance (equal to the weight). The mechanical advantage is the ratio of resistance to effort. Therefore,

$$\mathbf{IMA} = \frac{F}{f} = \frac{A}{a} = \frac{(\text{diameter of large piston})^2}{(\text{diameter of small piston})^2}$$

Example: The areas of the pistons of a hydraulic press are 50 in^2 and 2 in^2, respectively. If a force of 3 lb is applied to the small piston, how large a weight can be supported by the large piston?

$$\frac{F}{f} = \frac{A}{a} \quad \text{or} \quad \frac{F}{3\text{ lb}} = \frac{50\text{ in}^2}{2\text{ in}^2} \quad \text{or} \quad F = 75\text{ lb} = W$$

ARCHIMEDES' PRINCIPLE

The apparent loss in weight of an object immersed in a fluid equals the weight of the displaced fluid. When an object is placed in a fluid, some of the fluid is pushed out of the way—that is, some of the fluid is displaced. This is so because no two objects can occupy the same space at the same time (impenetrability of matter). In the case of an object that is fully *submerged,* like object *A,* the *volume* of fluid displaced equals the volume of the object. A stone held in water appears lighter than when held in air. The apparent loss in *weight* equals the weight of the displaced fluid (Archimedes).

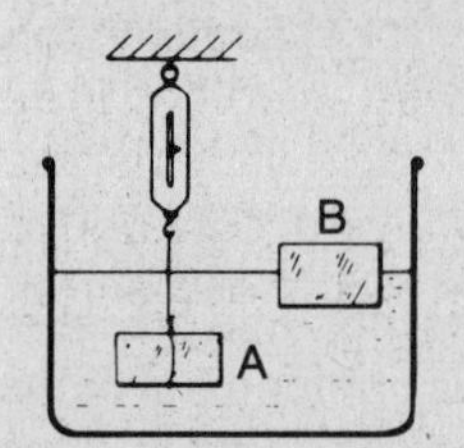

Assume that in the above diagram object *A* is submerged in water and that object *B* floats in the water. The pull that object *A* exerts on the spring is less than if *A* were in air. We can think of the water as exerting an upward push on the object. This upward push of a fluid on an object immersed in it is the buoyant force of the fluid or its *buoyancy.* The apparent loss in weight equals the buoyancy.

Example: Object *A* weighs 50 gm in air and 30 gm in water. Since the apparent loss in weight is 20 gm, the weight of the displaced water is 20 gm. Since the density of water is 1 gm/cm^3, the volume of water displaced is 20 cm^3. Therefore the volume of object *A* is 20 cm^3.

If an object, like *B, floats* in water, it appears to weigh nothing. The *weight* of the displaced fluid equals the weight of the object in vacuum. (If the fluid is a liquid, instead of using the object's weight in vacuum we use the object's weight in air.) Notice that the *volume* of the displaced liquid is less than the volume of the object.

Archimedes' principle can be applied to finding the specific gravity of solids and liquids. The specific gravity of solids and liquids equals their weight divided by the weight of an equal volume of water. A solid submerged in water displaces a volume of water equal to the volume of the solid; the weight of this equal volume of water is equal to the apparent loss in weight of the solid. Therefore, for a *solid that sinks in water*

$$\textbf{sp. gr.} = \frac{\textbf{weight in air}}{\textbf{apparent loss of weight in water}}$$

The specific gravity of liquids can be found by using a solid that sinks in the unknown liquid as well as in water. The solid must be insoluble in both. The object displaces the same volume in water as in the unknown liquid. The solid's apparent loss of weight in water equals the weight of the displaced water; the solid's apparent loss of weight in the unknown liquid equals the weight of the same volume of the liquid. Therefore, for a *liquid,*

$$\textbf{sp. gr.} = \frac{\textbf{apparent loss in weight of solid in liquid}}{\textbf{apparent loss in weight of solid in water}}$$

The specific gravity of liquids can also be found by using a *hydrometer.* It floats higher in denser liquids and has a scale on it that gives specific gravity directly. Another instrument for measuring specific gravity of liquids is the pycnometer or bottle of known volume. (A *hygrometer* is used for measuring relative humidity of the atmosphere.)

The apparent loss in weight due to the buoyant effect of gases can usually be neglected. However, the rising of balloons depends on this. The lifting force of the gas in a balloon is equal to the weight of the air displaced by the balloon minus the weight of the gas in the gas bags. For example, if 20 ft^3 of helium in a balloon weigh 4 ounces, the balloon will be pushed up by force equal to the weight of the 20 ft^3 of displaced air, about 28 ounces. The lifting force of the helium is (28 — 4) ounces. This 24-ounce force can be used to counteract the weight of the gas bags and anything else attached to the balloon.

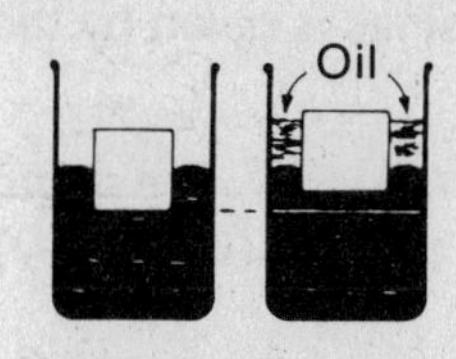

Now let us consider the question asked at the beginning of the chapter. The iron (sp. gr. about 7.6) will float in the mercury (sp. gr. about 13.6). Therefore the iron will sink until the weight of the mercury it displaces equals its own weight. (Also, the volume of the displaced mercury equals the volume of the iron which is below the level of the mercury. If the iron has the shape of a rectangular solid, such as a cube, this volume equals the area of the base times the depth.)

When oil is poured on top of the mercury, the iron rises somewhat out of the mercury as shown in the diagram on the right. The reason for this is that the oil (sp. gr. usually less than one) floats on top of the mercury and pushes down on it. According to Pascal's Principle this push is transmitted in all directions and therefore results in an upward push on the underside of the iron. Note that this is true even when enough oil is poured in to cover the iron. The iron will rise until the buoyant force due to the displaced oil plus the buoyant force due to the displaced mercury equal the weight of the iron.

BERNOULLI'S PRINCIPLE

If the speed of a fluid is increased, its pressure is decreased. This principle is made use of in the design of airplane wings to give the plane lift. The wing is designed so that the air will move faster over the top of the wing than across the bottom. As a result, the air pressure on top of the wing is less than on the bottom, with the consequence that the upward push of the air on the wing is greater than the downward push. (In applying the principle, it usually helps to think of *sidewise* pressure as decreased when the speed of the fluid is increased.)

QUESTIONS Chapter 4

Select the choice which fits the question best.

For 1-2. A rectangular piece of metal, whose dimensions are 5 cm × 20 cm × 30 cm, is submerged in a liquid. The sp. gr. of the metal is 7.0. The sp. gr. of the liquid is 2.0.

1. The mass of the block is **(A)** 3000 gm **(B)** 3000 lb **(C)** 21,000 gm **(D)** 21,000 lb **(E)** 210,000 lb

2. The apparent loss in weight of the block in the liquid is **(A)** 3000 gm **(B)** 6000 gm **(C)** 15,000 gm **(D)** 21,000 gm **(E)** 21,000 lb

3. Two insoluble objects seem to lose the same weight when completely submerged in a liquid. The objects must have the same **(A)** weight in air **(B)** weight in the liquid **(C)** weight in water **(D)** density **(E)** volume

4. A block of wood having a volume of 90 cm^3 and a sp. gr. of 0.80 is held under water. The force one must exert to keep it from rising to the surface is, in grams, **(A)** zero **(B)** 18 **(C)** 72 **(D)** 80 **(E)** 90

5. If an object weighs 90 gm in air, 60 gm in water, and 20 gm in an unknown liquid in which it does not dissolve, the sp. gr. of the unknown liquid is **(A)** 1.3 **(B)** 1.5 **(C)** 2.3 **(D)** 3.0 **(E)** 4.5

For 6-7. A cube, 6 inches on edge, has a specific gravity of 4.

6. Its weight in air, in pounds, is approximately **(A)** 2 **(B)** 4 **(C)** 30 **(D)** 240 **(E)** 900

7. Its apparent weight in water, in pounds, is approximately, **(A)** 8 **(B)** 23 **(C)** 30 **(D)** 38 **(E)** 58

8. A cube of iron whose sides are of length h is put into mercury. The weight of the iron cube is W. The density of iron is d_1; that of mercury is d_m. The depth to which the cube sinks is given by the expression **(A)** $W/(h^2 d_m)$ **(B)** $W/(h^2 d_1)$ **(C)** $Wh^2 d_m$ **(D)** $Wh^2 d_1$ **(E)** W/h^2

9. The density of sea water is 64 lb/ft^3. The water pressure on a submarine is 5120 lb/in^2. The depth of the submarine below the surface of the water, in ft, is approximately **(A)** 20 **(B)** 40 **(C)** 60 **(D)** 80 **(E)** 11,000

10. The force exerted by the fluid in a hydraulic press on the piston is 2000 lb. If the area of the piston is 10 in.^{2} the fluid pressure on the piston is, in lb/in^2 **(A)** 20 **(B)** 200 **(C)** 2000 **(D)** 20,000 **(E)** 200,000

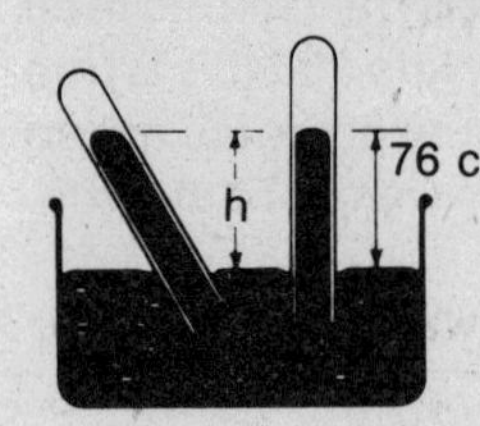

11. The column of mercury in a mercury barometer is 76 cm high on a certain day when the barometer is vertical. If the barometer is tilted a few degrees, the vertical height of the mercury column, h, is **(A)** still 76 cm **(B)** 76 cm minus the angle **(C)** 76 cm minus a function of the angle **(D)** 76 cm plus the angle **(E)** 76 cm plus a function of the angle

EXPLANATIONS TO QUESTIONS Chapter 4

Answers

1. (C)	4. (B)	7. (B)	10. (B)
2. (B)	5. (C)	8. (A)	11. (A)
3. (E)	6. (C)	9. (E)	

Explanations

1. **(C)** A simple exercise! Since density = mass/volume, the mass = density × volume. The volume of the block = length × width × height = (5 × 20 × 30) cm^3 = 3000 cm^3. Since the sp. gr. of the metal is 7.0, its density is 7.0 gm/cm^3. Therefore, mass = 7.0 gm/cm^3 × 3000 cm^3 = 21,000 gm.

2. **(B)** Apparent loss in weight = wt. of displaced fluid. Volume of displaced fluid = vol. of submerged block = 3000 cm^3.
Wt. of 3000 cm^3 of water = 3000 gm
Wt. of displaced fluid = sp. gr. × wt of equal vol. of water = 2.0 × 3000 gm = 6000 gm

3. **(E)** Archimedes' principle: The apparent loss in weight is equal to the weight of the displaced fluid. Both displace equal weights of the same fluid; therefore, both displace equal volumes of the same fluid. But the volume each displaces is equal to the volume of the object. Therefore, both objects must have the same volume.

4. **(B)** The buoyant force on the object = wt. of displaced water. Since the block is fully submerged, it displaces 90 cm^3 of water, which weigh 90 gm. The weight of the block acts downward on the block. The weight of the block = sp. gr. × weight of equal vol. of water = 0.8 × 90 gm = 72 g. An additional 18 gm of force downward is required to equal the (upward) buoyant force of the water.

5. **(C)** sp. gr. of liquid

$$= \frac{\text{apparent loss of weight in liquid}}{\text{apparent loss of weight in water}} = \frac{90\text{ gm} - 20\text{ gm}}{90\text{ gm} - 60\text{ gm}}$$

$$= \frac{70\text{ gm}}{30\text{ gm}} = 2.3$$

6. **(C)** Vol. of cube = $(\text{length})^3 = (\tfrac{1}{2}\text{ ft})^3 = \tfrac{1}{8}\text{ ft}^3$. Mass = density × vol. = (sp. gr. × density of water) × vol. = (4 × 62.5 lb/ft^3) × ⅛ ft^3 = 31.2 lb. Here, the weight is expressed in the same unit as the mass. The closest choice is 30 lb.

7. **(B)** Since the sp. gr. is greater than 1, the object sinks in water and displaces a volume of water = vol. of obj. = ⅛ ft^3. Weight of displaced water = vol. × density = ⅛ ft^3 × 62.5 lb/ft^3 = 7.8 lb. Apparent loss in weight = wt. of displaced water = 7.8 lb. Its apparent weight is (31.2 − 7.8) lb = 23.4 lb.

8. **(A)** A little difficult. Let y be the depth. The volume of the iron below the level of the mercury = area of base × depth = h^2y. This also equals the vol. of the displaced mercury. Wt. of displaced mercury (in mass units) = vol. of mercury × density of mercury = h^2yd_m. Therefore, $y = W/h^2d_m$

9. **(E)** pressure = depth × density;

$$\text{depth} = \frac{p}{\text{density}} = \frac{(5120 \times 144)\text{ lb/ft}^2}{64\text{ lb/ft}^3} = (80 \times 144)\text{ ft}$$

This is equal to 11,520 ft; but it isn't necessary to multiply out since the other answers are obviously much too small.

10. **(B)** $p = F/A = (2000\text{ lb})/(10\text{ in.}^2) = 200\text{ lb/in.}^2$

11. **(A)** Recall Pascal's vases? The liquid pressure depends on the vertical depth below the surface; this must remain unchanged to balance the pressure due to the atmosphere.

5

Molecular Activity

THE KINETIC THEORY

The kinetic theory of matter states that all matter is composed of molecules which are in constant random motion. In a gas the molecules are relatively far apart and exert little force on each other except when they collide. These molecules move freely and quickly between collisions. The pressure of a gas on its container is due to the collisions of the gas molecules with the walls of the container. In solids the molecules are much closer together than in gases, and there are considerable forces between the molecules and little freedom of motion. The molecules of the solids vibrate and therefore possess kinetic energy. The molecules of liquids are also quite close to each other but can move past each other readily. When a liquid is "still" its molecules are still vibrating. This is indicated by Brownian movements in which large particles are seen to move because of bombardment by invisible molecules. (Also see Chapter 6.)

ELASTICITY AND HOOKE'S LAW

When a force is exerted on a solid its dimensions change—often only a little bit. When the distorting force is removed the solid may return to its original dimensions; if it does it is said to be perfectly elastic. The force tending to restore the solid to its original dimensions is due to the forces between the molecules. For elastic materials *Hooke's law* applies: *The deformation or distortion of an elastic object is proportional to the distorting force* ($F = kx$). For example, if a 10-gm weight suspended from a spring stretches a spring 2 cm, a 20-gm weight will stretch it 4 cm.

SURFACE TENSION AND CAPILLARY ACTION

If water is poured out of a beaker, the inside of the beaker remains wet even if we try to shake the excess out of the beaker. This indicates that the molecules of water are attracted by the molecules in glass. The force of attraction between molecules of different substances is called *adhesion. Cohesion* is the force of attraction between molecules of the same substance. Cohesion in solids is often quite great—it is very difficult to pull steel apart. Since water wets glass we deduce that the adhesion between water and glass is greater than cohesion of water.

If we gently place a double-edge razor blade on top of some water so that the flat surface of the blade is in contact with the surface of the water, we notice that the blade "floats" in spite of the high density of steel. Careful examination shows that the blade does not actually penetrate the surface of the water. The surface of the water acts as though it were a thin elastic membrane; this phenomenon is referred to as *surface tension.* It is explained by noting that the molecule at the surface has neighbors different from the molecule in the body of the liquid. The latter is surrounded by molecules of its own kind, which attract it equally in all directions. The molecule at the surface of the water is not attracted by any water molecules above it (attraction by molecules in the air is negligible); it is attracted by molecules below and to the side of it. This leads to an unbalanced force tending to pull the molecule down. This contracting force on the surface of the liquid tends to keep the molecules together, thus opposing penetration.

If we examine the surface of the water in a glass, we find that it is not perfectly flat. We may notice

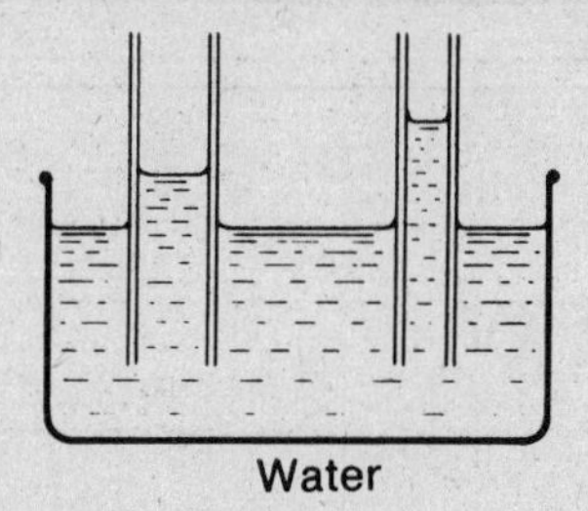

Water

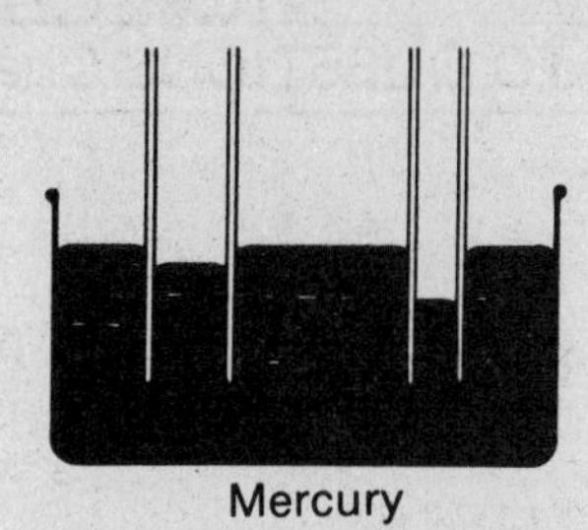

Mercury

that at the edges the water is slightly curved upwards. If we use mercury instead of water we notice that the edges are slightly curved downwards. The slight lifting of the water at the edges is due to the fact that the adhesion between water and glass is greater than the cohesion of the water. In the case of the mercury the cohesion is greater than the adhesion.

If we insert tubes of small diameter into the liquids, we notice that the level of the liquid inside the tube is different from the level outside. If the liquid wets the tube, the liquid rises in it; otherwise the liquid is depressed in it. *The smaller the diameter of the tube, the greater is the rise* (*or depression*) *of the liquid in it.* This phenomenon is known as *capillary action.* (Capillary means hair-like; tubes of small diameter are known as capillary tubes.) Capillary action is not fully explained. Adhesion, cohesion, and surface tension must play a part. The curved surface of a liquid column is called *meniscus.*

QUESTIONS Chapter 5

Select the choice which fits the question best.

1. Pressure exerted by a gas on the walls of its container is due to **(A)** adhesion between the gas molecules and the container **(B)** cohesion between the gas molecules and the container **(C)** collision between the gas molecules and the container **(D)** surface tension of the gas **(E)** gravitational pull on the gas

2. If water rises 4 cm in a long, thin tube because of capillary action, then, under corresponding conditions of use, the rise (in the tube) of a liquid whose density is 2 gm/cm^3 will be **(A)** 1 cm **(B)** 2 cm **(C)** 8 cm **(D)** 16 cm **(E)** unpredictable with the given information

3. Within normal limits of use, if a 100-gm object stretches a spring 3 cm, a 200-gm object will stretch the spring, in cm, **(A)** $3 \times \sqrt{2}$ **(B)** 3×2 **(C)** 3^2 **(D)** 3×2^2 **(E)** $3^2 \times 2^2$

4. The term *Brownian movement* refers to **(A)** irregular motions of small particles suspended in a fluid **(B)** convection currents in a liquid or gas **(C)** convection currents in a gas but not in a liquid **(D)** the stretching of a body beyond its elastic limit **(E)** the sinking of mercury in capillary tubes

5. A container has a small hole in the bottom. Air can go through this hole, but water cannot. This can be best explained by the statement that **(A)** water contains hydrogen atoms, air does not **(B)** water molecules are larger than molecules in the air **(C)** water molecules are smaller than molecules in the air **(D)** water is more viscous than air **(E)** surface tension of the water prevents it from escaping.

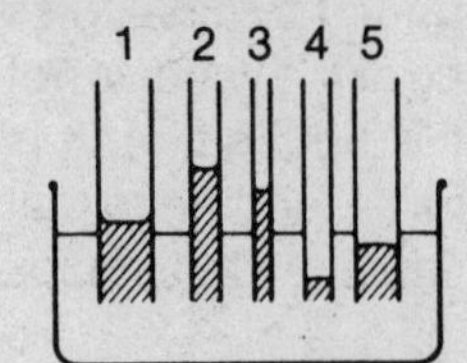

6. The two thin tubes which show most correctly the behavior of water are: **(A)** 1 and 2 **(B)** 2 and 3 **(C)** 3 and 4 **(D)** 4 and 5 **(E)** 1 and 4

EXPLANATIONS TO QUESTIONS Chapter 5

Answers

1. **(C)** 2. **(E)** 3. **(B)** 4. **(A)** 5. **(E)** 6. **(A)**

Explanations

1. **(C)** According to the kinetic theory the molecules of a gas collide with the walls of the container. Each collision exerts a push on the container. The effect of so many collisions is to produce a rather constant force. This force, divided by the area of the container on which it acts, produces the pressure of the gas.

2. **(E)** Not enough information is supplied. As stated in the chapter, the rise or fall also depends on the surface tension of the liquid. We are not told how the surface tension of the new liquid compares with that of water. Instead of rising, the level in the new liquid may actually fall.

3. **(B)** According to Hooke's law, the distortion is proportional to the distorting force. Since the distorting force is doubled, the distortion is also doubled.

4. **(A)** As described in the chapter, Brownian movement refers to the erratic movement of small particles suspended in a fluid. The movement results from the collision with the unseen but constantly moving molecules of the fluid.

5. **(E)** This is a famous demonstration of surface tension. If the water is poured in carefully, the surface of the water at the hole acts like a stretched membrane and does not let the water escape.

6. **(A)** Water wets glass; therefore the water should rise in the capillary tube. This rules out tubes 4 and 5. The narrower the tube, the higher the liquid should rise. This fits tubes 1 and 2, not tubes 2 and 3. (It also fits tubes 1 and 3, but that was not a choice.)

Heat, Temperature, Thermal Expansion

A brass washer with a hole-diameter of one inch is transferred from ice water to boiling water. What happens to the size of the hole? What happens to the energy contents of the washer? Discussion of these questions requires an understanding of heat and its effects.

TEMPERATURE AND HEAT

One way of defining *temperature* is to say it is the degree of hotness or coldness of an object. If we think in terms of the *kinetic theory* we get another definition. The molecules of a substance are in constant random motion. If we heat a gas, its molecules move faster; that is, when the temperature of a gas goes up, the average speed of its molecules goes up. When the speed of motion of the molecule goes up, its kinetic energy goes up. The molecules in a substance are not all moving with the same speed and therefore do not all have the same kinetic energy. We can speak of the average kinetic energy of these molecules. When we heat a gas, the average kinetic energy of its molecules goes up. This leads to thinking of *temperature* as a measure of the average kinetic energy per molecule of the substance. If we take a cupful of water from a bathtub full of water, the *average* kinetic energy per molecule, and therefore the temperature, is the same in the tub as in the cup.

If we have two different gases at the same temperature, perhaps a mixture of oxygen and hydrogen at 0°C, the average kinetic energy of the oxygen molecules is the same as that of the hydrogen molecules. However, on the average, the oxygen molecules will move more slowly than the hydrogen molecules because the oxygen molecule has the greater mass. (Remember, $E_K = \frac{1}{2} mv^2$.)

We have seen that the molecules of a substance have kinetic energy. In the ideal gases to be described on page 46, we assume that this is kinetic energy of translation only. This is accurate enough for monatomic gases like helium and diatomic gases like oxygen and nitrogen at moderate temperatures and pressures. In actual gases the molecules may also have kinetic energy of rotation. They may also have potential energy with respect to each other. For example, when a solid expands, work is done to pull the molecules away from each other against the cohesive force. Until recently *heat* was thought of as the sum total of the energy of its molecules. We now say that a substance has *internal energy* as a result of the kinetic and potential energy of its molecules. In Chapter 16 we shall see that some of the internal energy is also due to potential energy of the atoms. In crystalline solids much of the internal energy is due to the vibration of the nuclei of the atoms. At the present time the physicist's knowledge of internal energy is incomplete.

When a hot piece of copper is brought in contact with a cold piece, the hot piece gets colder and the cold piece gets hotter—both pieces end up at the same temperature. The hot piece lost some internal energy and the cold piece gained some. We define *heat* as the form of energy which flows between two bodies because they are at different temperatures. It is energy in transit.

THERMOMETERS

When an object is heated, many of its characteristics may change. For example, when a solid gets hotter there may be a change in its color, electrical resistance, dimensions. We may use these changes as means for measuring temperature. Many thermometers depend on the fact that substances tend to expand when heated.

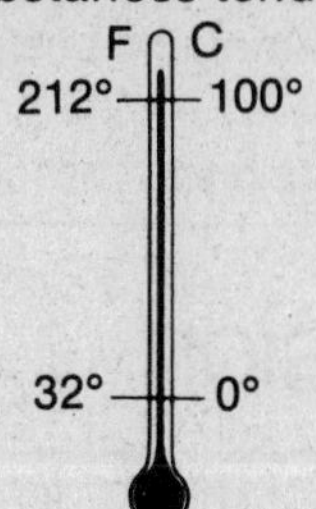

Let us describe briefly the making and calibration of a mercury thermometer. A uniform, thick-walled capillary tube with a glass bulb at one end is partly filled with mercury. The bulb is heated; the mercury expands and fills the tube, driving out all the air. Then the top of the tube is sealed. The selection of temperature scales requires the choice of a reference point and of the size of a unit. In practice two reference points are usually used for the mercury thermometer. If we insert the bulb of a mercury thermometer in water and heat the water steadily, the mercury level in the thermometer keeps rising until the water starts boiling. As long as the water keeps boiling, the level of the mercury in the thermometer remains constant, no matter how vigorously the water is boiled. This is, therefore, a readily obtained reference point. As the water is cooled, the level of the mercury keeps dropping until ice starts forming; as the water is cooled still further more ice keeps forming, but the level of the mercury remains the same as long as both ice and water are present and are mixed. This is the second reference point and is known as the *ice point*–the temperature at which ice melts or water freezes. On the *Celsius* or centigrade scale the ice point is marked 0°; on the *Fahrenheit* scale 32°.

The temperature at which water boils is affected significantly by variation in atmospheric pressure. Therefore in specifying the first reference point we emphasize the pressure to be used: the *standard pressure* is that which would be produced by a column of mercury 76 cm high. For short, we speak of a pressure of 76 cm of mercury, or 30 in. of mercury, or 760 mm of mercury, or of one (standard) atmosphere. The first reference point is known as the *steam point:* the temperature at which water under standard pressure boils. On the Celsius scale the steam point is marked 100°; on the Fahrenheit scale 212°.

Therefore, (212—32) Fahrenheit degrees equal (100—0) centigrade degrees: i.e., 180 F deg = 100 C deg. (N.B. In this book 180 F deg means 180 Fahrenheit degrees; 180° F is an actual temperature: 180 degrees Fahrenheit.)

change of temperature in Fahrenheit degrees = $^9/_5$ change in centigrade degrees

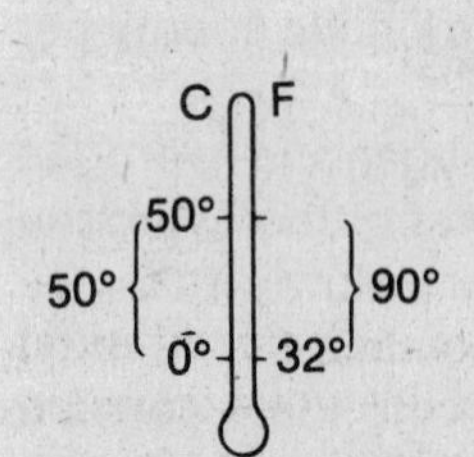

For example, if the temperature goes up 20 C deg (from 30° to 50°), the corresponding change on the Fahrenheit scale will be $^9/_5 \times 20$, or 36 F deg. What will the actual Fahrenheit temperature be? Fifty centigrade degrees correspond to ($^9/_5 \times 50$) Fahrenheit degrees, or 90 F deg. To get the actual Fahrenheit temperature we must add 32 F deg to the change of 90 F deg, or 122° F.

This indicates the formula:

$$F = {}^9/_5\,C + 32° \qquad \text{or} \qquad C = {}^5/_9\,(F - 32°)$$

for obtaining *actual temperatures.*

The *Kelvin scale* is used as an *absolute temperature* scale. This scale has no negative temperatures. *Absolute zero* $\doteq -273°C = 0°K$. $273°K = 0°C$. Each degree on the Kelvin scale has the same size as each degree on the Celsius scale. We shall say more about absolute temperature when we describe the behavior of gases.

EXPANSION AND CONTRACTION

Solids

Most solids expand when heated and contract when cooled. The same length of different solids expands different amounts when heated through the same temperature change. For example, brass expands more than iron. This difference is made use of in thermostats, where a bimetallic strip is used. The bimetallic strip (sometimes known as a *compound bar*) may be made by welding or riveting together brass and iron strips. When heated the bimetallic strip bends with the brass forming

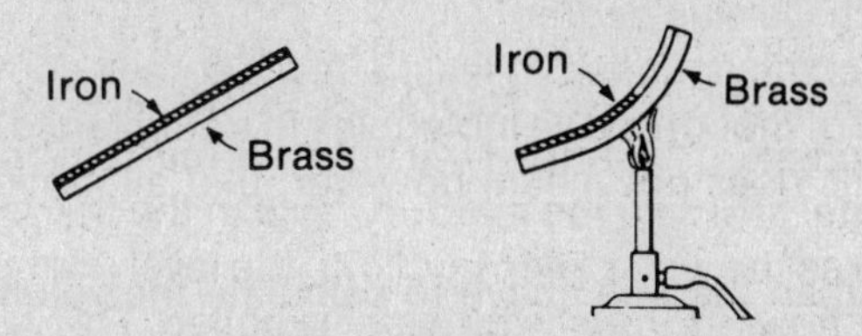

the outside of the curve. When the strip bends it can make or break an electric contact. In cooling the reverse happens; making a circuit can result in the operation of a heater.

When a solid is heated, it expands not only in length but also in width and thickness. Therefore, thermal expansion results in an increase in length, area, and volume. If a hollow solid, like a glass flask, is heated, the empty space increases in volume just as though it were made of the same material as the walls.

The *coefficient of linear expansion* tells what change in length takes place per unit length when the temperature of the solid goes up 1 degree. For example, the coefficient for brass is 1.9×10^{-5} per C deg. This means that for each centigrade degree rise in temperature: 1 cm will increase in length 1.9×10^{-5} cm; 1 ft will increase in length 1.9×10^{-5} ft; 1 yd will increase in length 1.9×10^{-5} yd. Since the temperature change indicated by the Fahrenheit degree is only $^5/_9$ as great as by the centigrade degree, the coefficients per Fahrenheit degree are $^5/_9$ as great as those per centigrade degree. For example, the coefficient for brass is $^5/_9 \times 1.9 \times 10^{-5}$ per F deg. When a solid is cooled the corresponding contraction takes place and the same thermal coefficient can be used. When a rod or wire is heated

change in length = original length × coeff. of expansion × temp. change

Liquids

Most liquids expand when heated, contract when cooled. Different liquids expand different amounts when heated through the same temperature change. The coefficient of volume expansion tells what change in volume takes place per unit volume when the temperature goes up 1 degree. This difference is greater for most liquids than for solids. For example, when the mercury thermometer is heated, the mercury expands more than the glass bulb. Therefore the level of the mercury rises to compensate for the difference in volume expansion.

Water behaves peculiarly in this respect. As water is cooled from 100°C it contracts until its temperature reaches 4° C. If it is cooled further the water will expand; that is, water is densest at 4° C. Therefore, as a lake is cooled at the surface by the atmosphere, the bottom of the lake tends to reach and maintain a temperature of 4° C. If the surface freezes, the resulting ice stays at the top since the specific gravity of ice is only about 0.9 that of water. Water molecules form hydrogen bonds with each other. These are abundant in ice and decrease in relative abundance as the temperature of water goes up.

Gases

If the pressure on a gas is kept constant, heating the gas will result in an increase in its volume. For all gases the coefficient of volume expansion is nearly the same. For example, if we start with a gas at

0° C, for each centigrade degree rise in temperature the volume increases by $^1/_{273}$ of whatever volume the gas occupied at 0° C, provided the pressure of the gas is not allowed to change. If a gas occupying a volume of 273 cm^3 at a pressure of 700 mm is heated from 0° C to 10° C, and the pressure is kept at 700 mm, the volume of the gas will become 273 cm^3 + 273 ($^{10}/_{273}$) cm^3 or 283 cm^3. This rule also operates when the gas is cooled, provided it is not cooled too much. For example, if the above gas is cooled to −10°C, its volume will become 263 cm^3. *If* gases continued to obey this rule even at very low temperatures, the gas would disappear at −273° C. This suggests using −273° C as the zero temperature on a gas scale of temperatures. No gas behaves this way. Fortunately, other theory indicates that no temperature below approximately −273° C can exist. Therefore, this is taken as the zero temperature on the absolute scales. 0° K ≐ −273° C. Each Kelvin degree measures the same temperature change as each Celsius degree.

Actual gases actually become liquefied at least a few degrees above absolute zero. The kinetic theory states that the speed of motion of the molecules in a substance decreases as the substance's temperature decreases. This does not mean, however, that all motion within the atom ceases at absolute zero.

A more convenient way than the one above to calculate the volume of a gas as the result of a temperature change is to use *Charles' law:* If the pressure on a gas is kept constant, its volume is directly proportional to its absolute temperature.

$$\frac{V_1}{V_2} = \frac{T_1}{T_2} \quad \text{or} \quad \frac{V_1}{V_2} = \frac{C_1 + 273°}{C_2 + 273°}$$

where V_1 is the volume of the gas at absolute temperature T_1 or temperature C_1 in centigrade degrees. The volume may be expressed in any volume units; of course, both volumes must be in the same units.

Another important gas law is *Boyle's law:* If the temperature of a gas is kept constant, the volume of the gas varies inversely with its pressure ($V = k/p$):

$$p_1V_1 = p_2V_2$$

(Note that the pressure may be in any unit and that the volume may be in any unit, but both pressures must be in the same units and both volumes must be in the same units.) For example, if the pressure on a gas is doubled without changing its temperature, its volume is reduced to half of its original volume.

Boyle's law and Charles' law may be combined to give the general gas law:

$$\frac{p_1V_1}{T_1} = \frac{p_2V_2}{T_2}$$

Note: Gauges used to measure air pressure in automobile tires do not give the true pressure that is required for the above gas laws. To get the true or absolute pressure we must add to this gauge pressure the pressure of the atmosphere, about 15 lb/in^2.

The above gas laws can be derived on the basis of the kinetic theory of gases.

ANSWERING THE QUESTIONS Let's look at the questions at the beginning of the chapter. The temperature of the brass washer is raised from 0° C to 100° C. The hole increases in size just as a brass disk of the same size would. The coefficient of linear expansion for brass is 1.9×10^{-5} per C deg. Therefore the increase in the diameter = original length × coeff. of expansion × temp. change = 1 inch × 1.9×10^{-5} per C deg × 100 C deg = 1.9×10^{-3} in. or 0.0019 inch. Since the temperature of the brass goes up, the kinetic energy of the particles in it goes up. We can say that the internal energy of the brass washer goes up. In the next chapter we'll find out how large this increase is.

QUESTIONS Chapter 6

Select the choice which fits the question best.

1. On the Fahrenheit scale a temperature of 50°C will be recorded as approximately **(A)** 40° **(B)** 10° **(C)** 40° **(D)** 105° **(E)** 120°

2. A temperature of 100°C is equivalent to a temperature on the Kelvin scale of **(A)** 0° **(B)** 173° **(C)** 212° **(D)** 273° **(E)** 373°

3. A change in temperature of 20 C deg is equivalent to a change in temperature on the Fahrenheit scale of approx. **(A)** 0.9° **(B)** 11° **(C)** 6.7° **(D)** 36° **(E)** 68°

4. An aluminum rod having a diameter of one inch is heated from a temperature of 27°C to a temperature of 54°C. As a result the diameter of the rod **(A)** does not change **(B)** increases to something less than two inches **(C)** doubles **(D)** shrinks to ½ inch to compensate for the increase in length **(E)** shrinks, but remains larger than ½ inch.

5. If the volume of a gas is doubled without changing its temperature, the pressure of the gas is **(A)** reduced to ½ of the original value **(B)** reduced to ¼ of the original value **(C)** not changed **(D)** doubled **(E)** quadrupled

6. The volume of a given mass of gas is doubled without changing its temperature. As a result the density of the gas **(A)** is reduced to ¼ of the original value **(B)** is halved **(C)** remains unchanged **(D)** is doubled **(E)** is quadrupled

7. The volume of a given mass of gas will be doubled at atmospheric pressure if the temperature of the gas is changed from 150°C to **(A)** 300° C **(B)** 423°C **(C)** 573°C **(D)** 600°C **(E)** 743°C

8. All of the following express approximately the same pressure *except* **(A)** 1 atmosphere **(B)** 15 lb/in^2 **(C)** 15 lb/ft^2 **(D)** 30 in of mercury **(E)** 760 mm of mercury

9. If 500 cm^3 of gas, having a pressure of 760 mm of mercury, is compressed into a volume of 300 cm^3, the temperature remaining constant, the pressure of the gas will be, in mm of mercury, approximately **(A)** 500 **(B)** 900 **(C)** 1100 **(D)** 1300 **(E)** 1500

10. How many cubic feet of air at atmospheric pressure must be pumped into a 20-ft^3 tank containing air at atmospheric pressure in order to raise the absolute pressure to 90 lb/in^2? Assume that the temperature remains constant. **(A)** 70 **(B)** 85 **(C)** 100 **(D)** 120 **(E)** 140

EXPLANATIONS TO QUESTIONS Chapter 6

Answers

1. **(E)**	4. **(B)**	7. **(C)**	10. **(C)**
2. **(E)**	5. **(A)**	8. **(C)**	
3. **(D)**	6. **(B)**	9. **(D)**	

Explanations

1. **(E)** $F = {}^9/_5C + 32° = {}^9/_5 \times 50° + 32° = 122°$

2. **(E)** $T = °C + 273° = (100 + 273)\ °\ K = 373°\ K.$

3. **(D)** change in F deg $= {}^9/_5$ change in C deg $= {}^9/_5 \times 20° = 36°$

4. **(B)** When a metal is heated, its dimensions increase in all directions. Therefore, the diameter of the rod will increase. However, the expansion per degree of temperature rise is a very small fraction of the original length; (check value of coefficient of expansion). The increase will be to something considerably less than two inches.

5. **(A)** Boyle's law applies. The pressure of the gas varies inversely with its volume. Since the volume is doubled, the pressure must be reduced to one-half of its original value.

6. **(B)** Density = mass/volume. The mass of the gas does not change in the problem; the volume is doubled. Therefore the density must go down to one-half of its original value.

7. **(C)** Charles' law applies: At constant pressure, the volume of a gas is proportional to its absolute temperature. Therefore, to double the volume of the gas we must double the absolute temperature. The starting temperature: $T = °C + 273° = (150 + 273)°\ K = 423°\ K$. The final absolute temperature is $2 \times 423°\ K = 846°\ K$. The final Celsius temperature is $(846 - 273)° = 573°$.

8. **(C)** You should know the equivalent values of atmospheric pressure. A simple recall question like this is not likely to be asked by CEEB.

9. **(D)** Boyle's Law applies.

$$p_1 V_1 = p_2 V_2$$

$$p_1\ (300\ \text{cm}^3) = 760\ \text{mm} \times 500\ \text{cm}^3$$

$$p_1 = \frac{760 \times 500}{300}\ \text{mm} = 1270\ \text{mm}$$

10. **(C)** Boyle's law applies, with a slight catch in the problem. First we calculate the volume which the 20 ft^3 of air at a pressure of 90 lb/in.^2 would occupy if its pressure were changed to atmospheric pressure (15 lb/in^2) without changing its temperature.

$$p_1 V_1 = p_2 V_2;$$

$$90 \frac{\text{lb}}{\text{in.}^2} \times 20\ \text{ft}^3 = 15 \frac{\text{lb}}{\text{in.}^2} \times V_2$$

$$V_2 = 120\ \text{ft}^3$$

of air at atmospheric pressure. However, 20 ft^3 of air at atmospheric pressure were in the tank originally. Therefore, only an additional 100 ft^3 have to be put in.

Measurement of Heat; Humidity

In the last chapter we stated that heat is a form of energy. When we want to measure the amount of heat which is transferred from one object to another, the laboratory setup includes an instrument called a *calorimeter.* This should be a container which is well insulated from the outside, but in a high school laboratory sometimes this is nothing more than a glass tumbler or a plastic cup. As with other measurements, we must establish units.

The calorie (cal) and kilocalorie (kcal) are often used as units of heat. They are now defined in terms of the joule. For ordinary calculations the old definitions simplify calculations and give sufficiently accurate results. We shall use them here.

The *calorie* is the amount of heat needed to raise the temperature of 1 gram of water 1 C deg. The *kilocalorie* is the amount of heat needed to raise the temperature of 1 kilogram of water 1 C deg. (This is the same as the calorie used to describe the nutritional value of foods.) The *British thermal unit (Btu)* is the amount of heat needed to raise the temperature of 1 lb of water 1 F deg.

$$1 \text{ kcal} = 1000 \text{ cal}$$
$$1 \text{ kcal} = 4190 \text{ joules}$$
$$1 \text{ Btu} = 252 \text{ cal}$$

Specific Heat

From the definition of the calorie we know that, if the temperature of 1 gm of water goes up from 10° C to 11° C, then one calorie was added to the water. However, we find that less heat is required to raise the temperature of 1 gm of most other substances one centigrade degree. We often define *specific heat* (*sp. ht.*) of a substance as the number of calories required to raise the temperature of 1 gm of the substance 1 C deg; this number is the same as the number of Btu required to raise the temperature of 1 lb of the substance 1 F deg. From the above we can see that the specific heat of water is 1 cal/gm-C deg, or 1 Btu/lb-F deg. The specific heat of copper is 0.09 cal/gm-C deg. Sometimes specific heat is defined in a slightly different way, so that it is not necessary to give any units. In that case we would say that the specific heat of copper is 0.09, and it will be understood that 0.09 cal is needed to raise the temperature of 1 gm of copper 1 C deg.

The specific heat of a substance is often obtained in the laboratory by using the method of mixtures. For example, a 100-gm block of iron is moved from boiling water (100° C) to a beaker containing 110 gm of water at 18° C. Assume that this raises the temperature of the water in the beaker to 25° C. We can now calculate the specific heat of iron:

heat lost by hot object = heat gained by cold objects

Also, for any object that loses or gains heat,

heat lost (or gained) = mass × sp. ht. × temp. change (see change of state below)

The heat lost by the iron = 100 gm × sp. ht. × (100° − 25°). Heat gained by the water = 110 gm × 1 cal/gm-C deg × (25° − 18°). If we assume that no other substances are involved in the heat exchange, 100 gm × sp. ht. × 75° = 110 gm × 1 cal/gm-C deg × 7° and sp. ht. = 0.10 cal/gm-C deg. (The accepted value for iron is 0.11.)

Change of State

This equation applies if there is no change of state. Earlier we defined solid, liquid, and gas. These are three states of matter. If we heat a solid, its temperature goes up until it starts melting. Further transfer of heat to the solid results in melting or *fusion* without change of temperature. Continued heating after all the solid has melted again leads to a temperature change—this time of the liquid. In the above equation the specific heat of the liquid must be used; this is usually different from the specific heat of the solid. For example, the specific heat of ice is 0.5; that for water is 1.

When a substance is changed from solid to liquid, or from liquid to gas, heat must be supplied to pull the molecules away from each other against cohesive forces. This represents an increase in potential rather than kinetic energy. Therefore the temperature remains the same. When the reverse change of state takes place, e.g., solidification, this heat is released. For a given crystalline substance (metals, ice, etc.), melting and freezing occur at the same temperature.

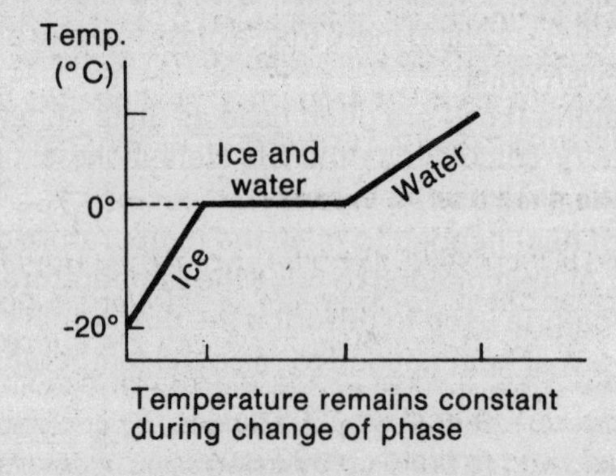

Temperature remains constant during change of phase

Heat of Fusion

Heat of fusion of a substance is the amount of heat required to melt a unit mass of the substance without change of temperature. For ice this is 80 cal/gm or 144 Btu/lb.

heat required for melting = mass × H_F

where H_F is the heat of fusion of the substance.

When we defined the ice point in the previous chapter we made use of this fact that the temperature does not change while melting takes place. It was also indicated there that this temperature is practically independent of ordinary changes in atmospheric pressure. If the pressure on ice is twice ordinary atmospheric pressure, its melting point becomes −0.0075° C. However, in ice-skating a person's weight is supported by a small area—the area of the bottom of the blades. This results in a high pressure (F/A) on the ice—enough to melt it; when the skater passes, the water freezes again. *Regelation* is this melting under increased pressure and refreezing when the pressure is reduced. This happens with substances like ice which expand on freezing. On the other hand, substances which contract on solidifying have a higher melting point at higher pressures. A lower melting point also results from dissolving solids in liquids. That is why salt sprinkled on an icy sidewalk tends to melt the ice.

Vaporization

Vaporization is the process of changing a substance to a vapor. It includes evaporation, boiling, and sublimation.

EVAPORATION As explained above, the molecules of a liquid are in constant motion. The speed of the different molecules is different. Some of the fastest moving molecules at the surface can escape from the liquid in spite of gravity and the attractive force between molecules. The molecules that escape are relatively far apart from each other, as in a gas, and form the vapor of the liquid. Since the molecules with the greatest kinetic energy escaped, the average kinetic energy of the remaining molecules in the liquid decreased; that is, unless the liquid is heated, the liquid cools when evaporation takes place. Evaporation is a cooling process. It occurs at the surface of the liquid. The rate of evaporation can be increased by making it easier for the molecules to escape from the liquid: heat the liquid to give more molecules higher speeds; spread the liquid over a greater area; decrease the air pressure on top of the liquid; blow away the vapor-filled air above the liquid.

BOILING In evaporation the liquid is converted to vapor at the surface of the liquid. In boiling the liquid is converted to vapor within the body of the liquid. We observe the vapor in the form of bubbles. Another way to define boiling is in terms of pressure. The molecules of the vapor move around freely like the molecules of a gas and exert pressure. We can speak of the pressure of a vapor as we speak of the pressure of the atmosphere. *Boiling* occurs when the pressure of the vapor forming inside the liquid equals the external pressure on the liquid; this is usually atmospheric pressure. During boiling the temperature of the liquid doesn't change. The temperature at which this occurs is the *boiling point* of the liquid. Although the temperature of the liquid does not change during boiling, energy must be supplied for the vaporization (to do work against the cohesive force between molecules). *Heat of vaporization* (H_v) of a substance is the amount of heat needed to vaporize unit mass of the substance without changing its temperature. For water at 100° C this is 540 cal/gm (or 972 Btu/lb).

heat required for vaporization = mass × H_v

When the pressure on a liquid is increased, its boiling point goes up; when the pressure on a liquid is decreased, its boiling point goes down. For example, at standard atmospheric pressure (760 mm) the boiling point of water is 100° C. In mountainous areas the atmospheric pressure is usually considerably below 760 mm and water boils at temperatures significantly below 100° C. Cooking at these lower temperatures proceeds rather slowly. To hasten the cooking, pressure cookers may be used; in these the pressure is allowed to build up by using special covers which do not let the vapor escape. The building up of the pressure raises the boiling point.

SUBLIMATION is the direct change from solid to vapor. For example, at ordinary room temperatures, solid carbon dioxide ("dry ice") changes directly to carbon dioxide gas. *Condensation* is the changing of a vapor to a liquid. This is accompanied by the release of the same amount of heat that would be required to vaporize the same quantity of the liquid. The term *condensation* is also used occasionally to refer to the reverse of sublimation; that is, it refers to the direct conversion of a vapor to a solid, as in the formation of frost on a cold night.

When there is a change of state, the method of mixtures described above has to be modified:

heat gained (or lost) = mass × sp. ht. × temp. change
+ mass melted × heat of fusion
+ mass vaporized × heat of vaporization

In this formula any term may be omitted that does not fit the situation. *For example,* to calculate the heat required to boil away completely 10 grams of water at 100° C if the starting temperature of the water is 60° C, we omit the second term since no melting is involved in the illustration. Heat gained = mass × sp. ht. × temp. change + mass vaporized × heat of vaporization = (10 × 1 × 40) cal + (10 × 540) cal = 5800 cal.

TABLE OF CONSTANTS FOR WATER

System	sp. gr.	density	sp. ht.	H_F	H_v
Metric (CGS)	1	1 gm/cm^3	1 (cal/gm-C deg)	80 cal/g	540 cal/g
Metric (MKS)	1	1000 kg/m^3*	1 (kcal/kg-C deg)	80 kcal/kg	540 kcal/kg
English (FPS)	1	62.5 lb/ft^3	1 (Btu/lb-F deg)	144 Btu/lb	972 Btu/lb

*This can also be expressed as 1 kg/liter.

HUMIDITY

The atmosphere contains water vapor, but there is a limit to how much water can be evaporated into a given volume of air, just as there is a limit to how much sugar can be dissolved in one cupful of coffee. More sugar can be dissolved in hot coffee than in cold A given volume of air can hold more water vapor at a higher temperature than at a lower temperature. The air is said to be *saturated* when it holds as much water vapor as it can at that temperature. At 20° C a cubic meter of air can hold about 17 gm of water vapor; at 30° C it can hold about 30 gm. Usually the atmosphere is not saturated. *Relative humidity* (expressed in percent) is the ratio of the mass of water vapor actually present in a given volume of air to the mass which would be present in it if it were saturated. For example, if a cubic meter of air at 20° C contains 12 gm of water vapor, the relative humidity is $\frac{12 \text{ gm}}{17 \text{ gm}} \times 100 = 71\%$. Hygrometers are instruments for measuring relative humidity. Readings on wet and dry bulb thermometers can be compared with the aid of a chart from which one can then read off the relative humidity. The basic principle of this is that evaporation is a cooling process. The rate of evaporation from the wet-bulb thermometer will be high when the relative humidity is low, and therefore on such a day the wet-bulb thermometer will read considerably below the dry-bulb one. There is no simple formula for converting this temperature difference to relative humidity, and therefore a chart is used.

If unsaturated air is cooled, its relative humidity goes up. If the temperature of the air drops sufficiently, saturation is reached and excess moisture precipitates out. The *dew point* is the temperature to which the air must be cooled so that it will be saturated and condensation will just form.

QUESTIONS Chapter 7

Select the choice which fits the question best.

1. One thousand cal of heat are added to 100 gm of water when its temperature is 40° C. The resulting temperature of the water is **(A)** 10° C **(B)** 32° C **(C)** 50° C **(D)** 80° C **(E)** 100° C

2. If 2 Btu of heat are added to 504 gm of water at 100° C and standard pressure, the temperature of the water will **(A)** not change **(B)** rise 1 C° **(C)** rise 1 F° **(D)** rise 2 C° **(E)** rise 2 F°

3. Fifteen hundred calories are added to 100 grams of gold when its temperature is 20° C. The specific heat of gold is 0.03. The resulting temperature of the gold is **(A)** 50° C **(B)** 120° C **(C)** 320° C **(D)** 520° C **(E)** 3020° C

4. A boy has 240 grams of water at 50° C. The number of grams of ice at 0° C which he must add to the water to lower the water temperature to 0° C is **(A)** 35 **(B)** 150 **(C)** 12,000 **(D)** 50,000 **(E)** infinite

5. A certain alloy has a melting point of 1000° C. The specific heat of the solid is 0.3; its heat of fusion is 20 calories per gram. The number of calories required to change 10 grams of the solid from

room temperature (20°C) to the liquid at 1000 C° is approximately **(A)** 240 **(B)** 440 **(C)** 3140 **(D)** 3740 **(E)** 29,600

6. When 10 grams of steam at 100°C condense to form water at 0°C, the number of calories liberated is **(A)** 10 **(B)** 800 **(C)** 1000 **(D)** 5400 **(E)** 6400

7. It is desired to change 10 gm of ice at 0°C to steam at 100°C. The number of Btu's required is, approximately, **(A)** 30 **(B)** 1800 **(C)** 2730 **(D)** 6400 **(E)** 7200

8. In an experiment performed at atmospheric pressure, 1 gm of ice at −20°C is changed to steam at 130°C. If the sp. ht. of both steam and ice is 0.5, then the number of calories required is, approximately, **(A)** 24 **(B)** 200 **(C)** 540 **(D)** 640 **(E)** 740

9. When one gram of atmospheric water vapor condenses in the air, it results in **(A)** always cooling the surrounding air **(B)** always heating the surrounding air **(C)** always leaving the surrounding air temperature unchanged **(D)** heating the surrounding air only if the change occurs below 100°F **(E)** cooling the surrounding air only if the change occurs below 100°F

10. If the temperature of the atmosphere rises above the dew point, **(A)** dew will form **(B)** precipitation will take place **(C)** the relative humidity will become 100% **(D)** evaporation is likely to take place **(E)** condensation will take place if nuclei such as dust are present.

EXPLANATIONS TO QUESTIONS Chapter 7

Answers

1. **(C)**
2. **(A)**
3. **(D)**
4. **(B)**
5. **(C)**
6. **(E)**
7. **(A)**
8. **(E)**
9. **(B)**
10. **(D)**

Explanations

1. **(C)** Heat gained = mass × sp. ht. × temp. change. 1000 cal = 100 gm × 1 cal/gm-C deg × temp. change. Temp. change = 10 C deg. Final temp. = (40 + 10)° C = 50°C.

2. **(A)** The given temperature of the water is its normal boiling point. Further addition of heat allows boiling to go on without a change in temperature, as long as the pressure remains the same. (Did you read the question too quickly?)

3. **(D)** Heat gained = mass × sp. ht. × temp. change. 1500 cal = 100 gm × 0.03 cal/gm-C deg × temp. change. Temp. change = 500 C deg. Final temp. = (500 + 20)° C = 520° C.

4. **(B)** Heat gained by ice = heat lost by water.

$$\text{Mass} \times H_F = \text{mass of water} \times \text{sp. ht.} \times \text{temp. change.}$$
$$\text{Mass}_i \times 80 \text{ cal/gm} = 240 \text{ gm} \times 1 \text{ cal/gm-C deg} \times 50 \text{ C deg}$$
$$\text{Mass}_i = 150 \text{ gm}$$

5. **(C)** Heat required = heat to raise temp. of solid to melting point + heat to melt solid

$$\begin{aligned}\text{Heat required} &= \text{mass} \times \text{sp. ht.} \times \text{temp. change} + \text{mass} \times H_F \\ &= 10 \text{ gm} \times 0.3 \text{ cal/gm-C deg} \times 980 \text{ C deg} + 10 \text{ gm} \times 20 \text{ cal/gm} \\ &= 2940 \text{ cal} + 200 \text{ cal} = 3140 \text{ cal}\end{aligned}$$

6. **(E)** Heat liberated = heat given off by steam when condensing at 100°C + heat given off by water when cooling from 100° C to 0°C.

$$\begin{aligned}\text{Heat liberated} &= \text{mass} \times H_v + \text{mass} \times \text{sp. ht.} \times \text{temp. change} \\ &= 10 \text{ gm} \times 540 \text{ cal/gm} + 10 \text{ gm} \times 1 \text{ cal/gm-C deg} \times 100 \text{C deg} \\ &= 5400 \text{ cal} + 1000 \text{ cal} = 6400 \text{ cal}\end{aligned}$$

7. **(A)** Heat required = heat to melt ice at 0° C + heat to raise temp. of resulting water from 0° C to 100° C + heat to change water at 100° C to steam at 100° C.
Heat required = mass × H_F + mass × sp. ht. × temp. change + mass × H_V
= 10 gm × 80 cal/gm + 10 gm × 1 × 100 C deg + 10 gm × 540 cal/gm
= 800 cal + 1000 cal + 5400 cal = 7200 cal
But 1 Btu = 252 cal; heat required = (7200/252) Btu = 28.5 Btu.

8. **(E)** Heat required = heat to raise temp. of ice to 0° C + heat to melt ice at 0° C + heat to raise temp. of water from 0° C to 100° C + heat to change water to steam at 100° C + heat to raise temp. of steam to 130° C. Heat required = 1 gm × 0.5 cal/gm-C° × 20C° + 1 gm × 80 cal/gm + 1 gm × 1 cal/gm-C° × 100C° + 1 gm × 540 cal/gm + 1 gm × 0.5 cal/gm-C° × 30C° = (10 + 80 + 100 + 540 + 15) cal = 745 cal.

9. **(B)** Heat is required to vaporize a liquid. Whenever the vapor condenses to a liquid, heat has to be given off again to the surroundings. In this case the heat is given off to the surrounding air and heats it.

10. **(D)** At the dew point the air is saturated with water vapor. When the temperature of the air goes up it can hold more water vapor; its relative humidity is less than 100%. If water is present, evaporation will take place.

Heat and Work; Heat Transfer

It is the English scientist James Prescott Joule who, probably more than anyone else, is responsible for our belief that heat is a form of energy. Since energy is the ability to do work, we should not be surprised that we can use hot objects to do work for us; we have already studied the reverse—when work is done with a machine, heat is produced since some energy is used to do work against friction. How much work can be done with 1 Btu of heat? How can a hot object be used to do work for us?

Imagine a gas in a cylinder with a movable piston. The molecules of the gas bombard the piston and produce a pressure on it. If we want to compress the gas, we must exert a force against the piston. We do work in compressing the gas. As we push the piston, we exert a force on the molecules of the gas thus giving them greater speed. The temperature of the gas goes up; if the cylinder is insulated, the work done in compressing the gas results in an equivalent increase in the internal energy of the gas. If the cylinder is not insulated, the temperature of the gas can be kept constant by letting this heat equivalent go instead to the material outside the cylinder. Mechanical energy can be converted completely to heat. How much heat is produced?

778 ft-lb = 1 Btu
4.19 joules = 1 cal

If the gas in the cylinder is allowed to expand, the gas does work as it moves the piston. If the cylinder is insulated, as the gas expands its temperature goes down; the energy required for the work comes from the internal energy of the gas. If the cylinder is not insulated, the temperature of the gas can be kept constant. Then, as the gas expands, heat flows into it from outside the cylinder; this heat equals the work done by the gas in expanding.

Recall that an *isothermal* process is one in which the temperature remains constant. Boyle's Law ($pV = k$) applies to a gas which expands or contracts while its temperature is kept constant. An *adiabatic* process is one in which no heat is allowed to enter or leave the system. This is true in the above process in which the cylinder is kept insulated. Boyle's Law does not apply to an adiabatic process.

HEAT ENGINES

Heat engines are used to convert heat to mechanical energy. Examples of heat engines are: gasoline engine, diesel engine, steam turbine, reciprocating steam engine. In these heat engines hot gases are allowed to expand; as they expand they do work. If the fuel is burned inside the cylinder of the engine itself, the engine is known as an *internal combustion engine;* this is true of the gasoline and diesel engines. If the fuel is burned in a separate chamber outside the engine proper, the engine is known as an *external combustion engine;* this is true of the steam engine and steam turbine. In these, the fuel, which may be coal, is burned in a separate furnace and is used to heat water in a boiler. The steam is then directed into the engine.

Steam Engine

Let us look at the diagram of a simple reciprocating steam engine. Steam under pressure enters the steam chest *C* from the boiler. With the valve *V* in the position shown, this steam has a path to the right side of the piston *P*. As a result the piston is pushed to the left. Rod *R*, pulled by the piston, in turn pulls the flywheel *F*, to which it is connected off center. (A *flywheel* is a wheel with a heavy rim; its inertia is large.) Since the steam does work, it loses some of its energy and becomes cooler.

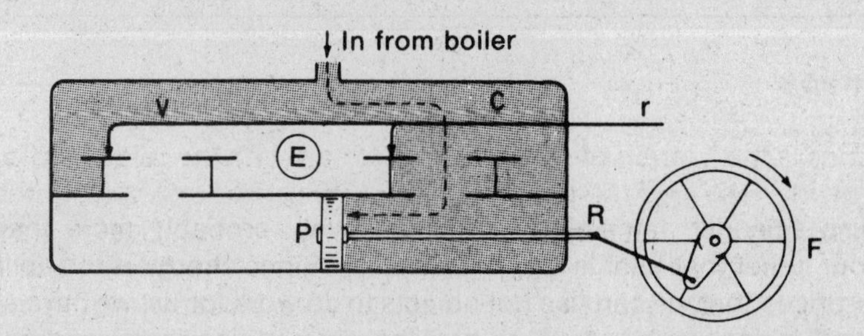

Steam remaining on the left side of the piston from a previous cycle is pushed out through the exhaust pipe *E* by the piston. The valve is connected to the flywheel by means of rod *r* (connection not shown). At the right time the valve is pulled to the right so as to provide a path for the steam from the *In* pipe to the left side of the piston; this reverses the motion of the piston. This back and forth motion is known as *reciprocating* motion. It can be converted to rotary motion by means of an *eccentric* (off-center device).

The *steam turbine* is more efficient than the reciprocating steam engine. In the turbine a set of cups is placed at the rim of a movable wheel. Steam is directed at these cups from a set of fixed nozzles. The force of the steam against the cups produces rotary motion.

Gasoline Engine

The *gasoline engine* is an internal combustion engine commonly used in automobiles. These engines frequently have six or eight cylinders connected to the same flywheel by means of a common shaft. One of these cylinders is shown. A complete cycle of operation consists of four strokes: Intake, compression, ignition or power, exhaust. The cylinder is shown at the beginning of the intake stroke. The exhaust valve *E* is closed. Gasoline and air mixture enter through the open intake valve *I*, and the piston moves down. Then both valves are closed and the mixture is compressed as the piston moves up during the second stroke. At the proper instant, high voltage is applied to the spark plug *S;* the resulting spark ignites the mixture. This produces an explosion which provides the power for pushing the piston and turning the crankshaft, flywheel, and ultimately the wheels of the car. During the last stroke the piston is pushed up again; this time the intake valve is closed and the exhaust valve is open. The waste gases are pushed out and the cycle can start again. Notice that there are two complete up and down motions during each cycle. The *carburetor* provides the proper mixture of air and gasoline vapor, which is then burned in the cylinders. The engine is kept running between power strokes by the flywheel *F*. The whole process is usually started by briefly switching a battery to a starting motor.

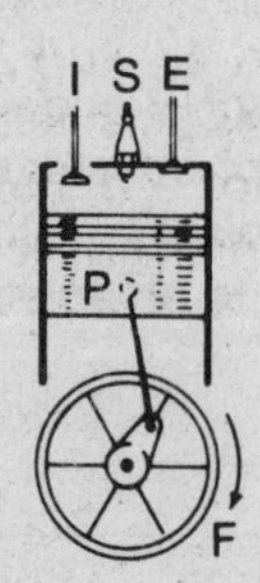

Diesel Engine

The *diesel engine* is also an internal combustion engine. Neither carburetor nor spark plug is needed. The air in the cylinder is compressed so much that its temperature rises sufficiently to ignite the fuel, which is then sprayed in. The fuel is denser and cheaper than gasoline. The efficiency of a diesel engine (about 40%) is higher than steam engines and gasoline engines.

Jet and Rocket Engines

Jet and rocket engines operate on the principle of Newton's third law. Gases escaping under pressure in one direction exert a push on the engine in the opposite direction. The jet engine takes from the atmosphere the oxygen needed for burning its fuel, while the rocket carries its oxygen as well as its own fuel.

Thermodynamics

Thermodynamics is the branch of science which deals with the relationship between changes in energy, the flow of heat, and the performance of work. For example, in the previous section we mentioned heat engines. Let us consider one in more detail. Heat is provided to a cylinder with a movable piston as shown for the gasoline engine. The heated gas pushes the piston which then does useful work in turning the wheel, and the piston returns to the starting position. During the cycle some of the heat supplied also goes into increasing the internal energy of the system, such as increasing the kinetic energy of translation and rotation of the molecules of the cylinder, piston, and waste gases.

The *first law of thermodynamics* states this quantitatively as a form of the conservation of energy: the heat supplied to a system is equal to the work done by the system plus the increase in the internal energy of the system. The *second law of thermodynamics* reminds us that in a closed system energy is always conserved, but you may not get exactly what you want. If all you want is heat, you can often make the energy change 100%, although part of the heat may escape the system. But getting work done with 100% efficiency is a different problem. The second law states: it is impossible to construct an engine which will extract heat from a source at constant temperature and use the heat completely to do work. This is indicated by the low efficiency given above for even the diesel engine.

METHODS OF HEAT TRANSFER

Sometimes we want heat to escape; sometimes we want to keep it from escaping. In the case of heat engines, such as the gasoline engine, only some of the heat can be used to do work. It is necessary to get rid of the excess heat or the engine would rapidly overheat. In the case of heating a home, we want the heat to get from the furnace to the apartment, but we may want to keep the heat from getting out of the apartment. The three methods of heat transfer are conduction, convection, and radiation.

Conduction

Conduction is the process of transferring heat through a medium which does not involve appreciable motion of the medium. For example, if we heat one end of a copper rod, the other end gets hot, too. The energy transfer in metallic conductors is chiefly by means of "free" electrons, but also involves bombardment of one molecule by the next. In general, metals are good conductors; good conductors of heat are also good conductors of electricity. Silver is the best metallic conductor of heat and electricity. Copper and aluminum are also very good.

Liquids and gases, as well as non-metallic solids, are poor conductors of heat. Poor conductors are known as *insulators.*

Convection

Convection is the process of transferring heat in a fluid which involves the motion of the heated portion to the cooler portion of the fluid. The heated portion expands, rises, and is replaced by cooler fluid, thus giving rise to so-called convection currents. This may be observed over a hot radiator: The heated air expands; since its density is different from the surrounding colder air it reflects some light and can be "seen" to move up. Radiators heat rooms chiefly by convection, and, therefore, are more effective when placed near the floor than when placed near the ceiling. Also note that when the heated fluid rises it mixes with the cooler fluid.

Radiation

Radiation is the process of transferring heat which can take place in a vacuum; it takes place by a wave motion similar to light. The wave is known as an *electromagnetic wave* and will be discussed further in a later chapter. The higher the temperature of an object, the greater the amount of heat it radiates. Black objects radiate more heat from each square centimeter of surface than light-colored objects at the same temperature.

The *vacuum bottle* (thermos bottle) is designed with the three methods of heat transfer in mind. The bottle is made of glass—a good insulator. The stopper is made of cork or plastic—also good insulators. This minimizes heat transfer by conduction. The space between the double walls is evacuated, minimizing heat transfer by conduction or convection. The inside surfaces (facing the vacuum) are silvered; shiny surfaces reflect electromagnetic waves. This minimizes heat transfer by radiation. As a result, hot liquids inside the bottle stay hot and cold liquids stay cold many hours.

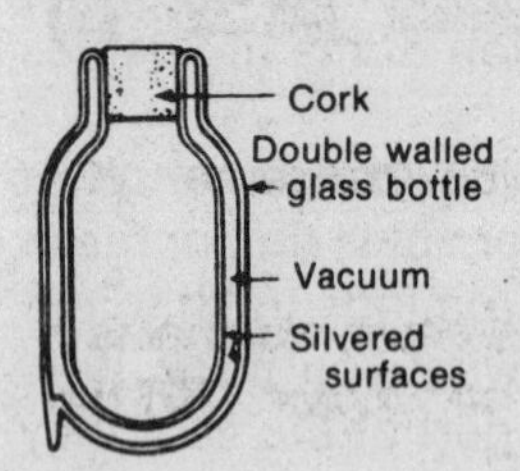

ENERGY SOURCES

For heating buildings and operating machines man has depended almost solely on fossil fuels such as coal and oil. Although new deposits keep being located and tapped, many people think that these fuels will be found inadequate in very few years. Waterfalls, of course, can be used in locations close enough to exploit economically this source of energy. Waterfalls and fossil fuels ultimately owe their energy to the sun. Energy keeps coming to the earth from the sun by radiation (about 2 calories per minute on each square centimeter in full sunlight). In some parts of the world large, curved mirrors are arranged to permit the use of this sunlight for cooking and heating. A tremendous amount of energy is available in the constant motion of the tides. However, it is believed that as the population grows and the standard of living goes up, all these sources of energy will prove inadequate. Some think the hope of the future lies in nuclear energy—energy released when certain changes take place in the nucleus of atoms. Man has already learned to obtain nuclear energy resulting from the splitting or fission of nuclei of heavy elements such as uranium. This is done in a controlled manner in the nuclear reactor (or atomic pile). Only relatively small quantities of such fissionable materials are available. Nuclear energy can also be obtained by the combining or fusion of the nuclei of light atoms, such as hydrogen. If this can be done in a controllable manner, a practically endless supply of energy will be available, because hydrogen, the necessary "fuel," is obtainable from the oceans in almost unlimited quantity. Nuclear energy will be discussed further in a later chapter. (Also see *solar cells.*) Some houses have roof-top installations with circulating fluids to absorb solar energy for producing hot water and heat.

QUESTIONS Chapter 8

Select the choice which fits the question best.

1. Radiation is the chief method of energy transfer **(A)** from the sun to an earth satellite **(B)** from a gas flame to water in a teakettle **(C)** from a soldering iron to metals being soldered **(D)** from water to an ice cube floating in it **(E)** from a mammal to the surrounding air

2. Of the processes below, the one in which practically all the heat transfer is by conduction is **(A)** from the sun to an earth satellite **(B)** from a gas flame to the top layer of water in a teakettle **(C)** from a soldering iron to metals being soldered **(D)** from the bottom of a glass of water to an ice cube floating in it **(E)** from a mammal to the surrounding air

3. A piece of bread can supply about 300 Btu. About how many pounds could be lifted 100 ft if it were possible to use this energy with 100% efficiency? **(A)** 3 **(B)** 600 **(C)** 750 **(D)** 2300 **(E)** 7000

4. A gallon of gasoline weighs about 6 lb. This gasoline can supply about 15,000 Btu/lb. If this gasoline is used with 30% efficiency in an 8-cylinder engine, the amount of work that can be done by the engine with each gallon of gasoline is, in ft-lb, approximately **(A)** 48 **(B)** 750,000 **(C)** 6×10^6 **(D)** 8×10^6 **(E)** 20×10^6

5. In a certain steam engine, the average pressure on the piston during a stroke is 50 lb/in^2 The length of each stroke is 12 in, the area of the piston is 120 in^2, and the diameter of the flywheel is 5 ft. The amount of work done on the piston during each stroke is, in ft-lb, approximately **(A)** 250 **(B)** 6000 **(C)** 30,000 **(D)** 72,000 **(E)** 96,000

6. Dark, rough objects are generally good for **(A)** conduction **(B)** radiation **(C)** convection **(D)** reflection **(E)** refraction

7. Dark plastic handles are often used on kitchen utensils because **(A)** the black material is a good radiator **(B)** the plastic is a good insulator **(C)** the plastic is a good conductor **(D)** the plastic softens gradually with excessive heat **(E)** the material is thermoplastic

8. A person seated in front of a fire in a fireplace receives heat chiefly by **(A)** convection of carbon dioxide **(B)** convection of carbon monoxide **(C)** convection of air **(D)** conduction **(E)** radiation

9. The unit which does not represent the same physical quantity as the others is the **(A)** British thermal unit **(B)** calorie **(C)** horsepower **(D)** joule **(E)** foot-pound

10. If friction is negligible, when a gas is compressed rapidly **(A)** its temperature remains the same **(B)** its temperature goes down **(C)** its temperature rises **(D)** it is liquefied **(E)** work is done by the gas

EXPLANATIONS TO QUESTIONS Chapter 8

Answers

1. **(A)**	3. **(D)**	5. **(B)**	7. **(B)**	9. **(C)**
2. **(C)**	4. **(E)**	6. **(B)**	8. **(E)**	10. **(C)**

Explanations

1. **(A)** The space between the sun and an earth satellite is an almost perfect vacuum. Conduction and convection cannot take place through a vacuum. Although some particles are shot out of the sun

and thus carry kinetic energy with them to the satellite, this energy is small in comparison with the amount constantly carried away by electromagnetic waves—by radiation. (Also see answer to the next question.)

2. **(C)** The soldering iron is in good contact with the metal to be soldered and thus provides excellent opportunity for heat transfer by conduction. The tip of a soldering iron is often made of copper to provide for better heat conduction. Although a teakettle, if made of metal, provides for some heating of the top layer of water in the kettle by conduction along the kettle, the top layer is heated chiefly by convection: the bottom layer gets heated, rises towards the top where it mixes with the colder top and also pushes the top layer toward the bottom. Similar reasoning applies to choice (D). Mammals lose heat by exhaling of warm air, evaporation of perspiration from the skin, convection, etc.

3. **(D)** 1 Btu = 778 ft-lb; 300 Btu = (300 × 778) ft-lb = 233,400 ft-lb.

$$\text{Work} = \text{wt} \times \text{height lifted}$$
$$233{,}400 \text{ ft-lb} = \text{wt} \times 100 \text{ ft};\ \text{wt} = 2334 \text{ lb}$$

4. **(E)** 1 lb of gasoline supplies 15,000 Btu; 6 lb of gasoline supply 90,000 Btu. 1 Btu = 778 ft-lb

$$90{,}000 \text{ Btu} \doteq 70.7 \times 10^6 \text{ ft-lb}$$
$$\text{output} = \text{efficiency} \times \text{input} = 0.30 \times 70.7 \times 10^6 \text{ ft-lb} = 20 \times 10^6 \text{ ft-lb}$$

5. **(B)** pressure = force/area; force = pA

$$\text{work} = \text{force} \times \text{distance} = \text{pressure} \times \text{area} \times \text{distance}$$
$$\text{work} = 50 \text{ lb/in.}^2 \times 120 \text{ in.}^2 \times 1 \text{ ft} = 6000 \text{ ft-lb}$$

Note that the distance the piston moves has to be converted to feet in order to have the unit of work be in ft-lb. Also note that the diameter of the flywheel is irrelevant.

6. **(B)** This question involves mere recall of a fact mentioned in the chapter which everyone is expected to know.

7. **(B)** Although the black material is a good radiator, this is irrelevant. The important part is that the plastic is a good insulator of heat. It is not desirable that the plastic soften. The term thermoplastic is not essential for elementary physics. It refers to the property of softening when heat is applied. Such a plastic is usually not desirable in handles of kitchen utensils; choice (E) is essentially the same as choice (D).

8. **(E)** Convection of any kind would lead to the rising of the heated gases, in this case up the chimney, not to the person. Very little conduction to the person can take place, since the intervening air is a poor conductor.

9. **(C)** All units are units of energy (heat or any other form) except horsepower, which is a unit of power.

10. **(C)** When a gas is compressed, work is done on the gas. This tends to heat the gas even if friction is absent. If the compression takes place rapidly, the heat produced can't escape completely and the temperature of the gas rises. Similarly, if a gas is allowed to expand rapidly its temperature goes down.

Wave Motion and Sound

What is the difference between ultrasonic and supersonic? What happens to the speed of sound under a bell jar from which the air is being exhausted? What happens to the loudness of sound when a second wave arrives at a point where before there was only one wave?

WAVE MOTION

Wave motion in a medium is a method of transferring energy through a medium by means of a distortion of the medium which travels away from the place where the distortion is produced. The medium itself moves only a little bit. For example, a pebble dropped into quiet water disturbs it. The water near the pebble does not move far, but a disturbance travels away from that spot. This disturbance is the wave. As a result some of the energy lost by the pebble is carried through the water by a wave, and a cork floating at some distance away can be lifted by the water; thus the cork gets some energy. We can set up a succession of waves in the water by pushing a finger rhythmically through the surface of the water. A vibrating tuning fork produces waves in air.

Waves can be classified in different ways. The wave produced by the pebble dropped into the water is a *pulse:* a single vibratory disturbance which travels away from the source of the disturbance through the medium. If we push our finger regularly and rapidly down and up through the surface of the water, we produce a periodic wave.

Periodic motion

Periodic motion is motion which is repeated over and over again. The motion of a *pendulum* is periodic. As long as its arc of swing is small, the time required for a back-and-forth swing is constant. This time is called its period and is given by the formula $T = 2\pi\sqrt{L'/g}$, where L' is the length of the string from which the bob is suspended, and g is acceleration due to gravity.

The motion of the tuning fork prongs is periodic. As the prongs vibrate back and forth they push particles in the air; these push other particles, etc. When the prong moves to the right, particles are pushed to the right; when the prong moves to the left, the particles move back into the space left by the prong. As the prong moves back and forth, the particles of the medium move back and forth *with the same frequency* as the prong. The energy travels away from the vibrating source, but the particles of the medium vibrate back and forth past the same spot (their equilibrium position).

Longitudinal and Transverse Waves

Two basic waves are the longitudinal wave and the transverse wave. The wave produced in air by a vibrating tuning fork illustrates the former. A *longitudinal wave* is a wave in which the particles of the medium vibrate back and forth along the path which the wave travels. Sound waves are longitudinal waves.

A *transverse wave* in a medium is a wave in which the vibrations of the medium are at right angles to the direction in which the wave is traveling. A transverse wave is readily set up along a rope. Either a longitudinal or transverse wave can be readily sent along a helical spring. A water wave is approximately transverse. The propagation of light can be described as a transverse wave. Since light can travel through a vacuum, this raises an interesting question regarding what is vibrating. This will be discussed further under the heading of electromagnetic waves.

Simple longitudinal and simple transverse waves can be represented by a sine curve. Let us first consider the production of a transverse wave set up in a rope. Imagine the rope stretched hori-

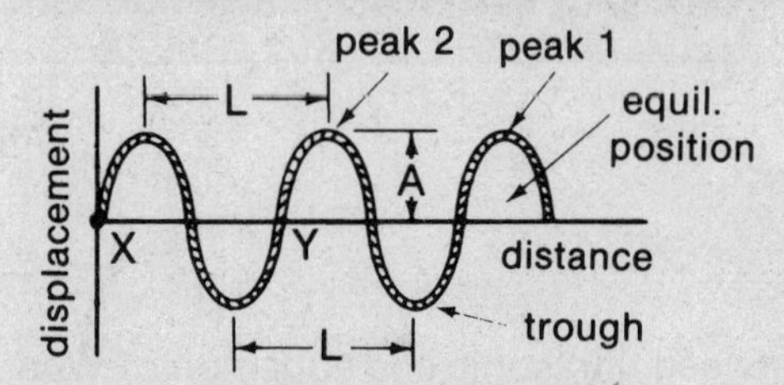

zontally with a prong of a vibrating tuning fork attached to the left end of the rope so as to produce a transverse wave in it. When the prong moves up, the rope is pulled up and the upward disturbance starts to travel to the right. As the prong moves down, the rope is pulled down and a downward displacement moves along the rope, following the upward displacement. When the prong moves up again, it produces the second peak to follow peak 1.

The whole pattern or wave shown can be imagined moving to the right with a certain speed, the speed of the wave in the medium (the rope). If we keep our eye on some specific spot on the rope, we see that part of the rope going up and down, transverse, or at right angles to the direction of motion of the wave as a whole.

The *frequency of vibration of each part of the rope is the same as the frequency of vibration of the source* (*the tuning fork*) *and the same as the number of peaks or troughs produced per second. The amplitude* of the wave is the maximum displacement or distance moved by each part of the medium away from its average or equilibrium position. In the diagram *A* represents the amplitude of the wave. *L* is the *wavelength*–the distance between any two successive peaks, or the distance between any two successive corresponding parts of the wave, such as *X* and *Y*.

As we watch the wave moving past a given spot of the medium, we can see peak after peak moving past the spot. The time required for two successive peaks to pass the spot is known as the *period* of the wave. It is the time required for the wave to move a distance equal to one wavelength. It is the same as the period of vibration of the source.

Since, for motion with constant speed, distance = speed × time, the speed of the wave equals the wavelength divided by the period. The *frequency* (*f*) of the wave is the number of complete vibrations back and forth per second. Since the period is the time for one vibration,

$$\mathbf{T = \frac{1}{f}}$$

Therefore, the speed of the wave equals the wavelength divided by one over the frequency. Instead of dividing, invert and multiply. The end result is the important relationship for a wave.

speed of a wave = frequency of wave × wavelength

$$\mathbf{v = f \times L}$$

The Greek letter *lambda* (λ) is frequently used for wavelength.

When something occurs over and over again, the term *cycle* is sometimes used to refer to the event which is repeated. We can speak of a wave having 100 vibrations per second or having 100 cycles per second. The term *wave* is sometimes used to refer to one cycle of the disturbance in the medium. In the sine curve shown above, the portion of the curve from X to Y would then be one wave.

Example: A wave having a frequency of 1000 cycles/sec has a speed of 1200 m/sec in a certain solid. Calculate the period and the wavelength.

The period $T = (1/f) = (1/1000)$ sec $= 0.001$ sec.

The wavelength $L = (v/f) = (1200 \text{ m/sec})/(1000 \text{ cy/sec}) = 1.2$ m/cy. Notice that, in the calculation of wavelength, *sec* cancels. Often we also drop *cycle* from the answer and give the wavelength as 1.2m.

The sine curve shown above can be used to represent a longitudinal wave as well as a transverse wave. This will be described under the discussion of sound. The above relationships and definitions apply to both types of waves. All waves also exhibit certain phenomena in common: reflection, refraction, interference, and diffraction. In addition, transverse waves exhibit the phenomenon of polarization. These phenomena will be described later.

SOUND

In physics, when we speak about sound, we usually mean the sound wave. Sound waves are longitudinal waves in gases, liquids, or solids. They are produced by vibrating objects. Sound cannot be transmitted through a vacuum. In order to describe the sound wave let us imagine the production of sound by means of a tuning fork vibrating in air. As the prong moves to the right, air is pushed to the right; since the air at the right is pushed closer together, a region of higher pressure is produced. This region of higher pressure is known as a *compression* or *condensation.* Because of the higher pressure, some air from the compression travels on to the right, thus producing another compression. In this way a compression travels away from the tuning fork. The speed with which it travels is the speed of the wave in air. In the meantime, because of its elasticity, the prong of the tuning fork starts moving back towards the left. As it does so air from the right rushes into the space left by the prong. The space from which the air rushes then becomes a region of lower pressure than normal. Regions of reduced pressure are known as *rarefactions*. Air at the right of the rarefaction is pushed towards the left, in turn producing a rarefaction further over to the right. Thus a rarefaction travels out to the right, following the compression. When the prong starts moving to the right again it starts another condensation, etc. The *longitudinal wave* is a succession of compressions and rarefactions, one complete vibration of the fork producing one compression and rarefaction.

If the wave hits another medium, such as a solid, compressions and rarefactions will be set up in the new medium. The frequency of the wave in the new medium will be the same as in air, but the speeds, and therefore the wavelengths, will be different.

If a sine curve is to be used to represent a longitudinal wave, the Y-axis may be used to represent the change in pressure of the medium as a result of the wave. A compression has an increase in pressure and can be represented as a positive number, while a rarefaction has a decrease in pressure, which would be represented by a negative number. The abscissa is used as for the transverse wave.

The abscissa may be the distance along the direction of wave travel. The curve then represents the change in pressure at a given moment plotted against distance along the medium. A crest then represents a compression, a trough a rarefaction. The wavelength is the distance between successive compressions or successive rarefactions. (It is interesting to note that a point of maximum pressure is also a point of zero displacement.)

A sine curve can also be drawn to represent the change in pressure at a *specific point* in the medium against *time.* The distance between successive compressions on such a graph would give the *period* of the wave as well as the period of vibration of a particle in the medium.

The *speed of sound* in air is approximately 1090 ft/sec or 331 meters/sec at 0° C. The speed goes up 0.6 m/sec (2 ft/sec) for each Celsius degree that the temperature goes up. Don't get mixed up in memorizing this fact—note that the change in speed is given per Celsius degree in both systems of units. The speed of sound in air is independent of the atmospheric pressure. In general, sound travels faster in liquids and solids than in air.

Musical Sounds

Sounds produced by regular vibration of the air are musical. Irregular vibrations of the air are classified as unpleasant sounds or noise. This is often demonstrated with a *siren* consisting of a wheel with concentric sets of holes. The wheel rotates at constant speed around an axis perpendicular to the wheel and passing through its center. The holes in the two innermost circles are evenly spaced. When air is blown gently at either of these circles, a pleasant, musical sound is heard. The wheel blocks the air at a constant rate and the blasts of air going through the holes produce compressions at a constant rate. The holes in the outermost circle are irregularly spaced. When air is blown gently at this circle, air allowed through the holes produces successive compressions which follow each other at irregular intervals. What a person perceives is noise.

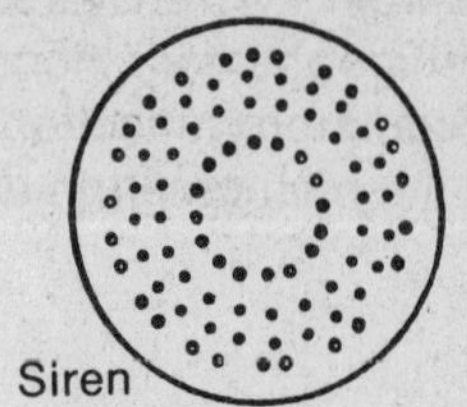
Siren

The *range in frequencies* of musical sounds is approximately 20-20,000 cycles per second (cy/sec). Some people can hear higher frequencies than others. Longitudinal waves whose frequencies are higher than those within the audible range are called *ultrasonic* frequencies. Ultrasonic frequencies are used in sonar for such purposes as submarine detection and depth finding. Ultrasonic frequencies are also being tried for sterilizing food since these frequencies kill some bacteria. Sound waves of all frequencies in the audible range travel at the same speed in the same medium. In the audible range, the higher the frequency of the sound the higher is the *pitch.* For example, in the above siren, the pitch produced with the middle circle of holes is higher than with the innermost circle. The term *supersonic* refers to speed greater than sound. An airplane traveling at supersonic speed is moving at a speed greater than the speed of sound in air at that temperature. *Mach 1* means a speed equal to that of sound; Mach 2 means a speed equal to twice that of sound, etc.

Musical sounds have three basic *characteristics:* pitch, loudness, and quality or timbre. As was indicated above, *pitch* is determined largely by the frequency of the wave reaching the ear. The higher the frequency the higher is the pitch. *Loudness* depends on the amplitude of the wave reaching the ear. For a given frequency, the greater the amplitude of the wave the louder the sound. To discuss quality of sound we need to clarify the concept of overtones. Sounds are produced by vibrating objects; if these objects are given a gentle push, they usually vibrate at one definite frequency producing a pure tone. This is the way a tuning fork is usually used. When objects vibrate freely after a force is momentarily applied, they are said to produce their *natural frequency.* Some objects, like strings and air columns, can vibrate naturally at more than one frequency at a time. The lowest frequency which an object can produce when vibrating freely is known as the object's *fundamental frequency;* other frequencies that the object can produce are known as its *overtones.* The *quality* of a sound depends on the number and relative amplitude of the overtones present in the wave reaching the ear.

When a wave reaches another medium, part of the wave is usually reflected. When a sound wave is reflected, a distinct *echo* is heard if the reflected sound wave reaches the ear at least 1/10 second after the sound traveling directly from the vibrating source to the ear. Echoes may be used to determine the distance of reflecting objects, as in sonar. The total distance traveled by the wave is equal to the speed of the wave in the medium multiplied by the time between the start of the sound and the arrival of the echo. If the reflected wave comes back along the same path as the incident wave, the distance to the reflecting surface, such as a wall, is one-half of the total distance.

Resonance and Interference

As was stated above, objects tend to vibrate at their natural frequency. A simple pendulum has a natural frequency. (Recall that

period of a simple pendulum $= 2\pi\sqrt{\textit{length/acceleration due to gravity}}$

and frequency equals one over the period: $f = 1/T$. Similarly a swing has a natural frequency of vibration. If given a gentle push the swing will vibrate at this frequency without any other push. Because of friction the amplitude of vibration decreases gradually; to keep the swing going, it is necessary to supply only a little push to the swing at its natural frequency or at a submultiple of this frequency. The swing can be forced to vibrate or oscillate at many other frequencies. All one has to do is to hold on to the swing and push it at the desired frequency. Such vibration is known as *forced vibration.* For example, our eardrum can be forced to vibrate at any of a wide range of frequencies by the sound reaching the ear. We can think of a distant vibrating tuning fork forcing the eardrum to oscillate by means of a longitudinal wave from the fork to the ear.

Resonance

Resonance between two systems exists if vibration of one system results in vibration or oscillation of the second system, whose natural frequency is the same as that of the first. In the case of sound the two systems may be two tuning forks of the same frequency. We mount the two forks on suitable boxes at opposite ends of a table. We strike one tuning fork, let it vibrate for a while, and then put our hand on it to stop the vibration. We will then be able to hear the second fork. It resonated to the first one. What caused the second tuning fork to vibrate? Compressions starting from the first fork gave the second fork successive pushes at the right time to build up the amplitude of its oscillations. Thus energy was transferred from the first fork to the second fork by means of the wave. When resonance occurs with sound, the term *sympathetic vibration* is also used: the second fork is said to be in sympathetic vibration with the first one. Later we shall discuss cases of resonance which do not involve sound waves.

One can think of an *air column* in a tube as having a natural frequency. The natural frequency of an air column depends on its length. If a vibrating tuning fork is held over an air column of the right length, resonance will occur between the fork and the air column; if the length of the air column is varied, a loud sound will be heard at resonance. The length of the closed resonant air column is about one-fourth of the wavelength, as will be discussed later.

length = L/4

Interference

Interference occurs when two waves go through the same portion of the medium at the same time. If, at a given time and place, the two waves tend to make the medium vibrate in the same direction, we have *reinforcement* or constructive interference. For example, waves I and II combine to give wave III.

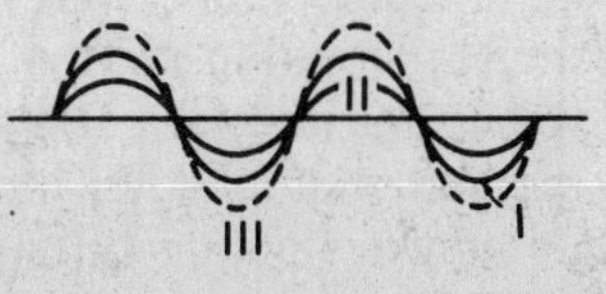

Note in the diagram that in waves I and II peaks coincide with peaks and troughs with troughs. Such waves are said to be in *phase.* They produce maximum reinforcement or *constructive*

interference. Sometimes we speak about *points* in a given wave as being in phase. For example, which point is in phase with point *B*?

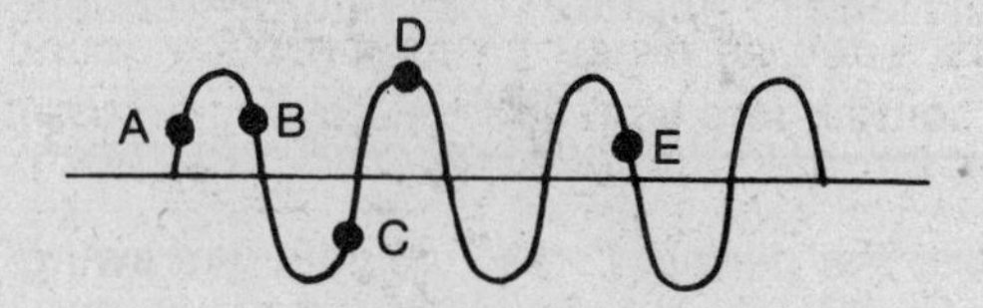

We can think of the diagram as representing a periodic wave moving to the right. The particles have been displaced from their equilibrium position represented by the horizontal line. Points on a wave which are a whole number of wavelengths apart are in phase; they are going through the same part of the vibration at the same time. Point *E* is two full wavelengths ahead of point *B*. They are in phase. Point *C* is about one-half of a wavelength ahead of point *B*. They are opposite in phase.

If two waves tend to make the medium vibrate in opposite directions, we have *destructive interference.* The two waves then tend to cancel each other. For example, waves IV and V together produce wave VI.

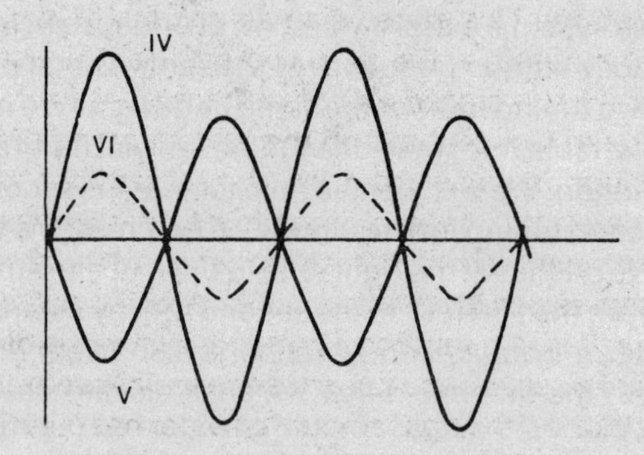

The peaks of one coincide with the troughs of the other. This is the condition for maximum destructive interference. Waves IV and V are said to be opposite in phase, or 180° apart, or 180° out of phase.

If two sounds reach the ear at the same time, interference takes place.

Beats are heard when two notes of slightly different frequencies reach the ear at the same time.

the number of beats = the difference between the two frequencies

A beat is an outburst of sound followed by comparative silence. Beats result from the interference of the two waves reaching the ear. When two compressions or two rarefactions combine, the eardrum vibrates with relatively large amplitude and a loud sound is heard. When a compression combines with a rarefaction, they tend to annul each other and relatively little sound is heard.

The resultant wave produced by two sources of slightly different frequencies sounded at the same time can be represented like this:

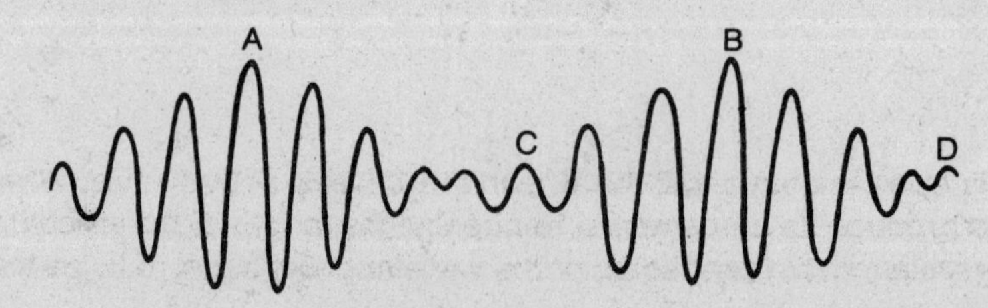

Notice that at *A* and *B* we have a relatively large amplitude corresponding to an outburst of sound and at *C* and *D* a rather small amplitude corresponding to comparative silence.

VIBRATING AIR COLUMNS

Some musical instruments, such as organ pipes, produce sound by means of vibrating air columns. We can set the air column into vibration by blowing across or into one end of the pipe.

Closed Pipes

A closed pipe has one end closed. The end we blow into is always considered open. As indicated before, one way to describe the vibrating air column is to say that it has natural frequencies of vibration. Its lowest frequency of vibration (fundamental) is one for which the wavelength equals approximately four times the length of the air column. *For example,* a closed air column which is one foot long produces a sound whose wavelength is 4 ft. If the speed of sound is 1100 ft/sec, the frequency of vibration = speed of sound divided by the wavelength, or 1100 ft/sec divided by 4 ft, which gives 275 vibrations per sec for the frequency (f). This is the fundamental frequency. The *overtones* which a closed air column can produce are odd multiples of the fundamental frequency ($3f$, $5f$, $7f$, etc.).

Another way to describe a vibrating air column is in terms of resonance. Imagine a narrow cylinder with some water at the bottom. This gives us an air column of length l_a. Since water is practically incompressible, the air in contact with the water cannot move down. This is, therefore, a closed air column. Imagine a tuning fork vibrating above the air column. Concentrate on the bottom prong of the fork. Its rest or equilibrium position is shown at *r;* its highest position at *4;* its lowest position during vibration at *2.* As the prong moves from *4* towards *2,* a compression forms at *r* and travels down the air column; when it reaches the water surface the compression is reflected. If the reflected compression reaches the prong just when it is at *r,* sending a compression in the direction away from the tube, the fork's vibration will be reinforced. The result is resonance between the vibrating tuning fork and air column—a loud sound is heard if the right amount of water has been poured in. How long should this air column be? The time required for the prong to go from *4* to *2* is one-half of the period of vibration. During this time the wave travels a distance equal to one-half of a wavelength (L). But in this experiment this distance traveled was up and down the air column. Therefore $2l_a = L/2$; or

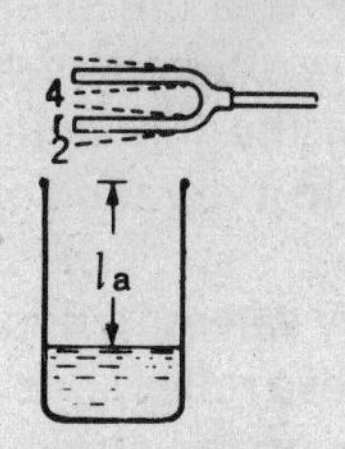

$$\boxed{L = 4l_a}$$

That is, for a resonant closed air column, the wavelength of the fundamental is equal to four times the length of the air column, as was stated before. However, suppose with the same air column we use a tuning fork whose frequency is three times higher than the one we used before. Its prong will vibrate from *4* to *2* in one-third the time, but the speed of the wave is not changed. Therefore, during this time the wave will have traveled only one-third of the total back-and-forth distance, which is two-thirds of the length of the air column. By the time the wave is reflected by the water and comes back to the tuning fork the prong has gone back to *4* and has just reached *2* again. Again the reflected air wave reinforces the vibration of the tuning fork. Again the air column resonated to the tuning fork. This time the closed air column vibrated at three times its fundamental frequency and produced its first overtone.

Open Pipes

An *open pipe* is open at both ends; the air can move freely at both ends. When an open pipe is vibrating so as to produce its fundamental frequency, the length of the air column (l_a) is equal to one-half of the wavelength *(L)* produced; or the wavelength is equal to twice the length of the air column.

$$\boxed{L = 2\,l_a}$$

An open pipe can produce not only odd multiples of its fundamental frequency, but also even multiples. The first overtone is twice the fundamental frequency; the second overtone three times, etc. ($2f$, $3f$, $4f$, $5f$, etc.)

VIBRATING STRINGS

If a string is fastened at both ends and plucked or bowed at some point in between, it will vibrate transversely. Some frequencies of vibration will be natural for the string and will persist. The *fundamental* frequency of the string is the lowest of these natural or free vibrations; for this frequency, the wavelength in the string is equal to twice the length of the string (l_s).

$$L = 2 l_s$$

Notice that this describes the wave in the *string.* The transverse vibration of the string sets up a longitudinal wave in the surrounding air: the sound wave. The frequency of the sound wave is the same as the frequency of vibration of the string; but since the speed of the wave in the string is different from the speed of the wave in the air, the wavelengths in the two media are different ($v = fL$). The *overtones* which a string can produce are the odd and even multiples of the fundamental frequency ($2f$, $3f$, $4f$, $5f$, etc.). Notice that in this respect a vibrating string is similar to a vibrating open air column.

Vibrating strings can also be described in terms of *standing waves* or stationary waves. When the string is plucked, the wave travels to both ends from which it is then reflected. These reflections can take place over and over again. As a result waves go in opposite directions through the string and interference takes place. At the ends, where the string is fastened, no vibration can take place. These places of no vibration are known as *nodes.* In between there may be one or more places where the waves reinforce each other completely. These places of maximum vibration are known as *antinodes.* When overtones are produced, there will also be one or more nodes along the length of the string. The distance between successive nodes is one-half a wavelength of the wave in the string. When a string produces its fundamental frequency, it is said to vibrate as a whole; when it produces overtones, it vibrates in parts.

Laws of Vibrating Strings

The fundamental frequency of vibration of a string depends on its length, its tension, and its mass per unit length. If one of these is varied at a time, the effect is as follows:

1. The frequency varies inversely as the length; i.e., making the string twice as long produces a frequency one-half as high.
2. The frequency varies directly as the square root of the tension; i.e., to double the frequency the force stretching the string must be quadrupled.
3. The frequency varies inversely as the square root of the mass per unit length.

DOPPLER EFFECT

If there is relative motion between the source of the wave and the observer, the frequency of vibrations received by the observer will be different from the frequency produced by the source. In the case of sound this will affect the pitch of the perceived sound, since pitch depends on the number of vibrations reaching the ear per second. When the source and observer approach each other, the pitch will be higher than if both were stationary; the pitch will be lower if source and observer are going further apart.

In the top diagram the source is stationary. The diagram shows compressions going to the right

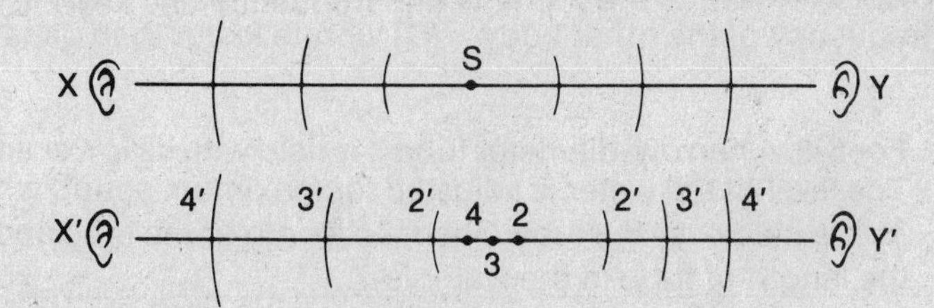

and left away from the source at *S*. The compressions are equally spaced, and the pitch will be the same for the stationary observers at *X* and *Y*.

The bottom diagram shows the compressions going to the right and left as a result of motion of the source at constant speed from point *4* to 2. Compression *4'* was produced when the source was at *4;* compression *3'* when the source was at *3*, etc. Notice that the compressions are more closely spaced in the direction towards which the source is moving (towards the right in this case). Therefore the pitch for observer *Y'* will be higher than for observer *X'*. Notice that the wavelength changed as a result of the source's motion. It decreased in the direction of motion and increased in the opposite direction.

Decibel

The human ear responds to an extremely wide range of sound intensities. It has therefore become customary to express the level of sound intensity or sound power by the logarithm of the ratio comparing the intensity or power to that of an arbitrary sound. The bel was orginally used as the unit to express this level; the decibel is now used. The intensity or power is said to be a number of decibels (db) above (or below) the reference level as given by the expression:

$$\textbf{decibels} = 10 \log \frac{P_1}{P_2}$$

where P_2 is the power or intensity level of the reference and P_1 is the power or intensity of the sound being investigated. The reference level (zero db) is usually taken as the one that is barely audible. The sound level of ordinary speech is about 20 db. The same system is used to describe the output of electronic amplifiers. The reference level in amplifiers is usually taken as 6 milliwatts of power.

QUESTIONS Chapter 9

Select the choice which fits the question best.

1. In order for two sound waves to produce beats it is most important that the two waves **(A)** have the same frequency **(B)** have the same amplitude **(C)** have the same number of overtones **(D)** have slightly different frequencies **(E)** have slightly different amplitudes

2. A boy produces a certain note when he blows gently into an organ pipe. If he blows into the pipe harder, the most próbable change will be that the sound wave **(A)** will travel faster **(B)** will have a higher frequency **(C)** will have a greater amplitude **(D)** will have a lower frequency **(E)** will have a lower amplitude

3. When the speed of sound in air is 330 meters per sec, the shortest air column, closed at one end, that will respond to a tuning fork with a frequency of 440 vibrations per sec has a length of approximately **(A)** 19 cm **(B)** 33 cm **(C)** 38 cm **(D)** 67 cm **(E)** 75 cm

4. A student wanted to make a pendulum whose period would be one second. He used a string of length *L* and found that the period was ½ sec. To get the desired period he should use a string whose length equals **(A)** ¼ *L* **(B)** ½ *L* **(C)** 2 *L* **(D)** 4 *L* **(E)** L^2

5. If a vibrating body is to be in resonance with another body, it must **(A)** be of the same material as the other body **(B)** vibrate with the greatest possible amplitude **(C)** have a natural frequency close to the natural frequency of the other body **(D)** vibrate faster than usually **(E)** vibrate more slowly than usually

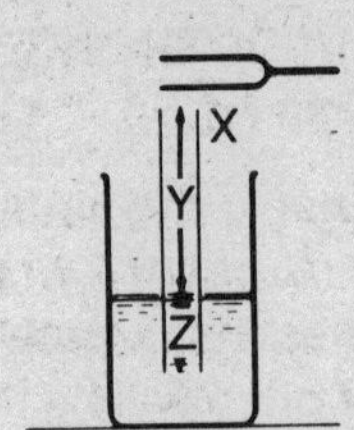

For 6-8. A narrow-diameter tube *X* is held with its lower end immersed in water. The level of the water is adjusted for maximum sound while a vibrating tuning fork is held over the tube. Then the length of the tube above the water is *Y* and the length of tube in the water is *Z*.

If fundamental frequencies only are involved,

6. the wavelength of the sound in the air will be approximately **(A)** 2 *Y* **(B)** 2 *Z* **(C)** 4 *Y* **(D)** 4 *Z* **(E)** *Y*—*Z*

7. the wavelength of the sound in water will be **(A)** less than 2 *Z* **(B)** between 2 *Z* and 2 *Y* **(C)** between 2 *Y* and 4 *Y* **(D)** 4 *Y* **(E)** greater than 4 *Y*

8. In the time required for the tuning fork to make one complete vibration, the wave in air will travel a distance equal to **(A)** wavelength/4 **(B)** wavelength/2 **(C)** wavelength **(D)** twice the wavelength **(E)** 1100 ft

For 9-10. A vibrating diaphragm sets up strong vibrations at the mouth of a horizontal tube containing air and a small amount of fine powder. If the powder becomes arranged in piles 1 cm apart and the speed of sound in air is 330 meters/sec,

9. the wavelength of this sound in air is **(A)** ¼ cm **(B)** ½ cm **(C)** 1 cm **(D)** 2 cm **(E)** 4 cm

10. the frequency of this sound is, in vibrations/sec, **(A)** 165 **(B)** 330 **(C)** 8,300 **(D)** 16,500 **(E)** 33,000

11. If a man moves, with a speed equal to 0.5 that of sound, away from a stationary organ producing a sound of frequency *f*, he would probably hear a sound of frequency **(A)** less than *f* **(B)** *f* **(C)** 1.5 *f* **(D)** 2.25 *f* **(E)** 2.5 *f*

12. A certain stretched string produces a frequency of 1000 per sec. If the same string is to produce a frequency twice as high, the tension of the string should be **(A)** doubled **(B)** quadrupled **(C)** reduced to ½ of the original value **(D)** reduced to ¼ of the original value **(E)** increased by a factor of $\sqrt{2}$

EXPLANATIONS TO QUESTIONS Chapter 9

Answers

1. **(D)**	4. **(D)**	7. **(E)**	10. **(D)**
2. **(C)**	5. **(C)**	8. **(C)**	11. **(A)**
3. **(A)**	6. **(C)**	9. **(D)**	12. **(B)**

Explanations

1. **(D)** The number of beats is equal to the difference in the frequencies of the two sounds. If the frequencies were the same there would be no beats. Even if the amplitudes are not the same, at times the two sounds will reinforce each other and at other times they will partially annul each other, giving rise to beats.

2. **(C)** The speed of all audible sound frequencies in air is independent of the frequency or the amplitude of vibration. However, the greater energy made available by blowing harder will lead to a greater amplitude of the wave. (It may also lead to the production of overtones, but these are higher in frequency than the fundamental.)

3. **(A)** speed of sound = freq. × wavelength; wavelength = v/f = (330 m/sec)/(440 vib/sec) = 0.75 m = 75 cm. The shortest closed air column for resonance = $L/4$ = 75 cm/4 ≐ 19 cm.

4. **(D)** For a simple pendulum, the period = $2\pi\sqrt{L'/g}$; the period is proportional to the square root of the length; if the length is multiplied by 4, the period is multiplied by the square root of 4, or by 2; i.e., if we want the period to be doubled, we must use a length of string four times the original length.

5. **(C)** This is described as the condition for resonance in the chapter: natural frequencies that are close to each other. The vibrations can be of rather small amplitude; the materials may be the same or different.

6. **(C)** This was described in the chapter as resonance for a closed air column. The resonant air column Y' is then one-fourth of the wavelength in air; or the wavelength in air = 4 Y.

7. **(E)** Z has nothing to do with the wavelength in water. However, the speed of sound is greater in water than in air; the frequency of the sound in water is the same as the original sound in air. Since speed of sound = freq. × wavelength, if the speed increases and the freq. remains the same, the wavelength has to increase also.

8. **(C)** This is described in the chapter as part of the discussion of the *period.*

9. **(D)** The problem describes the production of a standing wave. The powder forms piles at the nodes. The distance between successive nodes is ½ wavelength. Therefore ½ wavelength = 1 cm; wavelength = 2 cm

10. **(D)** freq = (speed of wave)/wavelength = (33,000 cm/sec)/2 cm = 16,500 per sec.

11. **(A)** This is an example of the Doppler effect. Since the man is moving away from the source of the sound, successive compressions will not reach him in such rapid succession as when he is stationary. (However, since he is moving at a speed less than sound, the compressions do catch up with him until the amplitude is too small to have any effect.) Therefore, the frequency he hears is less than f.

12. **(B)** The frequency of vibration of a string is proportional to the square root of its tension. The tension must be increased fourfold in order to have the frequency be doubled.

10

Nature of Light; Reflection

In many ways light behaves like a wave. For example, it is possible to have two light beams interfere with each other so that their combination is dimmer than either beam alone. In fact, some experiments suggest that light is a transverse wave. (Later we shall consider other experiments which the wave theory does not explain.) Since light comes to us from the sun through an excellent vacuum, one would like to know what vibration represents the wave. Until about 1900 it was customary to talk about some ether filling all of space and to describe light as a wave in this ether. Now when we think of light as a wave, we describe it as an *electromagnetic wave;* we think of electric fields and magnetic fields which vibrate (fluctuate in magnitude) at right angles to the direction in which the light travels. These fields can exist in a vacuum. Light is one type of electromagnetic wave; its wavelength is rather short: about 5×10^{-5} cm. The exact wavelength depends on the color of the light. Xrays are also electromagnetic waves—their wavelength is about 10^{-8} cm or 10^{-10}m. (One angstrom unit $= 10^{-10}$m.) Radio waves are electromagnetic waves with rather long wavelengths; they may be many meters long. (Electric and magnetic fields will be discussed later.)

Some phenomena, like the photoelectric effect, cannot be explained satisfactorily with the wave theory of light. We then turn to the *quantum theory* of light, which states that light is emitted and absorbed in little bundles or lumps of energy. These bundles of energy are known as *photons* or quanta (singular is *quantum*). The amount of energy in each quantum depends on the frequency of the light when described as a wave:

$$\boxed{E = hf}$$

where *f* is the frequency of the light and *h* is Planck's constant, named after Planck, who originated the quantum theory. (This theory will be discussed further in Chapter 16.)

The *speed of light* is fantastically high. In vacuum it is about 186,000 miles per sec, or about 3×10^{10} cm/sec, or about 3×10^{8} meters/sec. In air the speed of light is only slightly less. In transparent liquids and solids the speed of light is considerably less than in air.

ILLUMINATION

A *luminous* body is one that emits light of its own. An *incandescent* object is one that emits light because it has been heated. The filament in our electric light bulbs is incandescent; the firefly is luminous, but not incandescent. An *illuminated* object is one that is visible by the light that it reflects.

The rate at which an object radiates energy is sometimes referred to as *radiant flux*. Since this is a rate of sending out energy, a unit for radiant flux is the watt, as for power. Usually only part of this radiant flux produces illumination. *Luminous flux* is that part of the radiant flux which affects the optic nerve and can be used for seeing. Its unit is the *lumen.* The size of this unit is defined in terms of a standard source of light; this used to be a special candle whose luminous intensity is stated to be one *candlepower* (cp). One standard candle is said to emit 4π lumens.

By international agreement, the unit of luminous intensity in the MKS system now is the *candela* (cd) based on glowing platinum under specified conditions.

Illumination

Illumination (illuminance) represents the rate at which light energy falls on a unit area of surface; i.e., it is the luminous flux on unit area. Its unit can therefore be lumen/m^2, lumen/ft^2, foot-candle, etc.

$$\textbf{1 foot-candle} = \textbf{1 lumen/ft}^2$$
$$\textbf{1 foot-candle} = \textbf{10.8 lumen/m}^2$$

If a point source of light is used, the illumination on a screen at right angles to the light varies inversely as the square of the distance from the source and is proportional to the luminous intensity of the source.

$$\text{illumination} = \frac{\text{intensity of source}}{(\text{distance})^2} = \frac{\text{cd}}{s^2}$$
$$\text{ft-cdle} = \frac{\text{cd}}{\text{ft}^2}$$
$$\text{lumen}/\text{m}^2 = \frac{\text{cd}}{\text{m}^2}$$

For most purposes the ordinary incandescent light bulb is considered a pretty good point source, and the above formula can be used.

A *photometer* can be used to measure illumination or to determine the intensity of a light source. One set-up uses the photoelectric effect and is similar to the exposure meter used in photography. A set-up often used in the laboratory is as follows: A screen is placed on a meter stick so as to get light on one side from a standard bulb and light on the other side from the bulb of unknown intensity. The

screen is then moved back and forth along the meter stick until the illumination on the two sides of the screen is judged to be the same. Then,

$$\frac{\text{cd}_1}{s_1^{\,2}} = \frac{\text{cd}_2}{s_2^{\,2}}$$

REFLECTION

In discussing illumination it was not necessary to use either the wave theory or the quantum theory of light. This will also be true when describing reflection. It is convenient to use the *ray* to represent the direction in which the light travels and to think of light as traveling in a straight line as long as the medium doesn't change. When light hits a surface, usually some of the light is reflected. The *normal* for this light is the line drawn perpendicular to the surface at the point where the light ray touches the surface. The light that goes towards the surface is known as *incident* light and is represented by the incident ray. The angle of incidence (i) is the angle between the incident ray and its normal. The reflected light is represented by the reflected ray; the angle of reflection (r) is the angle between the reflected ray and its normal.

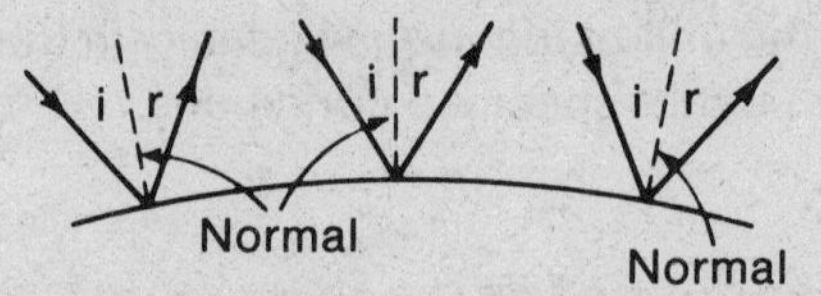

The *law of reflection* states that, when a wave is reflected, the angle of incidence equals the angle of reflection, and the incident ray, the normal, and the reflected ray lie in one plane. Notice that this applies not only to light but also to other waves and that this is true of smooth and rough surfaces.

If a surface is smooth and flat, we get *regular reflection.* Parallel incident rays are reflected as parallel rays. Most surfaces are rough to a degree. This is true even of this paper. From it we get

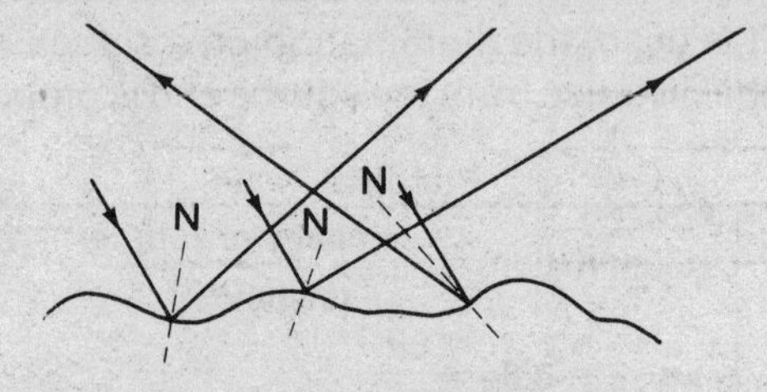

Diffuse Reflection

diffuse reflection: reflected light goes off in all directions even when the incident light is parallel. Non-luminous objects are visible because diffusely reflected light from their surfaces enters our eyes. A perfectly smooth reflecting surface cannot be seen. Luminous objects are seen because light they emit enters our eyes.

Image in Plane Mirror

A plane mirror is a perfectly flat mirror. The characteristics of an image produced by such a mirror can be determined by means of a *ray diagram:*

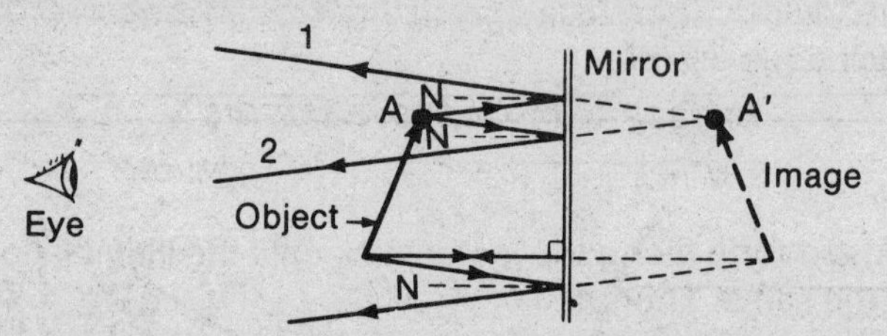

We can represent the object in some suitable way; often an arrow is used. We take two rays from the top of the object, point *A,* and draw the corresponding reflected rays, making the angle of reflection equal to the angle of incidence. The observer into whose eye these reflected rays go imagines them to come from some point behind the mirror: the point where the extensions of the reflected rays meet, point *A*. This is the image point corresponding to the top point on the object. The extensions of the reflected rays are shown dotted because the rays of light, and the image, are not actually behind the mirror. We do the same with the other end of the object. If we take a ray that is perpendicular to the surface, after reflection it will go right back along the same path. From the diagram we can determine the *characteristics* of an image in a plane mirror:

1. The image is the same size as the object.
2. The image is erect (if the object is erect).
3. The image is as far behind the mirror as the object is in front. (Every point is as far behind the mirror as the corresponding object point is in front.)

4. The image is *virtual.* This means that the image is formed by rays which do not actually pass through it; if a screen were placed at the position of the image, no image would be seen.
5. The image is laterally reversed. This can be thought of in terms of two people shaking hands: their arms extend diagonally between them. When a person holds his hand out to the mirror, the arm of his image comes out straight at him.

Image in Curved Mirrors

Spherical surfaces are commonly used as the reflecting surfaces of curved mirrors. A *convex* mirror is one whose reflecting surface is the outside of a spherical shell. A *concave* mirror is one whose reflecting surface is the inside of a spherical shell. The *center of curvature* (*C*) is the center of the spherical shell. The *principal axis* is the line through the center of curvature and the midpoint of the mirror. Incident rays parallel to the principal axis pass through a common point on the principal axis after reflection by a concave mirror. This point is the *principal focus* (*F*) of the mirror.

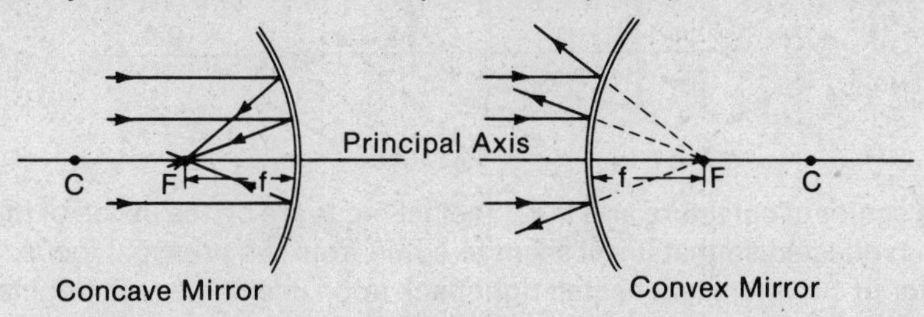

A convex mirror also has a principal focus. Incident rays parallel to the principal axis are reflected by a convex mirror so that the reflected rays seem to come from a common point on the principal axis. This principal focus is a virtual focus because the rays don't really meet there. The *focal length* (*f*) is the distance from the principal focus to the mirror. For a spherical mirror the focal length is equal to one-half of the radius of the spherical shell.

$$\mathbf{f = R/2}$$

We need one more idea in order to draw ray diagrams of spherical mirrors: a radius is perpendicular to its circle and is the direction of the normal. Therefore a ray directed along the radius of the spherical shell is perpendicular to the spherical mirror. The angle of incidence is then equal to zero and so is the angle of reflection. Therefore a ray going through the center of curvature (*C*) of the mirror is reflected right back upon itself.

Images with Concave Mirrors

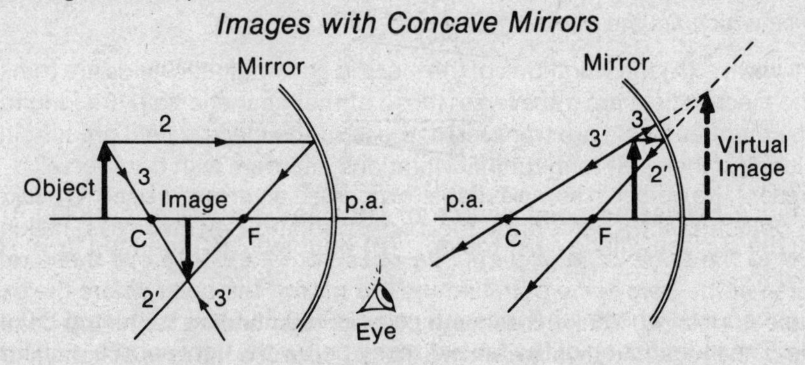

The technique for drawing the ray diagram follows: We try to find the points of the image that correspond to selected points of the object. We often use an arrow as the object, and if we place the arrow so that its tail touches the principal axis, we need to use rays only from the head of the arrow. The tail of the image will also touch the principal axis. We select two special rays from the head of the arrow; we already know their paths after reflection. Ray 2, parallel to the principal axis, after reflection, passes through the principal focus. Ray 3, whose direction passes through the center of curvature, is normal to the mirror and is reflected right back upon itself. The reflected rays, 2′ and 3′, intersect in the left diagram and produce a *real image* on a screen placed there.

In the diagram on the right the reflected rays don't meet and therefore do not produce a real image. However, a person into whose eye the reflected light goes will see an image behind the mirror, at the place where the extended reflected rays meet. This is a virtual image, as in the plane mirror. Rays extended behind the mirror are always drawn dotted. The virtual image is often shown dotted, too.

Any other ray starting from the head of the arrow, after reflection, will pass through the head of the image.

Notice that with the concave mirror we can get both real and virtual images. With the convex mirror we can get only virtual images. The procedure for the ray diagram is as above.

Images with Convex Mirrors

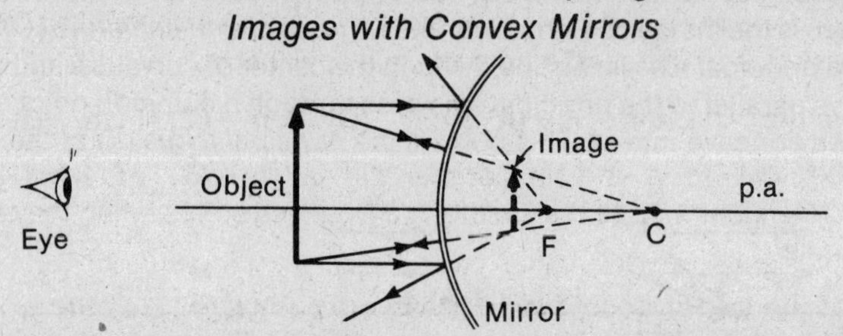

Notice that the center of curvature and the principal focus are on the inside of the spherical shell. The parallel ray is reflected so that it will seem to come from the principal focus. The ray directed towards the center of curvature is reflected right back upon itself. The rays originally starting from the same point on the object don't meet after reflection and therefore cannot produce a real image. The observer sees a virtual image behind the mirror. The virtual image formed with a convex mirror is smaller than the object; the virtual image formed with a concave mirror is larger than the object.

Careful analysis shows that spherical mirrors don't work precisely as described above. The difference is usually negligible. However, such aberrations can cause problems in astronomical telescopes using large mirrors. Instead of spherical mirrors they usually use parabolic mirrors, which are much more expensive.

Numerical calculations will be discussed in the next chapter under the heading of lenses.

QUESTIONS Chapter 10

Select the choice which fits the question best.

1. In the light wave **(A)** the vibrations of the electric and magnetic fields are transverse **(B)** the vibrations of the electric field are transverse; those of the magnetic field are longitudinal **(C)** the vibrations of the magnetic field are transverse; those of the electric field are longitudinal **(D)** all vibrations are longitudinal **(E)** longitudinal vibrations alternate with transverse.

2. If the distance between a point source of light and a surface is tripled, the intensity of illumination on the surface will be **(A)** tripled **(B)** doubled **(C)** reduced to ½ **(D)** reduced to $^1/_3$ **(E)** reduced to $^1/_9$

3. A small light source whose luminous intensity is 160 candela is located 2.0 meters above a horizontal table. The illumination of the table directly below the light source is, in lumens/m^2, **(A)** 20 **(B)** 40 **(C)** 80 **(D)** 320 **(E)** 640

For 4-5. A page in a book receives uniform illumination of 10 foot-candles from a point source of light placed 5 ft from it. The luminous flux is perpendicular to the page. The size of the page is 8″ x 6″

4. The intensity of the light source is, in candle-powers **(A)** 0.4 **(B)** 2 **(C)** 50 **(D)** 250 **(E)** 500

5. The number of lumens on the surface of the page is approximately **(A)** 3 **(B)** 10 **(C)** 35 **(D)** 100 **(E)** 480

For 6-8. *X* and *Y* are each 6 ft tall. *X* stands 4 ft from a vertical plane mirror, and *Y* 8 ft from the same mirror.

6. The size of *X's* image as compared with *Y's* is **(A)** 4 times as great **(B)** 2 times as great **(C)** the same size **(D)** ½ as great **(E)** ¼ as great

7. It will seem to *X* that *Y's* image, as compared with his own, is **(A)** 4 times as great **(B)** twice as great **(C)** the same size **(D)** smaller **(E)** beyond compare

8. The distance between *X's* image and *Y's* image **(A)** is 4 ft **(B)** is 8 ft **(C)** is 12 ft **(D)** is 16 ft **(E)** can't be calculated because of insufficient data

For 9-11. A candle 6 in. long is placed upright in front of a concave spherical mirror whose focal length is 10 cm. The distance of the candle from the mirror is 30 cm.

9. The radius of curvature of the mirror is **(A)** 5 cm **(B)** 10 cm **(C)** 15 cm **(D)** 20 cm **(E)** 6 in.

10. The image will be **(A)** virtual and smaller than the candle **(B)** virtual and larger than the candle **(C)** virtual and the same size as the candle **(D)** real and smaller than the candle **(E)** real and larger than the candle

11. If the top half of the mirror is covered with an opaque cloth, **(A)** the whole image will disappear **(B)** the whole image will be approximately one-half as bright as before **(C)** the top half of the image will disappear **(D)** the bottom half of the image will disappear **(E)** one cannot predict what will happen.

EXPLANATIONS TO QUESTIONS Chapter 10

Answers

1. **(A)**	4. **(D)**	7. **(D)**	10. **(D)**
2. **(E)**	5. **(A)**	8. **(E)**	11. **(B)**
3. **(B)**	6. **(C)**	9. **(D)**	

Explanations

1. **(A)** Light is an electromagnetic wave. In all electromagnetic waves the electric and magnetic fields vibrate at right angles (transverse) to the direction in which the wave is traveling.

2. **(E)** (The intensity of) illumination from a point source varies inversely as the square of the distance. As the distance from the light source increases, the illumination decreases. When the distance is 3 times as great, the illumination is $(^1/_3)^2$ (or $^1/_9$) as much as before.

3. **(B)**

$$\text{illumination} = \frac{\text{intensity of source}}{(\text{distance})^2}$$

$$\text{lumen}/\text{m}^2 = \frac{\text{cd}}{\text{m}^2} = \frac{160}{(2\text{m})^2} = 40$$

4. **(D)** $\text{ft-cdles} = \dfrac{\text{cd}}{(\text{ft})^2}$

$$\text{cd} = \text{ft-cdles} \times (\text{ft})^2 = 10 \times 5^2 = 250$$

5. **(A)** one foot-candle = one lumen/ft²
number of lumens on the surface = lumens/ft² × area of surface
area = 8″ × 6″ = 48 in² = (48/144) ft² = $^1/_3$ ft²
number of lumens = 10 lumen/ft² × $^1/_3$ ft² = 3.3 lumens

6. **(C)** The size of the image in a plane mirror is the same as the size of the object. Therefore, *X's* image is 6 ft tall and so is *Y's.*

7. **(D)** Although the actual size of the images is the same (see question 6), their apparent size will depend on the distance of the image from the observer, just as the apparent size of real objects will depend on their distance from the observer. The further an object or image is from the observer, the smaller it will *seem.* In this problem, since the image in a plane mirror is as far behind the mirror as the object is in front, *X's* image is 4 ft behind the mirror and 8 ft from *X.* For the same reason, *Y's* image is 8 ft behind the mirror and therefore at least 12 ft from *X,* who is 4 ft in front of the mirror. (Also see the next question.) Therefore to *X, Y's* image will seem smaller.

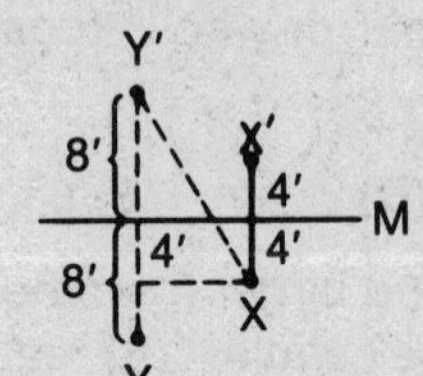

8. **(E)** The distance between the two images is the same as the distance between *X* and *Y.* This may be 4 ft or anything greater than 4 ft, as illustrated in the diagram, depending on how far over to one side *Y* stands.

9. **(D)** For a spherical mirror, the focal length = ½ of the radius of curvature. $R = 2f = 2 \times 10$ cm = 20 cm.

10. **(D)** The situation described in this problem is similar to the one for which the ray diagram is drawn in the chapter and a real image is obtained. Be able to draw such a ray diagram quickly and to deduce pertinent information from it. In this diagram the image is definitely smaller than the object. In the next chapter you will see how numerical calculations can be made rather quickly.

11. **(B)** All the rays starting from one point on the object will, after reflection from a spherical mirror, converge towards one point (for real image), or will diverge from one common point (for virtual image). This is the way points on the object are reproduced as points on the image, producing a sharp image. If part of the mirror is covered, less light will be available for each point reproduction, making the image dimmer.

Refraction of Light, Interference, Diffraction

Why is it that images seen in plane mirrors seem smaller than the corresponding objects, although "the size of the image in a plane mirror is the same as the size of the object"? Why is it that we can take a photograph of our image in a plane mirror, although "the image is virtual and cannot be seen on a screen placed at the position of the image"? We need to understand refraction and the operation of lenses and the eye to answer these and similar questions.

REFRACTION

Refraction is the bending of a wave on going into a second medium. Refraction depends on the difference in speed of the wave in the two media. The medium in which the light travels more slowly is known as the optically denser medium or, for short, the *denser medium.* This should not be confused with mass density (ratio of mass to volume of a substance). The medium in which the light travels faster is known as the (optically) *rarer* medium. Light travels faster in air than in liquids and solids. Light travels faster in water than in glass.

Law of Refraction

A ray of light passing obliquely into a denser medium is bent towards the normal; a ray of light entering a rarer medium obliquely is bent away from the normal; a ray entering a second medium at right angles to the surface goes through without being bent or deviated. The angle formed by the refracted ray and its normal is known as the angle of refraction.

The *index of refraction* (n) of a substance is defined as the ratio of the sine of the angle of incidence in vacuum to the sine of the angle of refraction in the substance. For many purposes we can use air instead of vacuum.

$$n = \frac{\sin \theta_1}{\sin \theta_2}$$

The index of refraction is also equal to the ratio of the speed of light in vacuum to the speed in the substance:

$$n = \frac{\text{speed of light in vacuum (or air)}}{\text{speed of light in the substance}}$$

The index of refraction is usually given as a number greater than one. This becomes awkward when the light goes from the denser substance into air, as in the right of the preceding diagram, where the angle of incidence is in glass and is smaller than the angle of refraction formed as the ray comes out into the air. Therefore, once we understand the principles involved, it is desirable to leave out of the definition the phrases "angle of incidence" and "angle of refraction." To calculate the index of refraction we divide the sine of the angle in vacuum (or air) by the sine of the angle in the denser medium. In the first figure shown in this chapter, for example,

$$n = \frac{\sin \measuredangle_3}{\sin \measuredangle_2}$$

In the above formulas for the index of refraction, vacuum is taken as the reference. These formulas give us the index of refraction with respect to vacuum, or the absolute index of refraction. If vacuum is not used, the relationship becomes:

$$\frac{n_2}{n_1} = \frac{\sin \theta_1}{\sin \theta_2}$$

Of course, this can be written as $n_1 \sin \theta_1 = n_2 \sin \theta_2$

If light goes through parallel surfaces of a transparent substance, such as a flat glass plate, the emerging ray is parallel to the entering ray, provided the two rays are in the same medium. If the two surfaces are not parallel, as in a triangular glass prism, the emerging ray is bent away considerably from its original direction. In drawing the ray diagram you should keep in mind the law of refraction.

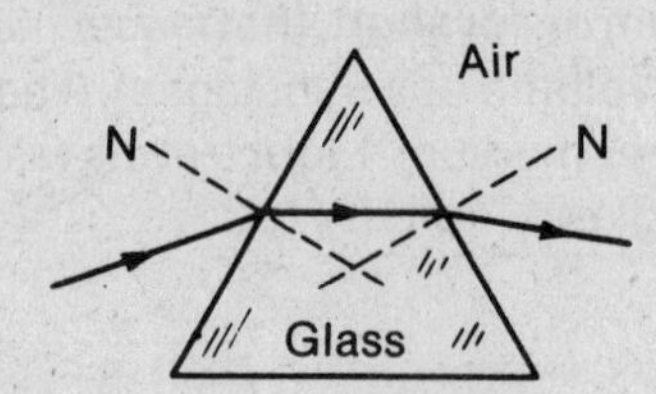

The diagram assumes that light of a single wavelength (*monochromatic light*) is used.

Critical Angle and Total Reflection

Whenever light enters a second medium whose index of refraction is different from that of the first medium, some of the light is reflected. The greater the angle of incidence, the greater is the amount of reflection. This is called *partial reflection* and was deliberately omitted in the above diagrams on refraction. In some cases all of the light is reflected. Consider this diagram, in which rays are shown making different angles of incidence; notice that they are all directed from the denser medium towards the rarer one.

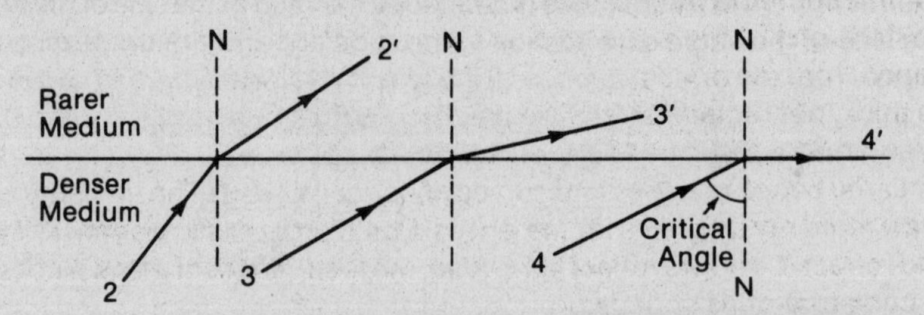

As we go from ray 2 to ray 3, etc., the angle (of incidence) in the denser medium keeps increasing, and therefore the angle (of refraction) in the rarer medium increases still more. For each ray the corresponding angle in the rarer medium is greater than the angle in the denser medium. As the angles of incidence increase, the refracted rays 2′, 3′, etc. lie closer and closer to the surface separating the two media. The *critical angle* is that angle of incidence in the optically denser medium for which the refracted ray makes an angle of 90° with the normal. *Total reflection* occurs when the angle of incidence in the optically denser medium is greater than the critical angle; then no light enters the rarer medium. Total reflection is made use of in prism binoculars and in curved transparent fibers and lucite rods: the light stays inside the curved rod because of successive internal reflections. For all reflections the angle of incidence equals the angle of reflection. For a common variety of glass the critical angle is about 42°. In the triangular glass prism shown, total reflection can take place. The acute angles are 45° and the entering ray shown is perpendicular to the glass surface and therefore is not refracted. The angle of incidence on the hypotenuse is therefore 45°, which is greater than the critical angle, and therefore there is total reflection from the hypotenuse.

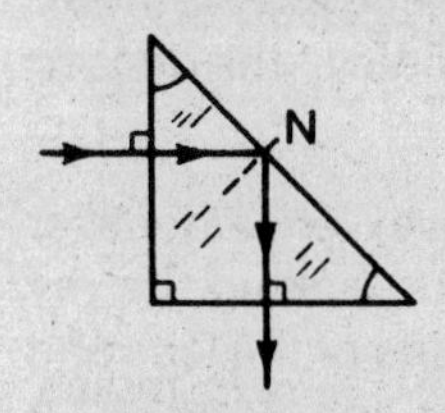

LENSES

A lens is a device shaped to converge or diverge a beam of light transmitted through it. Lenses described here are thin spherical lenses; one or both of their surfaces are spherical. If there is a nonspherical surface, it is plane. The material is transparent, usually glass. Unless stated otherwise in this description of lenses, the lens is surrounded by air. A *convex* lens is thicker in the middle than at the edges; it is also called a *converging* lens. A *concave* lens is thinner in the middle than at the edges; it is also called a *diverging* lens. The *principal axis* of a lens is the line connecting the centers of its spherical surfaces. A narrow bundle of rays parallel to the principal axis of a convex lens is refracted by the lens so that these rays, after refraction, pass through a common point on the principal axis known as the *principal focus* (*F*).

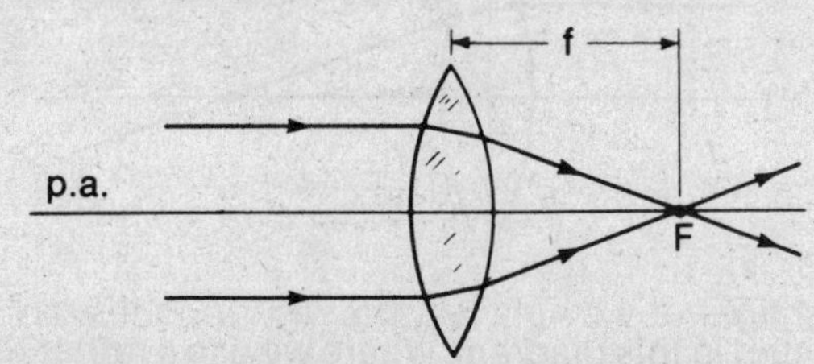

For the concave lens, rays originally parallel to the principal axis diverge after refraction and

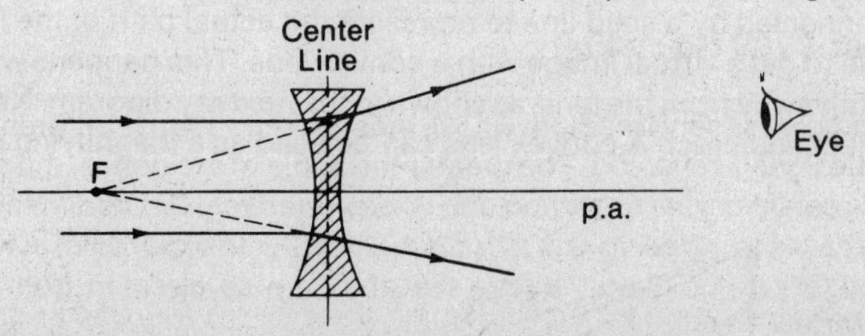

seem to come from a common point on the principal axis known as the *principal focus*. It is a virtual focus (*F*). The extensions of the actual rays are shown dotted and are called *virtual rays*. For both lenses, the distance from the principal focus to the lens is the *focal length* (*f*) of the lens. The lenses are drawn quite thick, but remember that we are discussing thin lenses only, and therefore we may show the focal length as a distance to the center line of the lens.

Notice the similarity between lenses and curved spherical mirrors. The similarity will become even greater as we draw more ray diagrams for the lenses. Don't forget one important difference: lenses let light through and refract it; mirrors reflect light. Also, the focal length of a lens is not so simply related to the radius of curvature of its surfaces.

Images Formed by Lenses

The *convex lens* is similar to the concave mirror in that it can be used to form both real and virtual images. We can show this by means of ray diagrams.

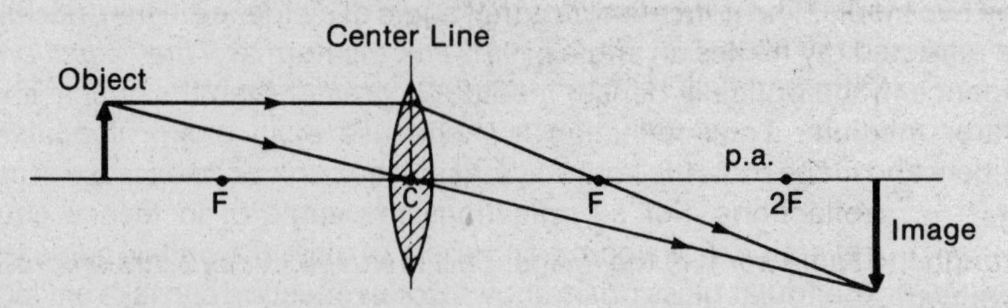

In the diagram the formation of a real image is shown. This happens with the convex lens whenever the object is more than one focal length away from the lens. The procedure for making the ray diagram is as follows. Draw the lens and its principal axis. Then draw the center line of the lens and note the center of the lens (*C*) where the principal axis intersects the center line. Mark off a convenient length along the principal axis to represent the focal length; this gives the location of the principal focus (*F*) on both sides of the lens. A point twice as far is represented by 2*F*. We usually measure this distance from the center line; since the lens is thin we can think of the center line as representing the lens with the arcs merely drawn in to remind us of the type of lens with which we are working. Place the object at some appropriate distance from the lens, depending on the situation. In this case the object was placed at a distance between one and two focal lengths away from the lens. We draw two special rays from the head of the object; one ray, parallel to the principal axis, is drawn to the center line and from there is drawn straight through the principal focus, *F*. The second ray is drawn as a straight line through the center of the lens. Where the two rays meet, the head of the image is formed. We draw in the rest of the image. A slight change is sometimes made in drawing of

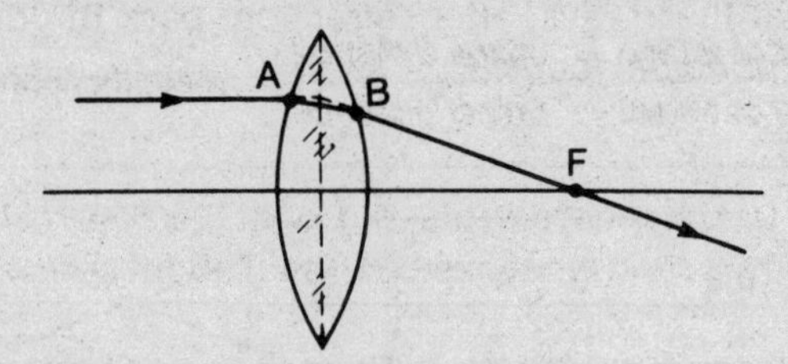

the first ray. This is exaggerated in this diagram, where we use a rather thick lens. First we proceed as above, but use broken lines inside the lens. The two points of intersection with the lens surfaces, *A* and *B*, are then connected by a solid line to represent the actual path of the light inside the lens.

It is also possible to get a virtual image with a convex lens. This happens when the object is less than one focal length away from the lens, as shown in the next ray diagram. Notice that the image is larger than the object and erect. A convex lens can be used as a magnifying glass.

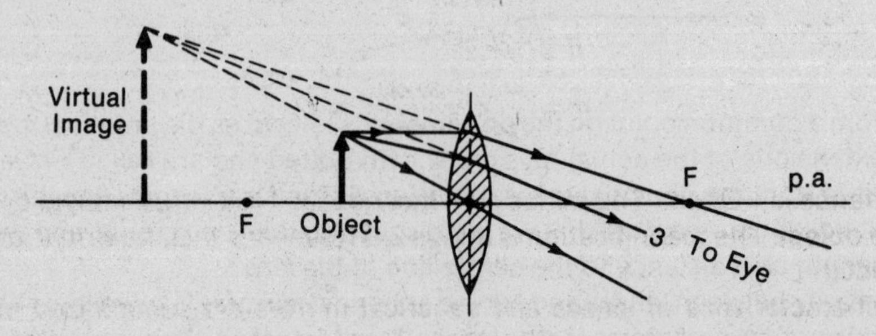

A *concave lens* always produces a virtual image which is smaller than the object and erect (if object is erect). This is illustrated in the next ray diagram. Remember, except for aberrations which will be mentioned later, all rays which start from the head of the object will, after refraction by the

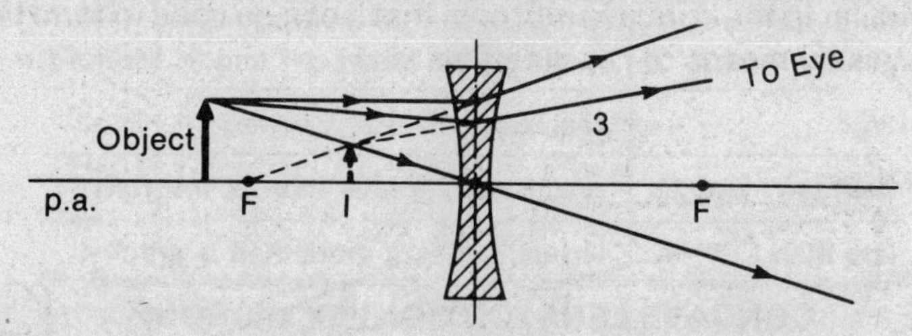

lens, pass through the same point of the image. This is shown by ray 3, drawn in after the rest of the diagram was drawn.

Numerical computations can be performed with the *lens equation.*

$$\frac{1}{\text{object distance}} + \frac{1}{\text{image distance}} = \frac{1}{\text{focal length}}$$

$$\frac{1}{p} + \frac{1}{q} = \frac{1}{f}$$

Rules

Rules in using lens equation: 1. Distances are measured from the lens; object distance is distance of object from lens, etc. 2. The object distance (p) is always positive. 3. The image distance (q) is positive for real images, negative for virtual images. 4. The focal length (f) is positive for convex lenses, negative for concave lenses.

The size (height) of the image can be calculated from this equation:

$$\frac{\text{size of image}}{\text{size of object}} = \frac{\text{image distance}}{\text{object distance}} = \text{magnification (m)}$$

The same equations apply to spherical mirrors, but what has been said in the rules about convex lenses applies to concave mirrors, and what has been said about concave lenses applies to convex mirrors.

Example: A convex lens whose focal length is 20 cm has an object placed 10 cm away from it. Describe the image.

$$\frac{1}{q} = \frac{1}{f} - \frac{1}{p}$$

$$= \frac{1}{20\text{ cm}} - \frac{1}{10\text{ cm}}$$

$$= -\frac{1}{20\text{ cm}}$$

$$\therefore q = -20\text{ cm}$$

$$m = \frac{q}{p} = \frac{-20\text{ cm}}{10\text{ cm}} = -2$$

The image distance is −20 cm. The minus sign means that it is a virtual image, on the same side of the lens as the object. The magnification is minus 2. This means that the virtual image is twice as large as the object.

Some of the characteristics of lenses and spherical mirrors are summarized in the following table.

Object Distance	Image Characteristics
CONVEX LENS (OR CONCAVE MIRROR)	
greater than $2f$	real, smaller, between f and $2f$, inverted
$2f$	real, same size, $2f$, inverted
between f and $2f$	real, larger, greater than $2f$, inverted
less than f	virtual, larger, q more than p, erect
CONCAVE LENS (OR CONVEX MIRROR)	
any distance	virtual, smaller, erect, q less than p

For example (referring to the first item in the table), when the distance of the object is more than two focal lengths away from the lens, the image is real, smaller than the object, between one and two focal lengths away from the lens, and inverted.

Note that, as the object gets closer to the lens, the image goes further away. An object "at infinity" will produce a point image a focal length away from the convex lens. A point source of light at the principal focus of a convex lens will produce light parallel to the principal axis after refraction by the lens.

Optical Instruments

The camera and eye have many points of similarity. The *camera* has a shutter to admit the light from the object at the right time. In the *eye* the eyelid corresponds to the shutter. In the camera the light then goes through an opening in a diaphragm, the aperture, through a convex lens (or combination of lenses) to the film with its surface sensitive to light. In the eye the light goes through the pupil, which is surrounded by the colored iris. The light then goes through a complicated lens system consisting of the "lens" and some liquids with curved surfaces. The light then falls on the retina where the image is formed. The image on the retina is real, inverted, reduced in size; appropriate electrical impulses are transmitted along the optic nerve to the brain. The image on the film of the camera is also real and inverted.

The refracting *astronomical telescope* and the *compound microscope* have some points of similarity. Both have convex lenses at opposite ends of a cylinder. The lens closer to the eye is known as the eyepiece; the other lens, closer to the object, is known as the objective. Keep in mind how these instruments are used. The telescope is used to look at very distant objects. Therefore the image produced by its objective lens is small in size and close to the principal focus of the objective. In order to make this image as large as possible, we examine it with the eyepiece, which is then used like a magnifying glass. The final image is virtual, if the telescope is used for visual examination.

$$\textbf{telescopic magnification} = \frac{\textbf{focal length of the objective}}{\textbf{focal length of the eyepiece}}$$

With the microscope we look at small things close by. Therefore, the object is placed at just a little more than one focal length away from the objective, to give an enlarged, real image to be examined by the eyepiece. This eyepiece is used like a magnifying glass, as in the telescope.

When we use a *projector,* we want an enlarged image on a screen. A convex lens will produce an enlarged, real image if the object is between one and two focal lengths from the lens. In a projector, the object (e.g., film) is a little more than one focal length away from a convex lens.

When a convex lens is used as a *magnifying glass,* the object is placed just a little less than a focal length away from the lens. The result is a virtual, erect, enlarged image. The magnification is, approximately,

$$\frac{25}{f\text{(in cm)}}$$

With *nearsighted* people the image of objects at ordinary distances is brought to a focus too soon in the eye. Therefore, by the time the light gets to the retina, the light has diverged again, giving a blurred image on the retina. This is corrected by eyeglasses with concave lenses, which make the light converge more gradually. With *farsighted* people the image of objects at ordinary distances would be in focus beyond the retina. To make the image come to a focus sooner, eyeglasses with convex lenses are used.

A lens is sometimes described by an *f-number,* such as *f*/8. This means that the diameter of the lens (measured along the center line) is ⅛ of the focal length. In a camera this *f*-number would mean that the aperture has been adjusted to ⅛ of the focal length. This is related to the "speed of the lens." If the *f*-number is changed from *f*/8 to *f*/4, the diameter of the aperture is doubled. Therefore, the area of the opening is quadrupled and, if the illumination of the scene is constant, four times as much light goes through the lens, and only ¼ of the exposure time is required.

COLOR AND LIGHT

In describing refraction by means of a triangular glass prism we specified the use of monochromatic light. If so-called *white light,* such as light from an incandescent tungsten filament bulb, is used, we get a beautiful display of colors. *Dispersion* is the breaking up of light into its

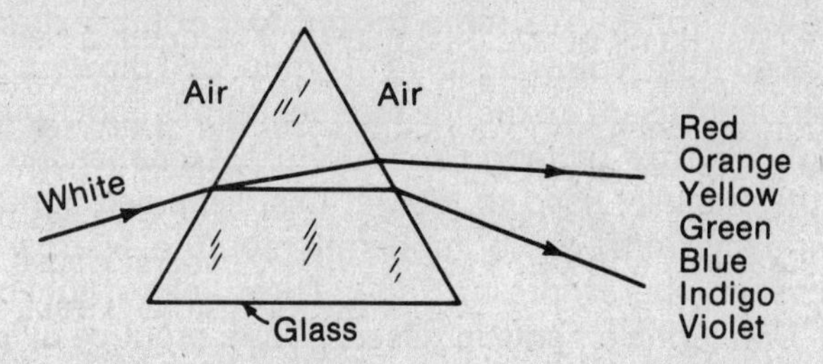

component colors. Seven distinct colors may be detected if white light is used. If a glass prism is used, the order of the colors is, from the one bent the least to the one bent the most: red, orange, yellow, green, blue, indigo, violet. Some students like to remember this sequence by using the first letters to spell a name. The term *spectrum* is used in many different ways. Here we shall define the *spectrum* of visible light as the array of colors produced by dispersion; or, the range of waves capable of stimulating the sense of sight. This range of wavelengths is approximately 4×10^{-5} cm for the violet to 7×10^{-5} cm for the red. *Infrared* "light" has a greater wavelength than red; *ultraviolet* a shorter one than violet. Both are invisible to the human eye. Dispersion is produced by the prism because the violet light is slowed up more than the red light in the glass (or water); the other colors travel at intermediate speeds. In vacuum all electromagnetic waves travel at the same speed.

The *color of an opaque object* is determined by the color of the light it reflects. A red object reflects mostly red; it absorbs the rest. If an object reflects no light, it is said to be black. If a red object is exposed only to blue light, which it absorbs, the red object would appear black. If an object reflects all the light, it is said to be white. A white object exposed to red light, only, will appear red. The color of a transparent object is determined by the light it transmits.

Primary and Complementary Colors

If the above seven colors of light are recombined, white light is produced. However, it is possible to produce white light with fewer colors. *Complementary colors* are two colors of light whose combined effect on the eye is that of white light. They are: red and blue-green; yellow and blue-violet; green and magenta (purple). With the proper choice of three colors, known as the *primary colors,* it is possible to reproduce nearly all colors. The three primary colors of light are red, green, and blue-violet. If we mix two primary colors, we get the complementary color of the third. For example, if we mix red and green light we get yellow, which is the complement of blue-violet, the third primary color. Don't confuse this with the mixing of *paints.* Paints, like other opaque objects, absorb light and are seen by the color of the light they reflect. If we mix two paints, each paint subtracts the color of light it absorbs, and the color of the mixture is the color of light neither absorbs. Color obtained by the mixing of *lights* is known as an additive process.

Aberrations

Even a perfectly made thin lens doesn't function exactly as described above, if white light is used. The operation of a lens depends on refraction. We saw that refraction of white light by the triangular prism resulted in dispersion. The different colors in white light are brought to a focus at different points by the lens. The focal length of a lens depends on the color of the light used, just as the index of refraction of a substance depends on the color of light used. The result is that images produced by lenses have a slight color fringe, and photographs with such lenses may be slightly fuzzy. This phenomenon is known as *chromatic aberration.* It can be minimized ("corrected") by using a suitable combination of lenses known as an *achromat.*

Types of Spectra

When white light was used above with the triangular prism, a *continuous spectrum* was produced: there was no gap as we went from one extreme, the red, to the other extreme, the violet. A continuous spectrum may be obtained from a glowing solid or liquid, or a glowing gas under high pressure. A *spectroscope* is an instrument used to examine the spectrum of light from any source. The light to be examined goes through a narrow rectangular slit before it reaches the prism which disperses the light. A telescope is then usually used to look at this dispersed light. A *bright-line spectrum* is obtained from a glowing gas under low or moderate pressure. Narrow rectangles or lines of light appear with gaps in between where there is no light. These lines of light can be thought of as images of the rectangular slit. Different elements in gaseous form produce different spectra. We say each element has a *characteristic spectrum* and use it for identification of the element. A *dark-line spectrum* is a continuous spectrum which is interrupted by thin dark lines. The dark lines result from the absorption of the energy of the missing wavelengths by gases that were between the hot source and the spectroscope. When a gas is cool, it can absorb some of the wavelengths which it emits when it is hot. When the sun's spectrum is examined superficially, it seems continuous. Actually it is a dark-line spectrum. The dark lines in the sun's spectrum are known as *Fraunhofer lines.* They are produced mostly by gases in the sun's atmosphere which absorb some of the energy emitted by the hotter core. (Spectra will be discussed further in Chapter 16.)

INTERFERENCE OF LIGHT

Coherent Sources

In the chapter on Wave Motion and Sound we noted that when two waves go through the same portion of the medium at the same time, interference occurs. The two waves are superposed. They may reinforce each other, giving constructive interference; or they may annul each other, giving destructive interference. This is readily observed in sound, where we can easily hear beats. It is also readily observed in standing waves in a string, or in water waves.

In order to observe interference we need two *coherent sources,* sources that produce waves with a constant phase relation. For example, two waves that are always in phase will always reinforce each other. If two waves are of opposite phase—that is, one wave is 180° out of step with the other—they will tend to annul each other. We have no trouble maintaining this constant phase relationship with mechanical vibrations. If we strike a tuning fork, both prongs will keep vibrating at the same frequency and in step. If we hold the vibrating fork vertically and rotate it around the stem as an axis, we can observe the interference produced by the waves produced by the two prongs. The two prongs are coherent sources.

Until recently it has been impossible to get two independent light sources which are coherent. The reason for this is that most sources of light emit it in short bursts of unpredictable duration. This would be like a tuning fork, momentarily stopping every few cycles. Two light sources might be out of phase during one burst but not during the next one. Annulment would be masked by the rapid change in phase relationship. It is now possible to get two lasers to be coherent. For most purposes, if we want interference of light, we take light from one source and have it travel by two different paths to the same place. This will be illustrated below. A *laser* produces a light beam which is monochromatic and coherent and diverges very little.

Interference with Thin Films

The beautiful colors of soap bubbles and of thin films of oil on a city pavement or highway are due to interference of light and give further evidence that light behaves like a wave. Let us assume that we have monochromatic light incident on a thin film. The film may be oil, soap solution, or even air trapped between two flat glass plates. Some of the light is reflected at the top surface of the film and is represented by ray 1. Most of the light goes through the film; some of this is reflected at the bottom surface of the film and comes out again. This is represented by ray 2. If we look down on the film, rays 1 and 2 may enter our eye and be focused on our retina. Coherent light is coming to our eye by two different paths. Ray 2 had to go back and forth through the film. The two rays (or light waves) arriving simultaneously on our retina will interfere with each other. (Refraction is not shown in the diagram because it is negligible in a thin film.)

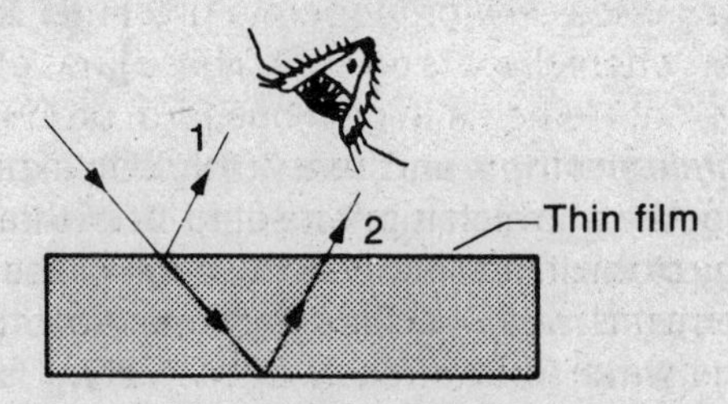

Ray 2 may be delayed a whole number of wavelengths by having to go through the film. If so, it will arrive at the eye in phase with ray 1, and the two waves will reinforce each other. The observer looking at the film in that direction will see a bright spot.

Ray 2 may be delayed an odd number of half-wavelengths—$\lambda/2$, $3\lambda/2$, $5\lambda/2$, etc. If so, it will arrive out of phase with ray 1, and the two waves will annul each other. The observer looking at the film in that direction will see a dark spot. In general, such patterns consist of bright fringes alternating with dark fringes.

(A complicated effect observed with all waves and pulses under certain conditions is phase reversal on reflection. This means that pulses and waves are reflected as though they were delayed one-half a wavelength when they are reflected from a medium with a greater index of refraction than the one they travel in. This change of phase on reflection will be ignored here. The phase difference in rays 1 and 2 will be assumed to be due only to the difference in path.)

If the film is ¼ wavelength thick, and if the incident ray is almost perpendicular to the surface, the path difference for ray 2 is ½ wavelength. The two waves cancel each other. This idea is applied in the coating of expensive photographic lenses to minimize undesirable reflections.

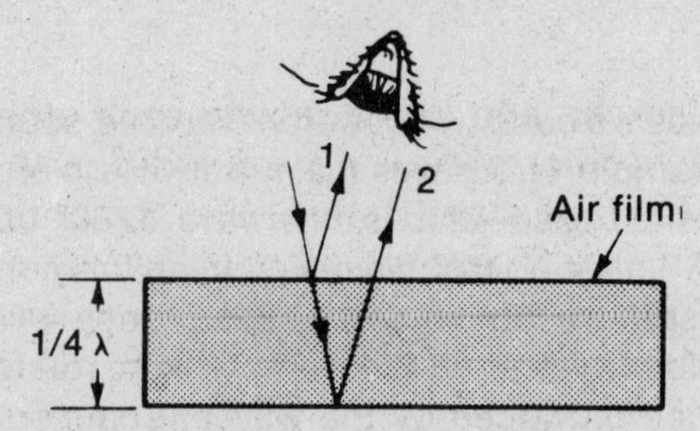

In ordinary daylight the film is illuminated by many different wavelengths. The thickness of the film may be just right in one direction to cause annulment of one color. All the other wavelengths will combine to give a colored spot. In another direction the thickness may be just right for the cancellation of a different wavelength, and the complement of the canceled color will be seen. As a result, when viewing a soap bubble or an oil film we see the full colors of the rainbow.

Interference with the Double Slit

In 1801 Thomas Young performed a famous experiment which suggested the interference of light. Its essential features are similar to the set-up described: Light from a small source *F* illuminates a barrier which has two narrow slit openings, S_1 and S_2, very close to each other. The light which gets through these two slits falls on a screen. The screen may be thought of as the retina of our eye, because we usually look at the light source through the two slits. Assume that the light source is monochromatic. What we see on the screen is surprising. We see a whole series of bright lines alternating with dark regions. The technique we used for drawing ray diagrams with lenses and mirrors is inadequate. On that basis we might have expected to see two bright lines, one for each slit which lets light through.

F
S_1 S_2
Barrier
Screen

We apply *Huygens' principle* and the wave theory of light to explain what we actually see. Imagine that we are looking down at the above setup. The light from the source spreads out and the wave fronts are represented by circles with source *F* at the center. The two slits are equidistant from the source. Each wave front reaches S_1 and S_2 at the same time. According to *Huygens' principle,* we can think of the portions of the wave front at each slit as acting like independent sources of light, producing their own wave fronts. These are represented as two sets of circles with S_2 being the center of the top set and S_1 the center of the bottom set. Since the light originally came from a single source, *F,* the waves produced by S_1 and S_2 are coherent.

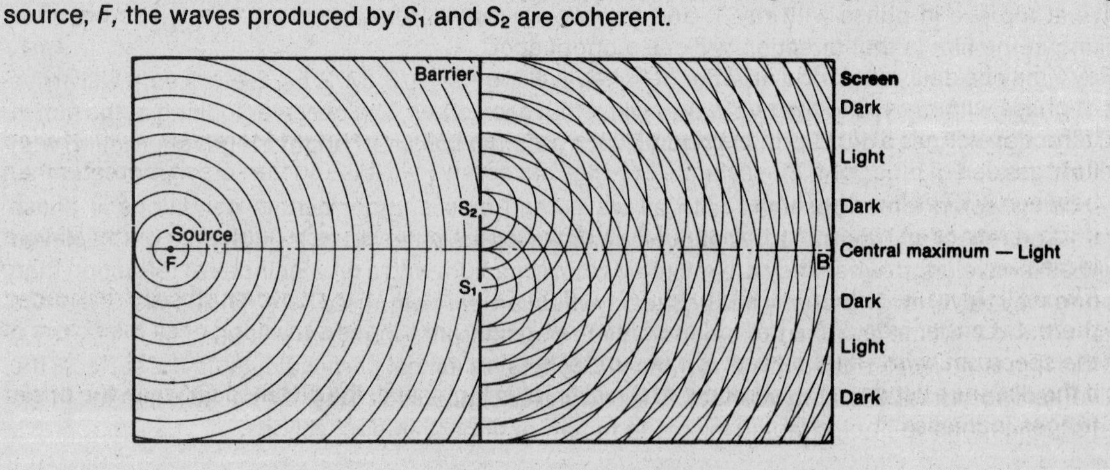

The waves overlap on the screen and interfere with each other. In some places the waves arrive in phase, reinforce each other producing antinodes, and give a bright region or line. One such place is B, which is equidistant from S_1 and S_2; at this place we have the central maximum, or the central bright line. There are other places, on either side of the central maximum, where we get bright lines, places where the two waves are in phase and reinforce each other. The waves are in phase, but one is a whole number of wavelengths behind the other. In between the bright lines we have dark regions. The two waves cancel each other, producing nodes, where they are out of step by ½ of a wavelength, $3/2$ wavelengths, or any other odd number of half-wavelengths.

Let us see if we can calculate the distance between bright lines or fringes. Let us look at the right half of the above diagram in a slightly different way. Let us think of the first bright line on one side of the central maximum and represent it as point P. It is bright because the wave from S_2, traveling in the

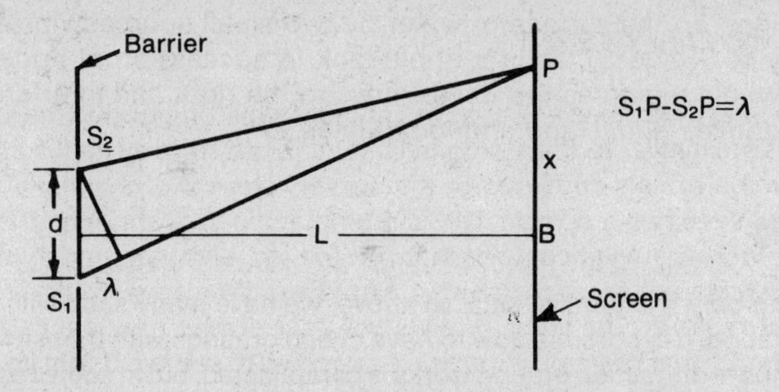

direction S_2P, arrives in phase with the wave from S_1 which travels in the direction S_1P. The length of path S_1P is greater than S_2P; how can the waves arrive in phase?—only if the difference in the length of the two paths is exactly one wavelength. For the second bright fringe, the difference in path is two wavelengths; for the third bright fringe, the difference in path is three wavelengths, etc.

If d is the distance between the two slits, L the distance between the barrier and the screen, λ the wavelength, and x the distance between the central maximum and the first bright fringe, it can be shown by selecting two triangles and making some approximations, that

$$\boxed{\frac{\lambda}{d} = \frac{x}{L}}$$

It can also be shown that the distance between any two adjacent bright fringes has the same value of x.

Complete annulment occurs midway between the bright fringes; the distance between adjacent nodes is also x.

Let us look at the important relationship shown above to see what predictions we can make. Since

$$\frac{\lambda}{d} = \frac{x}{L}$$

1. if the wavelength of light is increased (λ), the distance between fringes increases; if we use red light instead of blue light, the distance between successive red lines on the screen is greater than between successive blue lines;
2. if the distance between the two slits (d) is decreased, the distance between the bright fringes increases;
3. if white light is used, the central maximum (at B) is white, since all the wavelengths are reinforced there. On either side of the central maximum we get bright fringes consisting of all the colors of the spectrum, with violet closest to the central maximum;
4. if the distance between the screen and the slits (L) is increased, the distance between the bright fringes increases.

Example: Calculate the distance between adjacent bright fringes if the light used has a wavelength of 6000 angstroms and the distance between the two slits is 1.5×10^{-3} meter. The distance between the screen and the slits is 1.0 meter.

$$\frac{\lambda}{d} = \frac{x}{L}$$

$$\frac{6000 \times 10^{-10}\ \text{meter}}{1.5 \times 10^{-3}\ \text{meter}} = \frac{x}{1.0\ \text{meter}}$$

$$x = 4000 \times 10^{-7}\ \text{m, or } 0.00040\ \text{m}$$

In the diagram on the bottom of page 88, the pattern to the right of the barrier is very similar to the pattern obtained on the surface of water by two small sources vibrating in phase with the same frequency at S_1 and S_2, as in a ripple tank. A succession of nodes (and antinodes) is produced in the whole region as the waves move to the right and interfere with each other. This succession produces *nodal* (and *antinodal*) lines.

Diffraction Grating

If instead of having only two parallel slits, as above, we have many such slits close together, we have a diffraction grating. It is possible now to have cheap gratings with thousands of slits per inch. The full analysis of how a diffraction grating works is complicated, but in some respects it is similar to the double slit. However, the more slits we have, the brighter is the pattern. For approximate calculations, the above formula (for the double slit) may be used. It provides a convenient method for measuring the wavelength of light. Instead of the triangular glass prism the grating may be used in the spectroscope.

As with the double slit, the pattern obtained on the screen with a grating has a central maximum which is the same color as the light used. On either side of the central maximum we get bright fringes; if white light is used, the bright fringes are continuous spectra. The spectrum closest to the central maximum is known as the first-order spectrum. Notice there is a first-order spectrum on either side of the central maximum. If the source of light is a luminous gas, each "bright fringe" is the characteristic spectrum of the element.

DIFFRACTION

Wave Characteristic

When we deal with a wave, we expect to find *diffraction:* the bending of a wave around obstacles. We know that this happens in the case of sound because, for example, we can hear a speaker even if our direct view of him is blocked by a crowd of people. In the case of a water wave we can see the wave going around objects on the surface of the water, so that after a short distance the wave looks again as though there had been nothing in the way.

If light is a wave, why do objects cast sharp shadows? Why doesn't the light bend around the object and illuminate the region behind it? Theory shows that the smaller the wavelength the smaller the diffraction effects. We already know that the wavelength of light is only a small fraction of a millimeter, while that of radio and sound waves is many centimeters (and even meters for radio waves). It is because the wavelength of light is so small that shadows are so sharp and that a beam of light tends to travel along straight lines.

However, diffraction of light does occur; because its wavelength is so small, we must look very carefully. For example, if we photograph the shadow of a small ball, using a suitable light source, we find a small illuminated spot in the center of the shadow.

Single Slit Diffraction

Suppose we shine a beam of monochromatic light on an opaque barrier which has a narrow rectangular opening cut into it, a single narrow slit. Let the light which goes through the slit fall on a screen held parallel to the barrier. On the basis of what we did with ray diagrams, we might expect that the part of the beam which gets through the slit will have the same cross section as the slit and that the illuminated part of the screen will look exactly like the slit in shape and size.

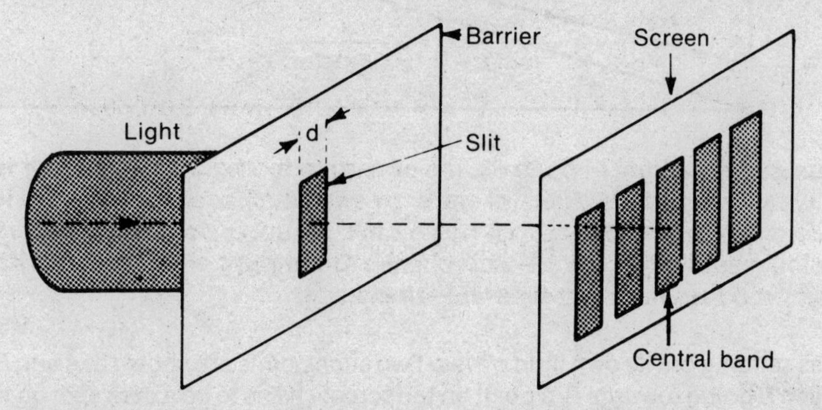

If the slit is narrow and we look carefully, what we actually see on the screen is many rectangular, parallel bands of light (*images of the slit*). Each band may be a little wider than the slit. The central band is very bright; to either side are bands of decreasing brightness. (The central band's intensity is about 20 times as great as the first bright band to either side.) Below is a rough graph of the intensity of the observed pattern plotted against distance along the screen.

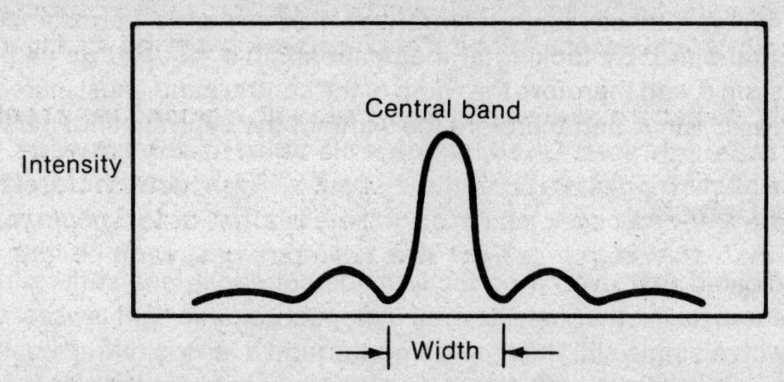

In the diffraction pattern, the central band is also called the central bright line or the central maximum. Careful analysis shows that

1. the width of the central maximum varies inversely as the width of the slit; that is, if we make the slit narrower, the central bright line becomes broader, or, in other words, there is more and more diffraction, more and more bending of the light around the edges of the slit;
2. the width of the central maximum varies directly as the wavelength; if we use red light, we get a wider or broader central line than if we use blue light;
3. the distance between successive dark bands is approximately the same as the distance between the first dark band and the central bright maximum.

If the opening is large compared with the wavelength, light practically travels along straight lines; diffraction is negligible.

The Wave Theory and Single Slit Diffraction

As in the case of the double slit interference pattern, we can apply Huygens' principle and the wave theory of light to explain the single slit diffraction pattern. Let us view the barrier and screen edge-on. Assume that the slit has a width d and that the barrier is a distance L from the screen. Think of a wave front arriving at the slit opening. According to *Huygens' principle*, every point on the wave front can

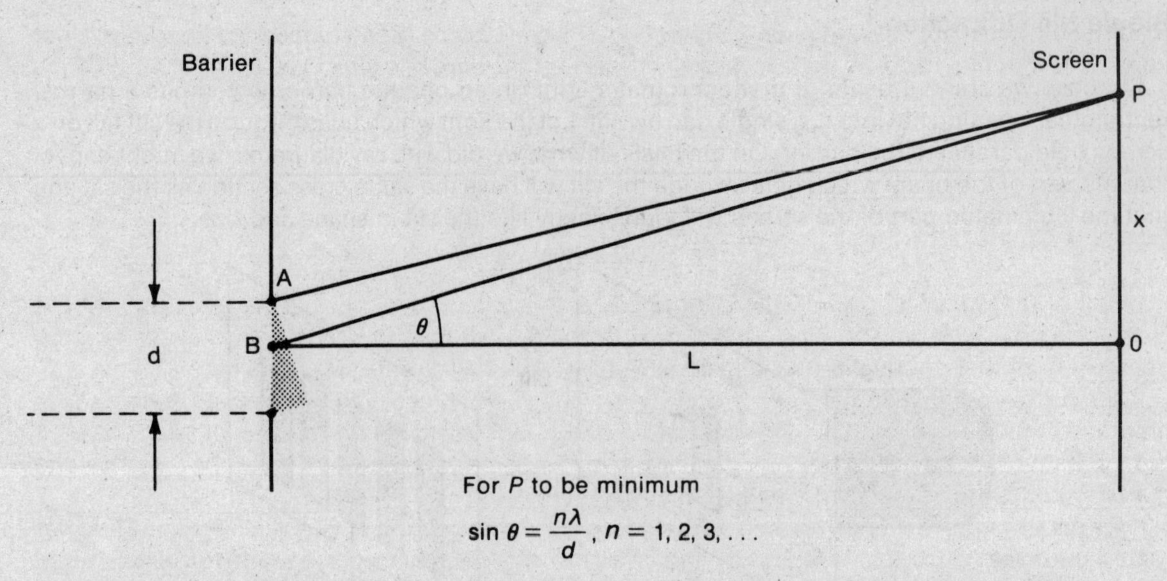

be thought of as sending out its own little waves. Two such points are shown at *A* and *B*. Think of the waves from *A* and *B* going towards *P*, a point on the screen. If *P* is to be a dark spot on the screen, the two waves traveling along *AP* and *BP* must annul each other; and, similarly, the waves from the two points just below *A* and *B* on the wave front at the slit, etc. Now path *BP* is larger than path *AP*. If this difference in path length is one-half wavelength, the two waves will annul each other.

With a little geometry it can be shown that for a dark point on the screen $\sin \theta = n\lambda/d$, where the letters have the values shown in the diagram, and λ is the wavelength of the light used. When $n=1$, *P* gives us the point of minimum intensity for the central line. In other words, the distance *OP* gives half of the width of the central band. By looking at the relation ($\sin \theta = \lambda/d$,), we can see that if the wavelength λ increases, $\sin \theta$, and therefore the width of the central band, must increase. Also, if the width of the slit *d* decreases, $\sin \theta$, and therefore the width of the central band, must increase.

Diffraction and Lens Defects

We have already indicated that even a perfectly made spherical lens suffers from chromatic aberration. Spherical lenses have other defects even if monochromatic light is used. One of these is due to diffraction. As with a single slit, the light going through a lens is diffracted. The amount of diffraction is usually negligible, but one effect is that parallel light entering the lens is not brought to a point focus, but to a disk focus. Around this bright diffraction disk we may be able to see bright diffraction rings. The fact that we get a disk diffraction pattern instead of a point focus with a lens sets a practical limit to the magnification we can get with a lens.

Resolving Power

The ability of a lens or optical instrument to distinguish between two point sources is its *resolving power.* Resolution or resolving power is increased by increasing the diameter of the lens; it works similarly for spherical mirrors. Suppose, for example, we look through a telescope at two stars that are close to each other in the sky. If the objective lens has a small diameter, the diffraction pattern of each star will be large and the two disks will overlap as in diagram *A*. The resolution is poor: it is hard

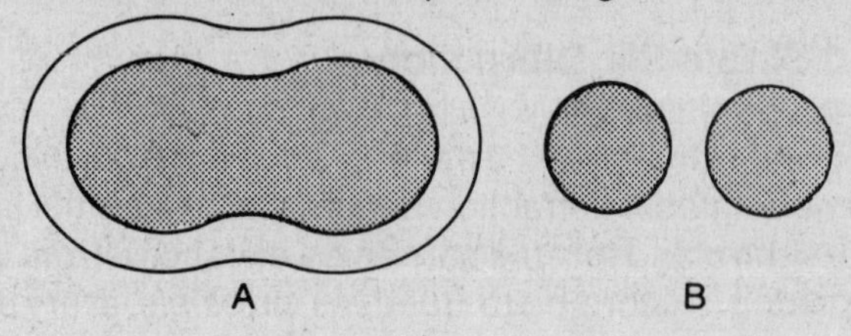

to tell whether we are looking at one star or two. If we try to use greater magnification we will not improve resolution, because we also magnify the size of the disk. If, instead, we use an objective lens with a larger diameter, we decrease the amount of diffraction, we decrease the size of the diffraction disk, and we increase the resolving power. The effect of using a larger diameter lens is shown in diagram *B*.

Doppler Effect with Light

When we studied sound, we noticed that the pitch we hear depends on the frequency of the vibration of the source. The pitch is affected by the relative motion between the source and the observer. A similar phenomenon is observed in the case of electromagnetic waves.

If the distance between the source and the observer is decreasing, the observed frequency is greater than that of the source. In the case of light, frequency determines the color of light we see. If the frequency is greater than the source frequency (wavelength is smaller), the light is more violet than it would have been if there were no relative motion.

If the distance between the source and the observer is increasing, the observed frequency is lower than that of the source. In the case of light, this means that the observed light would be shifted towards the red.

There are some interesting applications of the Doppler effect with electromagnetic waves. Radar may be used to measure the velocity of approach of airplanes, automobiles, and similar objects. At a radar installation a transmitter sends out an electromagnetic wave in a narrow beam. The beam is reflected back by many objects, especially metallic ones. The reflected wave is picked up by a radar receiver. If the object is stationary, the reflected wave has the same frequency as that sent out by the transmitter. If the object's distance from the radar is increasing, the reflected frequency is less than the radar frequency. If the object's distance is decreasing, the reflected frequency is greater. The change in frequency depends on the velocity of approach. The greater the velocity, the greater the change in frequency.

The famous *red shift* is explained by the *Doppler principle* as being due to an expanding universe. What does that mean? The light from the sun, stars, and other heavenly objects has been analyzed with a spectroscope. Present are the characteristic spectra of many elements we know on earth. By comparing the spectrum of an element on earth with the corresponding spectrum from stars, it has been found that the spectrum from some stars is shifted towards the blue end of the spectrum; this means that these stars must be approaching the earth. The spectrum from some other stars has been found to be shifted towards the red end of the spectrum. This means that these stars are moving away from the earth. There are some very distant objects in the sky known as *nebulae* (singular is nebula). Some are galaxies similar to our Milky Way galaxy. The spectra from these nebulae are shifted towards the red. The further the nebulae are, the greater the shift. This has been interpreted to mean that the distant parts of the universe are moving further and further away from us and from each other: the universe is expanding.

Polarization of Light

We have been saying that light behaves like a transverse wave. What is the evidence? We have already shown how light behaves like some kind of wave. We explained interference and diffraction of light by making use of wave theory. We can readily see how two waves can get to the same place at the same time and annul each other. We cannot explain this annulment if we think of the energy as arriving as streams of particles. However, this did not tell us whether the wave is longitudinal or transverse. We observed interference and diffraction with sound and decided that the sound wave is longitudinal.

An effect observed with light and not with sound is polarization. *Polarized light* is light whose direction of vibration has been restricted in some way. The direction of a longitudinal wave cannot be restricted; either it vibrates in the direction in which the energy is traveling (the direction of propagation), or it doesn't vibrate at all and we have no wave. On the other hand, in a transverse wave there are an infinite number of ways in which we can have vibrations at right angles to the direction of propagation of the wave.

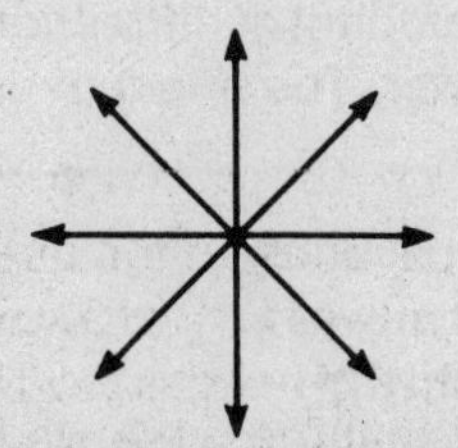

In the diagram, imagine a ray coming out of the paper. Vibration can be towards the top and bottom edges of the paper, towards the right and left, and in all other directions in between. We think of ordinary natural light, such as we get from our tungsten filament bulbs, as a transverse wave with all directions of vibrations present. Some crystals, natural and synthetic, allow only that light to go through whose vibration is in the direction of the axis of the crystal. Such light is then polarized:

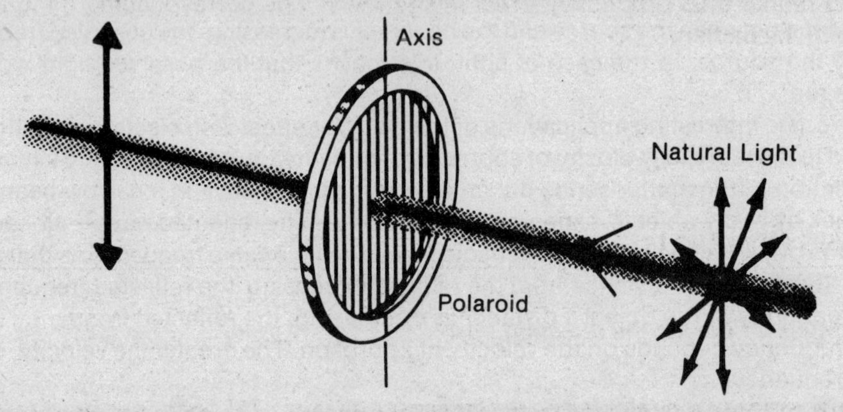

Polaroid is a synthetic material that does this. It consists of plastic sheets with special crystals embedded in them. If we let ordinary light shine through one polaroid disk, the light is polarized but the human eye can't tell the difference. If we allow this polarized light to shine through a second polaroid disk kept parallel to the first one, the amount of light transmitted by the second disk depends on how we rotate the second disk with respect to the first one. If we rotate the second disk so that its axis is at right angles to the axis of the first disk, practically no light is transmitted. The maximum amount of light is transmitted when the two axes are parallel to each other.

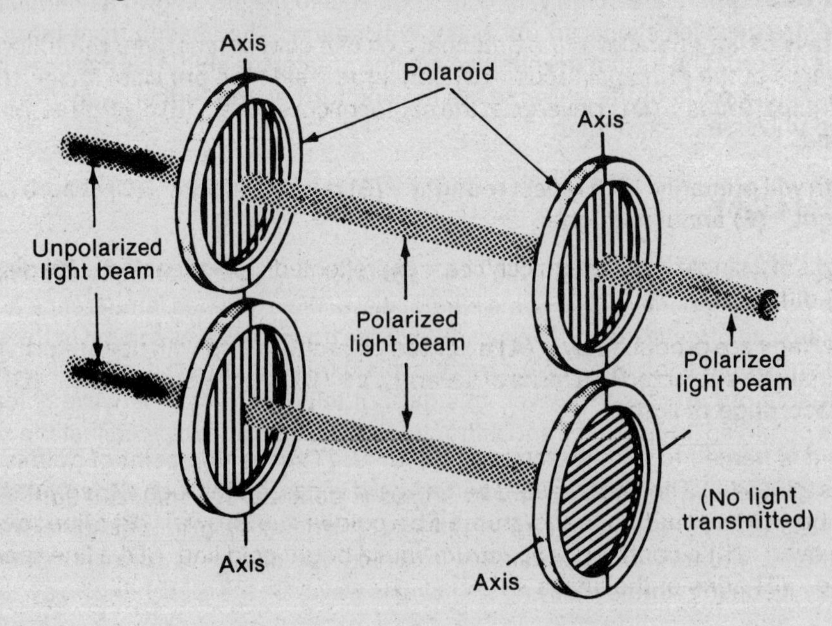

According to *Maxwell's theory,* light is a transverse wave in which electric and magnetic fields vibrate or fluctuate at right angles to the direction of propagation. Electric and magnetic fields will be discussed further in later chapters.

ANSWERING THE QUESTIONS Let us return to the questions at the beginning of the chapter. The size of an object doesn't change as it moves further away from us, but it looks smaller to us. The explanation for this lies in the characteristics of images produced by convex lenses, of which there is one in the eye. The further the object, the smaller the image on the retina. This is also true of the plane mirror image. Its size is the same as the size of the object, but the further it is from the eye, the smaller the image formed on the retina.

Incidentally, if you look at the diagram of the plane mirror image in the preceding chapter, you will observe that we see the image by means of diverging rays which come to our eye. (This is also true of rays coming to us from any point of a real object.) The convex lens in the eye brings these rays to a focus on the retina, thus producing a real image there. The corresponding thing is done by the convex lens in the camera.

QUESTIONS Chapter 11

Select the choice which fits the question best.

1. Real images formed by single convex lenses are always **(A)** on the same side of the lens as the object **(B)** inverted **(C)** erect **(D)** smaller than the object **(E)** larger than the object

2. A virtual image is formed by **(A)** a slide projector **(B)** a motion-picture projector **(C)** a duplicating camera **(D)** an ordinary camera **(E)** a simple magnifier

3. A person on earth may see the sun even when it is somewhat below the horizon primarily because the atmosphere **(A)** annuls light **(B)** reflects light **(C)** absorbs light **(D)** refracts light **(E)** polarizes light

4. Incident rays of light parallel to the principal axis of a convex lens, after refraction by the lens, will **(A)** converge at the principal focus **(B)** converge inside the principal focus **(C)** converge outside the principal focus **(D)** converge at the center of curvature **(E)** diverge as long as they are close to the lens.

5. A red cloth will primarily **(A)** reflect red light **(B)** refract red light **(C)** absorb red light **(D)** transmit red light **(E)** annul red light

6. Yellow light of a single wavelength can't be **(A)** reflected **(B)** refracted **(C)** dispersed **(D)** polarized **(E)** diffused

7. Newton's rings are produced by **(A)** a lighted cigarette falling with nonuniform acceleration **(B)** a lighted cigarette subjected to a force of several *g's* **(C)** interference of light **(D)** polarization of light **(E)** absorption of light

8. Some gold is heated to a temperature of 2000° C. (The melting point of gold is 1063° C; its boiling point is 2600° C.) The light emitted by this gold is passed through a triangular glass prism. The result will be **(A)** a continuous spectrum with a golden hue all over **(B)** a line spectrum with a golden hue all over **(C)** a continuous spectrum with a bright gold line **(D)** a line spectrum with a bright gold line **(E)** none of the above

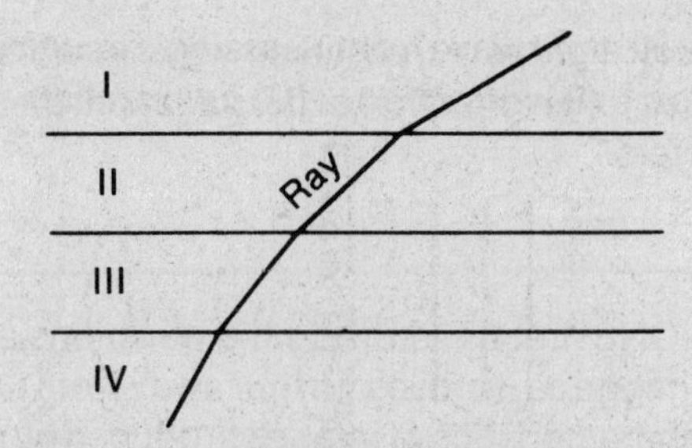

9. The diagram shows 4 layers of different transparent liquids on top of each other. The liquids do not mix. The path of an oblique ray of light through the liquids is shown. The medium in which the speed of the wave is lowest is **(A)** I **(B)** II **(C)** III **(D)** IV **(E)** indeterminate on the basis of the given information

10. The speed of light in a certain transparent substance is two-fifths of its speed in air. The index of refraction of this substance is **(A)** 0.4 **(B)** 1.4 **(C)** 2.0 **(D)** 2.5 **(E)** 5.0

11. A camera of 6-in. focal length is used to photograph a distant scene. The distance from the lens to the image is approximately **(A)** 0.5 ft **(B)** 1 ft **(C)** 2 ft **(D)** 4 ft **(E)** 6 ft

12. A film 2.0 cm wide is placed 6.0 cm from the lens of a projector. As a result a sharp image is produced 300 cm from the lens. The width of the image is **(A)** 12 cm **(B)** 50 cm **(C)** 100 cm **(D)** 300 cm **(E)** 600 cm

13. If an object is placed 30 cm from a convex lens whose focal length is 15 cm, the size of the image compared to the size of the object will be approximately **(A)** twice as large **(B)** more than twice as large **(C)** 1.5 times as large **(D)** smaller **(E)** the same size

14. A small beam of monochromatic light shines on a plate of glass with plane parallel surfaces. The index of refraction of the glass is 1.60. The angle of incidence of the light is 30° The angle of emergence of the light from the glass is **(A)** 18° **(B)** 30° **(C)** 45° **(D)** 48° **(E)** 60°

15. In an experiment, an object was placed on the principal axis of a convex lens 25 cm away from the lens. A real image 4 times the size of the object was obtained. The focal length of the lens is **(A)** 20 cm **(B)** 25 cm **(C)** 33 cm **(D)** 50 cm **(E)** 100 cm

16. Which will be produced when blue light with a wavelength of 4.7×10 meters passes through a double slit? **(A)** a continuous spectrum **(B)** two narrow bands of blue light **(C)** alternate blue and black bands **(D)** bands of blue light fringed with green **(E)** bands of blue light fringed with violet

17. Two pulses approach each other in a spring, as shown.

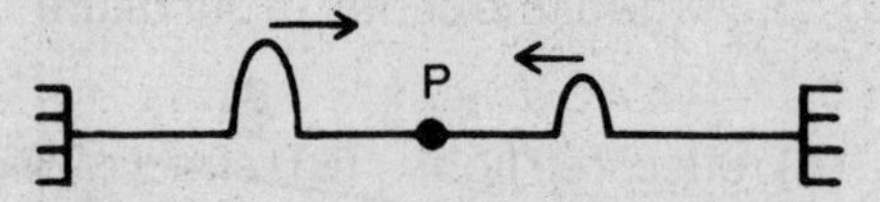

Which of the following diagrams best represents the appearance of the spring shortly after the pulses pass each other at point *P*?

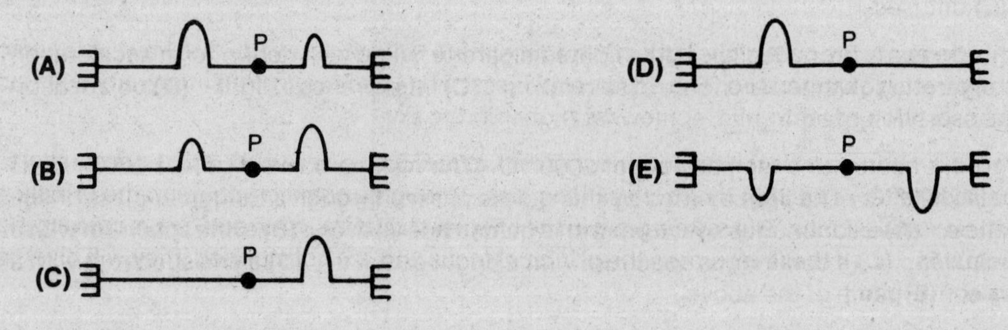

18. The diagram represents straight wave fronts passing through a small opening in a barrier. This is an example of **(A)** reflection **(B)** refraction **(C)** polarization **(D)** dispersion **(E)** diffraction

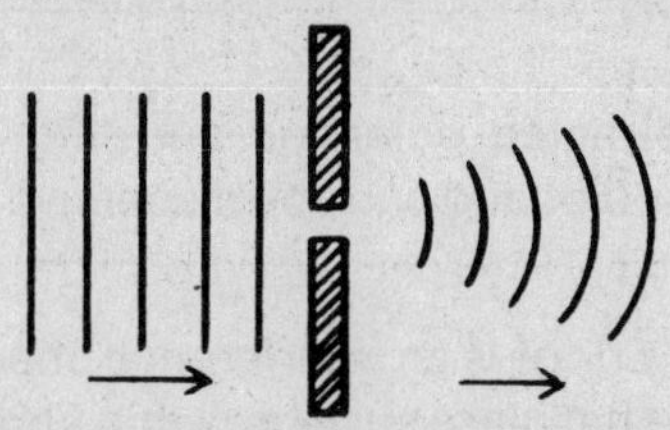

19. A ray of light goes obliquely from water into glass. The angle of incidence in the water is θ_1, the angle of refraction in the glass is θ_2, the index of refraction of water is n_1, and the index of refraction of the glass is n_2. Which of the following is a correct relationship for this case? **(A)** $n_1 \sin \theta_1 = \sin \theta_2$ **(B)** $n_1 \sin \theta_2 = \sin \theta_1$ **(C)** $n_2 \sin \theta_1 = \sin \theta_2$ **(D)** $n_2 \sin \theta_2 = \sin \theta_1$ **(E)** $n_2 \sin \theta_2 = n_1 \sin \theta_1$

EXPLANATIONS TO QUESTIONS Chapter 11

Answers

1. **(B)**	6. **(C)**	11. **(A)**	16. **(C)**
2. **(E)**	7. **(C)**	12. **(C)**	17. **(B)**
3. **(D)**	8. **(E)**	13. **(E)**	18. **(E)**
4. **(A)**	9. **(D)**	14. **(B)**	19. **(E)**
5. **(A)**	10. **(D)**	15. **(A)**	

Explanations

1. **(B)** Note from the ray diagrams in the chapter that real images formed by single convex lenses are always inverted and on the other side of the lens from the object. The real images may be larger, smaller, or the same size as the object, depending on the distance of the object from the lens.

2. **(E)** Projectors and cameras are used to put images on screens or films; images that appear on screens are real images. The simple magnifier (also called magnifying glass or simple microscope) is used to view something without projecting an image on a screen (other than the eye's retina); it produces a virtual image as shown in the ray diagram in the chapter.

3. **(D)** Light travels somewhat more slowly in air than in vacuum. Sunlight entering the atmosphere obliquely is refracted towards the normal. Since the atmosphere's density increases as it

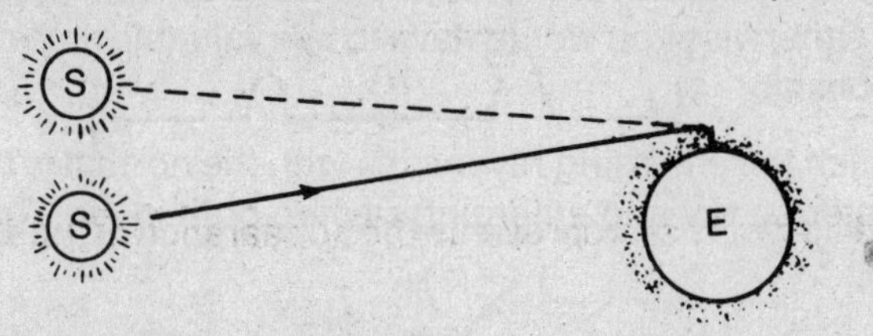

gets closer to the earth, the path of the light in the atmosphere is slightly curved. The drawing is not to scale and the effect is exaggerated. The observer projects a line backward to an imaginary position S' where the sun will appear to him: somewhat higher in the sky.

4. **(A)** This situation gave us the definition for the principal focus of a convex lens: rays parallel to the principal axis, after refraction by a convex lens, pass through a common point on the principal axis called the *principal focus.* The ray diagram in the chapter shows these refracted rays converging on the principal focus. (If these rays pass the principal focus and are not stopped, they will diverge *after* having converged.)

5. **(A)** Opaque objects are seen by the light they reflect. We say a cloth is red if it appears red when viewed in natural white light; it must, therefore, reflect red light, which is a component of the white light.

6. **(C)** All light can be reflected, refracted, polarized, and diffused. However, if it contains only one wavelength, it can produce only one color and can't be dispersed. (Definition of dispersion: breaking up of light into its component colors.)

7. **(C)** The production of Newton's rings is an interference phenomenon. They may be observed if you place a curved glass surface on a flat glass surface; a thin film of air is wedged between the glass surfaces. If you look down on this film you may be able to see several rings around the point where the two glass surfaces touch. The rings are produced by interference between the light reflected from the top surface of the airfilm and the light reflected from the bottom surface of the airfilm.

8. **(E)** At the temperature given, gold is an incandescent liquid. The light from incandescent solids or liquids produces a continuous spectrum without any special bright lines due to the color of the cold solid or liquid. The various colors, such as yellow and orange, gradually change from one to the other.

9. **(D)** A ray passing obliquely into a medium is bent towards the normal if it travels more slowly in the new medium; it is bent away from the normal if it travels faster in the new medium. In the diagram, if the ray is directed from medium I toward medium IV, it is bent closer to the normal in each successive medium. Therefore the ray travels most slowly in medium IV. If the ray is directed from medium IV toward medium I, we may notice that the ray is bent further away from the normal in each successive medium, again indicating that the wave travels fastest in medium I, slowest in medium IV.

10. **(D)** Index of refraction $= \dfrac{\text{speed of light in air}}{\text{speed of light in medium}} = \dfrac{v}{2/5\,v} = 5/2 = 2.5$

11. **(A)** You may just recall that, if the object is very far from a convex lens, the image is formed practically one focal length away from the lens. Or, you may use the lens equation: $1/p + 1/q = 1/f$; the object distance p is very large, so we'll use the infinity symbol (∞) for it.

$$1/\infty + 1/q = 1/f;\ 1/\infty = 0$$
$$1/q = 1/f;\ \text{and } q = f$$

That is, the image distance equals the focal length, in this case 6 in., which equals 0.5 ft.

12. **(C)** $s_1/s_0 = q/p$

$$s_1 = \frac{q}{p} \times s_0 = \frac{300\text{ cm}}{6\text{ cm}} \times 2\text{ cm} = 100\text{ cm}$$

13. **(E)** You should know that, if an object is 2 focal lengths away from a convex lens, a real image is formed 2 focal lengths away from the lens (on the other side from the object) and the image is the same size as the object. Otherwise you would have to use valuable time on calculations with the lens equation and size relationship.

14. **(B)** The angle which the emerging ray makes with the normal is the angle of emergence. We assume here that the entering ray and emerging ray are both in air. Then the two rays are parallel

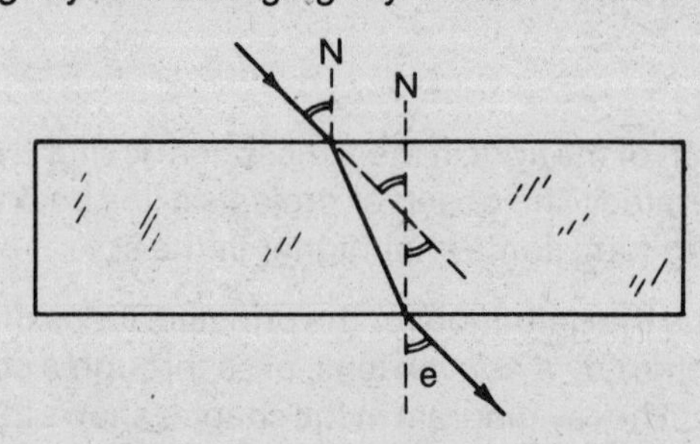

since the glass surfaces are parallel. Therefore, the angle of incidence equals the angle of emergence.

15. **(A)**

$$s_i/s_o = 4 = q/p$$
$$q = 4p\text{; but } p = 25 \text{ cm}$$
$$q = 100 \text{ cm}$$
$$1/f = 1/p + 1/q$$
$$1/f = 1/25 + 1/100 = 5/100$$
$$f = 100/5 \text{ cm} = 20 \text{ cm}$$

Also *note* that the object distance is between f & $2f$, since a real, enlarged image was produced. The only choice that fits this is (A).

16. **(C)** When light passes through a double slit, the pattern which forms on the screen where the light falls consists of alternating bands of light separated by black regions. If white light is used, the central band is white and on either side the bright bands consist of the spectrum of colors of which the white light is composed. In this question blue light of a single wavelength is used. Such light cannot be dispersed. Therefore all bands of light on the screen, including the central band, consist of this blue light. These bands are separated by bands of black regions.

17. **(B)** When two pulses or waves pass each other in a medium, each continues on as though there had been no interference from the other. In this question, therefore, the pulse with the larger amplitude will appear on the right side of P and continue traveling towards the right. The pulse with the smaller amplitude appears on the left and continues traveling towards the left.

18. **(E)** *Diffraction* is the bending of a wave around an obstacle. The shaded parts of the diagram represent the obstacle. As the wave goes through the small opening in the obstacle, it spreads behind the obstacle. The small opening acts like a point source of a wave.

19. **(E)** In refraction, if one of the two media is vacuum, we need only one index of refraction, that of the substance used. This is its absolute index of refraction, or its index of refraction with respect to vacuum. This is also usually adequate if the substance is solid or liquid and we use air instead of vacuum. However, for the general case we have to use the index of refraction of both substances, as pointed out in the text.

12

Static Electricity—Electric Circuits

Does a short circuit between *A* and *B* or *C* and *D* blow the fuse? What is the potential difference between *A* and *B* in the circuit as drawn? What is the potential difference when there is a short circuit?

To answer these questions you obviously must know and understand Ohm's law and associated definitions and concepts. Don't assume that you do. Carefully study the material below and check your real understanding. The problem will be considered in detail later.

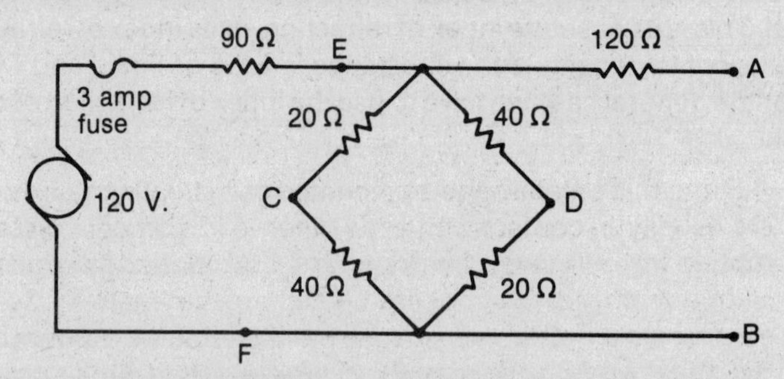

KINDS OF ELECTRIC CHARGE AND FORCE BETWEEN CHARGES

You should remember that a rubber rod rubbed with a piece of fur or wool acquires a negative charge; a smooth glass rod rubbed with silk becomes positively charged. Of course, the other material in the pair acquires an opposite charge, i.e., the silk becomes negative and the fur or wool becomes positive. Lucite may be used instead of glass; polystyrene instead of rubber.

Like charges repel; unlike charges attract. *Coulomb's law* applies to this force of attraction and repulsion if the charges are concentrated at points or small objects; the force is proportional to the product of the two charges and varies inversely as the square of the distance between them.

$$F = \frac{kQ_1Q_2}{d^2}$$

In the MKS system, the unit of charge *Q* is the coulomb, the distance *d* is in meters, and *k* is a proportionality constant which has the value 9×10^9 Nt-m²/coulomb² in vacuum or air. It is sometimes called the electrostatic or Coulomb's law constant.

METHODS OF CHARGING AND ELECTROSTATIC INSTRUMENTS

Charging an object usually results in a gain or loss of electrons. In a solid, the positive charges usually do not move readily. A gain of electrons results in making an object more negative; a loss of electrons in making it more positive. A neutral object usually acquires the same kind of charge as that of the charged object it touches. In *charging by induction,* a charged object is held near (but not touching) a neutral conductor while the latter is grounded. The latter thus acquires a charge opposite to that of the original charged object. The ground connection is broken before the charged object is removed. (See illustration below.)

The *electroscope* is an instrument used to determine the presence and sign of small electrical charges by the deflection of very thin metal leaves; in this type, gold leaf is frequently used. *The electrophorus, the Wimshurst machine,* and the *Van de Graaff* generator are devices for producing electrical charges by induction.

Nearly all the mass of an atom is concentrated in the *nucleus* which ordinarily contains protons and, except in the case of the common hydrogen atom, neutrons. The *electrons* move in orbits around the nucleus and are imagined to form "shells." Electrons in all atoms are alike: they have the same negative charge and small mass. The *proton* has a positive charge equal in magnitude to that of an electron; its mass is approximately 1836 times as much as the mass of the electron. The *neutron* is electrically neutral; its mass is slightly greater than that of the proton. In the neutral atom the number of protons equal the number of electrons. An *ion* is a charged atom (lost or gained electrons) or group of atoms. Metals are good conductors of electricity because their outermost electrons move readily.

Illustrations and Additional Facts

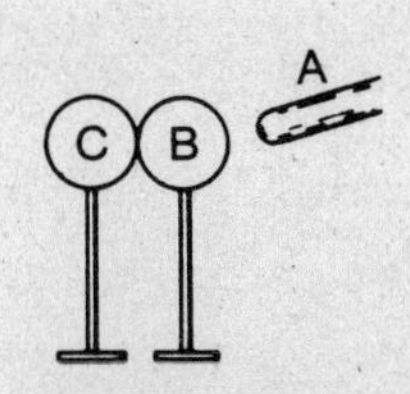

B and *C* are metal spheres mounted on insulating stands. The metal portions are initially in contact with each other. *A* is a smooth glass rod that has been rubbed with silk and is held near *B.* If *C* is touched by hand momentarily and *A* is then removed, *C* and *B* are both charged negatively by induction because electrons were attracted from the hand due to the nearness of the positive glass rod. What would have been the charge on *C* and *B* if the procedure had been to separate *C* and *B* slightly without touching the metallic portion and then removing *A?* Electrons would have been attracted to the end of *B* closest to *A,* leaving *C* positive. After separation *B* remains negative, *C* positive.

If we move a charged metallic sphere so that it will not be affected by any external charges, the excess charge, positive or negative, will be distributed uniformly on the outside of the sphere. In general, if the metallic object is not a sphere, the excess charge will distribute itself so that the greatest charge will be on the most pointed portion.

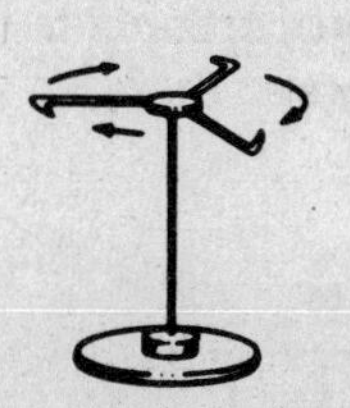

Electric discharge takes place readily from sharp points. This is often demonstrated with the *electric whirl,* which consists of a wire with sharp points bent as shown in the diagram and pivoted at its center of gravity so that it is free to rotate in a horizontal plane. No matter which terminal of a static machine the whirl is connected to, it always rotates in the direction shown, away from its points. The air always contains some ions. When the whirl is positively charged, its points are strongly repelled by positive charges in the air; when negatively charged, by negative charges in the air. Why isn't this effect neutralized by ions in the air of opposite charge? Opposite charges are attracted to the points and by successive collisions overcome the friction and inertia of the whirl. This produces rotation in the same direction as before.

ELECTRIC FIELDS

We noticed above that two charged objects exert a force on each other without touching. Physicists have found it desirable to introduce the concept of an electric field in order to visualize better and to describe more precisely this force acting at a distance. An *electric field* is said to exist wherever an electric force acts on an electric charge.

Electric Field Intensity

The *electric field intensity* at a certain point is the force per unit positive charge placed there. It is a vector quantity. Its direction is the direction of the force on the *positive* test charge; if this is to the right, the direction of the field intensity at that point is to the right. If we happen to have used a negative test charge, the direction of the force on it is opposite to the direction of the field intensity.

$$E = F/Q$$

where E is the electric field intensity (Nt/coul), F (Nt) is the force exerted on positive charge Q (coul).

We can represent the electric field intensity around a charge by drawing vectors with suitable direction and length. Frequently we represent the field by merely drawing lines whose direction is the same as the direction of the field intensity at the various points along the lines, but whose length is independent of the magnitude. These lines are called *lines of force* or *lines of flux*. The stronger the field, the more the number of lines drawn in that area.

A few patterns representing electric fields are shown below.

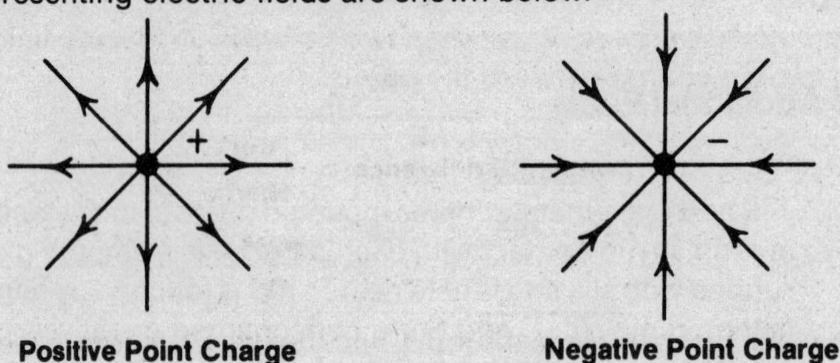

Positive Point Charge

Negative Point Charge

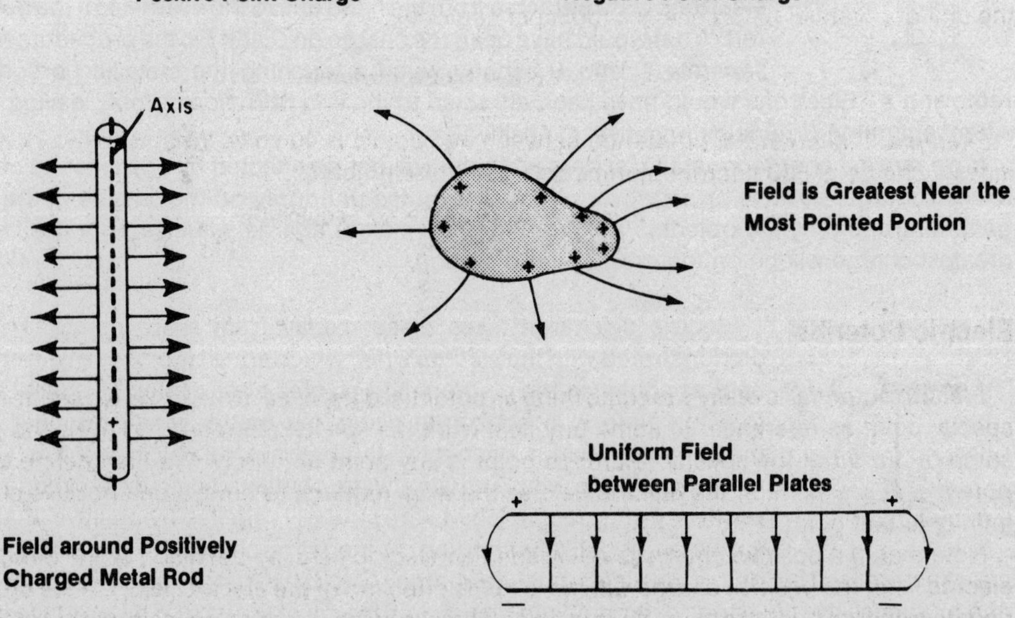

Field around Positively Charged Metal Rod

Note that, for metallic objects having a static electric charge, the lines of force are perpendicular to the surface at the points where they touch the surface.

For a point charge the field intensity varies inversely as the square of the distance from the charge. For a charged metallic sphere the field intensity varies inversely as the square of the distance from the center of the sphere, but there is no field inside the sphere. The excess charge is on the outside surface of the sphere. The field intensity around a long charged rod varies inversely as the distance (not the square of the distance) from the rod. The field between two parallel oppositely charged metallic plates is uniform; the intensity is the same everywhere between the plates.

ELECTRIC POTENTIAL AND POTENTIAL DIFFERENCE

In the diagram below an electric field is represented by three curved lines. Let *A* and *B* be two points in the field, as shown. If we put a positive charge *Q* at point *A*, the field will exert a force *F* on the

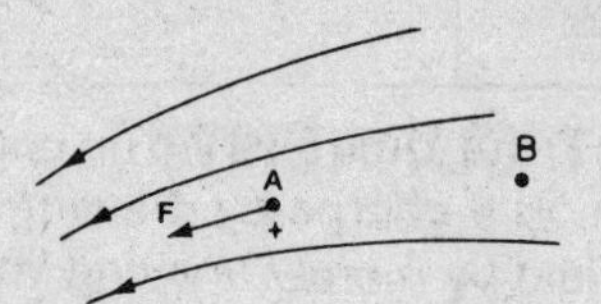

Work Is Required To Move a Charge from A to B

charge in the direction of the field. If we want to move the charge from *A* to *B*, we must do work. (Recall that work done equals the product of the force and distance moved in the direction of the force.)

We define the *potential difference, V,* between two points in an electric field as the work per unit charge required to move a charge between the points.

$$\textbf{potential difference} = \frac{\textbf{work}}{\textbf{charge}}$$

$$\mathbf{V} = \frac{\textbf{work}}{\mathbf{Q}}$$

In the MKS system, the unit of work is the joule, and the unit of charge is the coulomb. Therefore the unit of potential difference is a joule per coulomb.

1 volt = 1 joule/coulomb

Example: The potential difference between two points is 40 volts. What is the work required to move a charge of 800 microcoulombs between these points?

$$40 \text{ volts} = \text{work}/(800 \times 10^{-6} \text{ coulomb})$$

$$\text{work} = 0.032 \text{ joule}$$

Electric Potential

Electric potential is really the same thing as potential difference, except that we arbitrarily select a special point as reference. In some practical work the special reference "point" is the ground. In some of our work the special reference point is any point at infinity. We then define the *electric potential* at any point in the electric field as the work required to bring a unit positive charge from infinity to that point.

Note that, if a positive charge is released in an electric field, by definition of the direction of the electric field, the positive charge will move in the direction of the electric field. On the other hand, a negative charge released in an electric field will move in the direction opposite to that of the field. (If this bothers you, think of the field as being produced by a positive charge.) However, all objects released in the earth's gravitational field move in the direction of the field, namely towards the earth.

As will be discussed in more detail under Electric Currents, the two terminals of a battery are oppositely charged; there is a definite difference of potential between them. If two parallel metallic plates of the same size are connected to these terminals, one plate to a terminal, the two plates will acquire opposite and equal charges. (Such an arrangement is known as a *capacitor* and is discussed in Chapter 14.) There will be a difference of potential between them, the same as the difference of potential between the two terminals of the battery. We already mentioned that there is a uniform field between *oppositely charged parallel plates.* The intensity of this field is proportional to the difference of potential between the plates and inversely proportional to the distance between them:

$$E = V/d$$

By definition the field intensity is also equal to the force per unit charge. From this we see that

1 newton/coulomb = 1 volt/meter

ELECTRIC CURRENTS—DC (DIRECT CURRENT) POTENTIAL DIFFERENCE

Potential Difference

Potential difference is also referred to as difference of potential or voltage. Electromotive force is a special kind of potential difference.

1. The concept—Batteries (also dynamos, electrostatic generators, etc.) have two terminals, one positive the other negative. The positive terminal has a deficiency of electrons; the negative terminal an excess of electrons. Energy (chemical energy in the case of the battery) had to be used to attain this charged condition of the terminals; e.g., work had to be done to push electrons onto the negative terminal against the force of repulsion of electrons already there. *Potential difference* (*V*) *is the work per unit charge* that was done to get the terminals charged and that is therefore potentially available for doing work outside the battery. *Electromotive force* (*emf*) is the potential difference between the terminals of the source when no current flows. When current flows, as when a lamp is connected to a battery, the potential difference between the terminals of a battery is sometimes significantly less than its emf because of the battery's internal resistance (see page 106).
2. Unit of potential difference—the *volt;* e.g., the common flashlight cell supplies 1½ volts, the outlet at home 115 volts.

$$V = \frac{\text{work}}{Q}, \quad \text{volts} = \frac{\text{joules}}{\text{coulomb}}$$

Current

1. The concept—*Current* is the rate of flow of electric charge. In a metallic conductor (e.g., copper wire) the motion of electrons is the electric current; in liquids the electric current consists of moving negative and positive ions (an *ion* is a charged atom or group of atoms); in gases the electric current consists of moving ions and electrons.
2. Direction—If a copper wire is connected to the terminals of a battery or generator, electrons will go through the wire from the negative terminal to the positive terminal. This is sometimes taken as the *direction of the current* in the wire, but sometimes the opposite direction is used. In this book the first definition will be used: the direction of the electron flow. Usually on nationwide examinations, questions have been avoided which depended on a specific definition of current direction. Your knowledge of a consistent set of rules based on one definition is expected (see left-hand rules, page 114).
3. Unit—the ampere; 1 amp = 1 coulomb/sec; 1 coulomb = 6×10^{18} electrons.

Resistance

1. The concept—The amount of current in a device when it is connected to a given source of potential difference depends on the nature of the device. Resistance of a device is its opposition to the flow of electric charges, converting electrical energy to heat because of this opposition. (Note that in ac (alternating current) *impedance* is the *total* opposition to the electric current, and that portion of it which converts electrical energy to heat is the resistance.)
2. Unit—the ohm (Ω).

FACTORS AFFECTING RESISTANCE These are temperature, length, cross section, and material.

1. Effect of temperature: The resistance of most metallic conductors goes up when the temperature goes up.
2. Effect of other factors: The resistance of a conductor is directly proportional to its length and inversely proportional to its cross-sectional area, at constant temperature. This is expressed by the formula:

$$R = \frac{kL}{A}$$

L = length in meters
R = resistance in ohms
A = cross-sectional area in meters^2
k = a constant for the material and is called resistivity; unit is ohm-meter

(*Conductivity* is the reciprocal of resistivity. The resistivity of good conductors is low; their conductivity is high.)

Conductance

Conductance is the reciprocal of resistance, and its unit is the *mho.* (The higher the resistance, the lower or poorer is the conductance.)

OHM'S LAW AND ELECTRIC CIRCUITS—DC

Ohm's Law

The current in a circuit (I_T) is directly proportional to the potential difference applied to the circuit (V_T) and inversely proportional to the resistance of the circuit (R_T).

Formulas:

$$I_T = V_T/R_T; \quad R_T = V_T/I_T; \quad V_T = I_T R_T$$

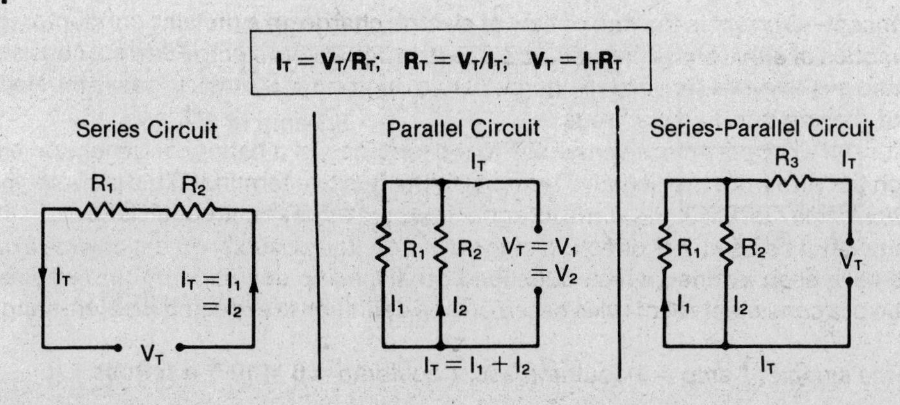

	Series Circuit	Parallel Circuit	Series-Parallel Circuit
Current	$I_T = I_1 = I_2$	$I_T = I_1 + I_2$	$I_T = I_3 = I_1 + I_2$
Resistance	$R_T = R_1 + R_2$	$\frac{1}{R_T} = \frac{1}{R_1} + \frac{1}{R_2}$ *	$R_T = R_3 + \frac{R_1R_2}{R_1 + R_2}$ *
Voltage	$V_T = V_1 + V_2$	$V_T = V_1 = V_2$	$V_T = V_1 + V_3 = V_2 + V_3$; $V_1 = V_2$
IR-drop	$V_T = I_TR_T$; $V_1 = I_1R_1$; $V_2 = I_2R_2$, etc.		
Symbols	I_1 = **current through** R_1; V_2 = **potential difference across** R_2, etc.		

*If only two resistors are connected in parallel, the combined resistance is equal to their product divided by their sum: $R_T = \frac{R_1R_2}{R_1 + R_2}$.

For a PORTION OF A CIRCUIT the same formulas apply, but I, V, and R have to be selected for that part of the circuit. The potential difference across a resistor is sometimes called *potential drop, IR-drop, or voltage drop.*

If n equal resistors are connected in parallel, their combined resistance $R_T = \frac{R}{n}$, *where* R is the resistance of one of them. The combined resistance of resistors in parallel is always less than the smallest resistance.

Notes: 1. V_T = potential difference at terminals of source = emf of source if internal resistance of source is negligible. 2. The *internal resistance* of the source can be treated as an external resistance in series with the rest of the circuit; to all of this is applied the emf of the source. 3. When the internal resistance is not negligible,

$$\boxed{V_T = \text{emf} - Ir,}$$

where I is the total current, r is the internal resistance of the source of electrical energy, V_T is the potential difference across the external circuit = voltage at terminals of source.

LINE DROP This is the voltage across wires connecting the source to devices getting energy. This may be negligible; if not, it can be represented and treated like R_3 in the above series-parallel circuit.

EXAMPLE 1 Series Circuit

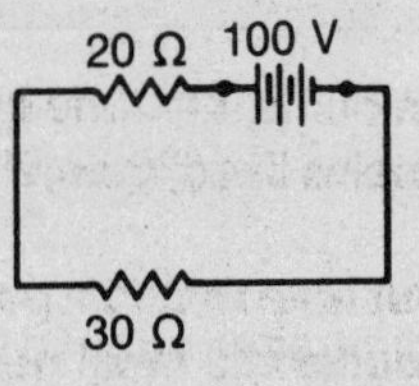

20 Ω 100 V

30 Ω

$V_T = 100$ V

$R_T = R_1 + R_2 = 20\ \Omega + 30\ \Omega = 50\ \Omega$

$I_T = \frac{V_T}{R_T} = \frac{100\text{ V}}{50\ \Omega} = 2.0$ amp

$I_1 = I_T = 2.0$ amp: this is current through each resistor

EXAMPLE 2 Parallel Circuit

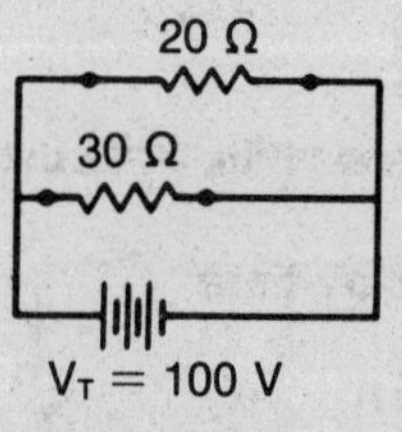

20 Ω

30 Ω

$V_T = 100$ V

$V_T = 100\text{ V} = V_1 = V_2$

$R_T = \frac{R_1R_2}{R_1 + R_2} = \frac{20\ \Omega \times 30\ \Omega}{20\ \Omega + 30\ \Omega} = 12\ \Omega$

$I_T = \frac{V_T}{R_T} = \frac{100\text{ V}}{12\ \Omega} = 8.3$ amp or $8^1/_3$ amp

$I_1 = \frac{V_1}{R_1} = \frac{100\text{ V}}{20\ \Omega} = 5.0$ amp through 20 Ω

$I_2 = \frac{V_2}{R_2} = \frac{100\text{V}}{30\ \Omega} = 3^1/_3$ amp through 30 Ω

N.B.: $I_1 + I_2 = (5.0 + 3^1/_3)$ amp $= 8^1/_3$ amp $= I_T$

EXAMPLE 3

Assume same circuit as for Example 2, except that battery has emf of 120 V and internal resistance of battery is 3 Ω.

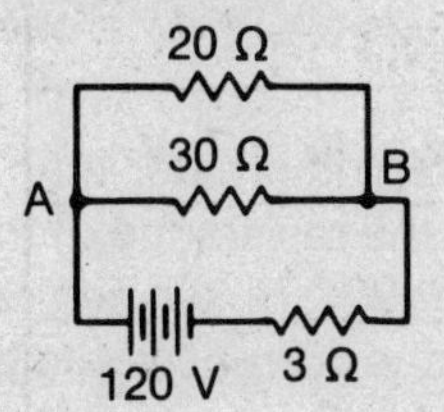

On the basis of calculations for Example 2, this circuit is equivalent to:

$$I_T = \frac{V_T}{R_T} = \frac{120\text{ V}}{(12+3)\ \Omega} = \frac{120}{15}\text{ amp} = 8.0\text{ amp}$$

$V_{AB} = IR = (8.0\text{ amp})\ (12\ \Omega) = 96\text{ V}$. the potential difference between *A* & *B* is 96 V. This is also the voltage across the 20 Ω resistor, as well as the voltage across the 30 Ω resistor. It is also the terminal voltage of the battery when used in this circuit.

SHORT CIRCUIT A low resistance path between two points where there is normally a relatively high resistance. This *may* result in an excessive current, known as an *overload.* If the circuit is overloaded, a fuse properly placed in the circuit will burn out.

In a parallel connection any device may be disconnected without breaking the circuit for the others.

ELECTRICAL ENERGY AND POWER

Joule's Law

The heat produced by current in a given resistor is proportional to the resistance, the time that the current flows through the resistor, and the square of the current through the resistor:

$$\mathbf{H = 0.24\ I^2Rt}$$

where H = heat in calories, I = current in amperes, R = resistance in ohms, t = time in seconds.
N.B. Heat produced is proportional to the square of the current if R remains fixed; then, if I is doubled, H is quadrupled, provided R and t remain the same.

In *Parallel Circuits* the same formula for heat produced applies, but note that in a parallel circuit the device with the least resistance draws the most current and produces the most heat.

Power

Power is the rate of using or supplying energy. The power consumption P of a resistor is:

$$\mathbf{P = VI; \quad P = I^2R; \quad P = V^2/R;}$$

where P = power in watts, V = potential difference across resistor in volts, I = current through resistor in amperes, R = resistance in ohms.

ADDITIONAL ENERGY AND POWER UNITS Since power = energy/time

$$\textbf{energy = power} \times \textbf{time,}$$

if power is expressed in watts, appropriate energy units are: watt-sec; watt-min; and watt-hr.

$$\textbf{1 watt-sec = 1 joule}$$

ANSWERING THE QUESTIONS Now let us return to the electric circuit at the beginning of the chapter. Note again that *some questions can be answered without detailed calculations.* You can't afford to take time for precise calculations when only an estimate is needed. QUICKLY EXAMINE THE SUGGESTED ANSWERS BEFORE MAKING CALCULATIONS. If there is a big difference between suggested answers, a rough estimate on your part may be adequate.

To determine whether the fuse will blow we need to know whether the current is greater than 3 amp. Any short circuit to the right of *E* or *F* in the diagram can at most reduce the resistance of the circuit to the right of *EF* to zero. Then the total resistance of the circuit would be 90Ω (IF we neglect the internal resistance of the generator,) and, using Ohm's law, $I_T = 120$ V/90 Ω, the current would be less than 2 amp. Therefore the fuse will not blow because of a short circuit from *A* to *B* or from *C* to *D*. To produce a *short circuit between A and B,* a copper wire may be connected between *A* and *B*. The resistance of such wires is usually assumed to be negligible; i.e., equal to zero. Therefore the IR-drop across the short circuit is zero. (Also note from above that the current through this short circuit would have to be something less than 2 amp.)

When there is *no connection from A to B,* no current flows through the 120-ohm resistor and the IR-drop across this resistor is therefore zero. The potential difference between *A* and *B* will then be the same as between *E* and *B*, or *E* and *F*. The simplified circuit will then be a fairly simple series-parallel circuit:

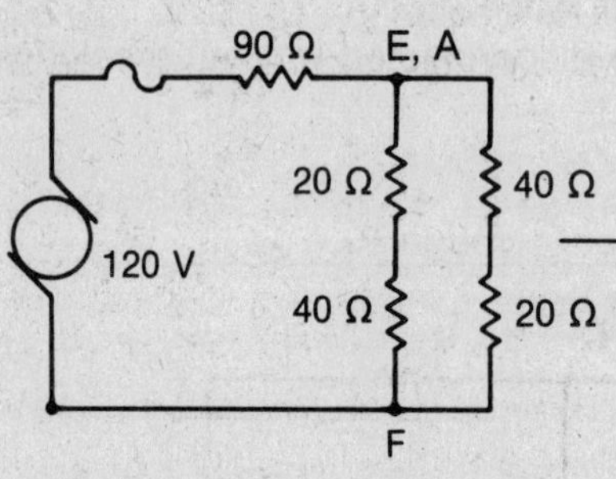

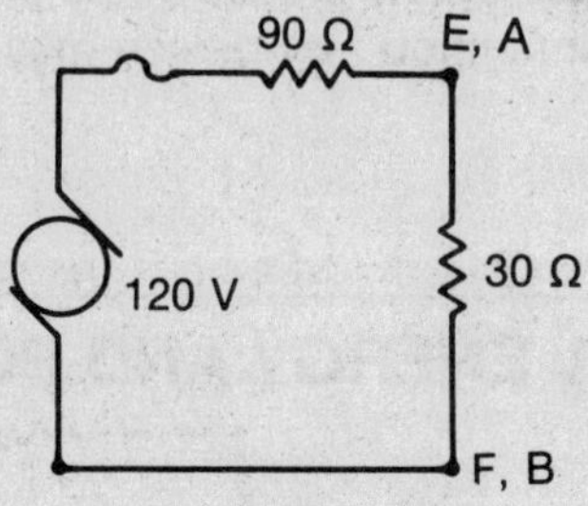

There are two parallel branches containing 60 Ω each. Their combined resistance is 60/2 or 30 Ω. R_T then is 120 Ω, $I_T = 120$ V/120 Ω = 1 amp, and the IR-drop across the 30-ohm resistance is 30 V. This is therefore also the voltage across each of the 40-ohm, 20-ohm branches; the current through each is 30 V/60 Ω = ½ amp. The IR-drop across the 40-ohm resistor is ½ amp × 40 Ω = 20 V, etc.

When there is a short circuit from A to B, the simplified circuit is indicated below, and the calculations can be carried out, if necessary, as indicated above, by using the appropriate relations as tabulated on p. 106.

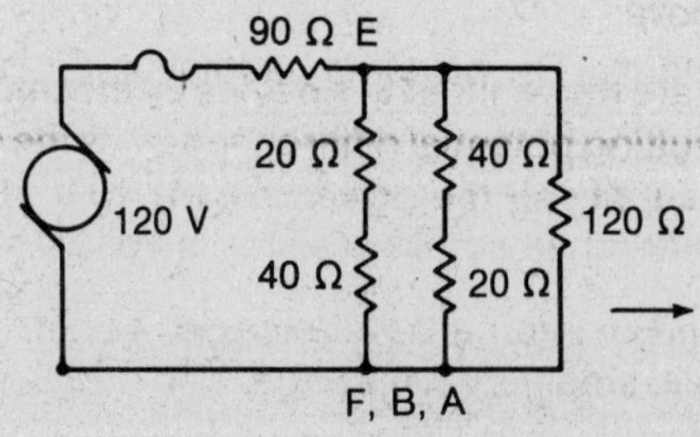

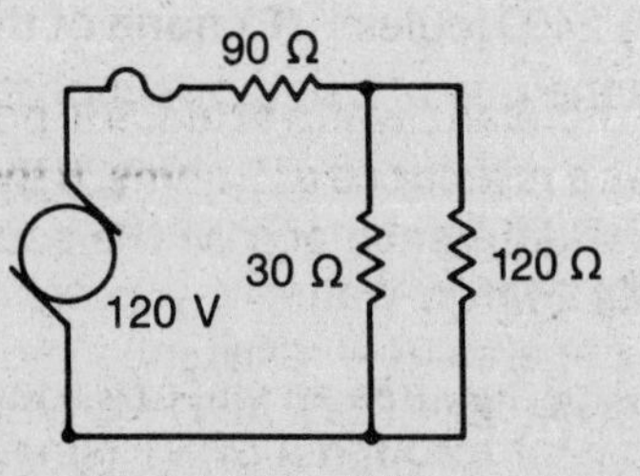

QUESTIONS Chapter 12

Select the choice which fits the question best.

1. The leaves of a negatively charged electroscope diverged more when a charged object was brought near the knob of the electroscope. The object must have been **(A)** a rubber rod **(B)** an insulator **(C)** a conductor **(D)** negatively charged **(E)** positively charged

For 2-3. Two point charges repel each other with a force of 4×10^{-5} newtons at a distance of 1 meter.

2. The two charges are **(A)** both positive **(B)** both negative **(C)** alike **(D)** unlike **(E)** equal

3. If the distance between them is increased to 2 meters, the force of repulsion will be, in newtons, **(A)** 1×10^{-5} **(B)** 2×10^{-5} **(C)** 4×10^{-5} **(D)** 8×10^{-5} **(E)** 16×10^{-5}

4. A proton (electric charge $= 1.6 \times 10^{-19}$ coul) is placed midway between two parallel metallic plates which are 0.2 meter apart. The plates are connected to an 80-volt battery. What is the magnitude of the electric force on the proton, in newtons? **(A)** 3.2×10^{-20} **(B)** 6.4×10^{-17} **(C)** 400 **(D)** 16 **(E)** 80

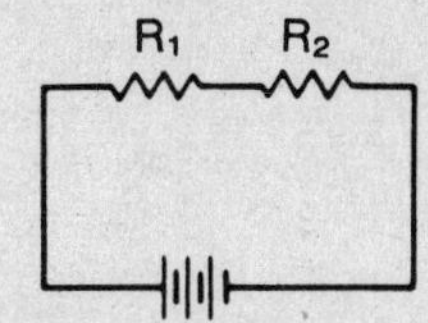

5. In the circuit shown, R_1 and R_2 are 30 Ω and 60 Ω, resp. $I_1 = 4$ amp. The potential difference across R_2 is equal to **(A)** 30 V **(B)** 60 V **(C)** 120 V **(D)** 240 V **(E)** a quantity which can't be calculated with the given information.

For 6-9. In the circuit shown, 4 amp is the current through R_1.

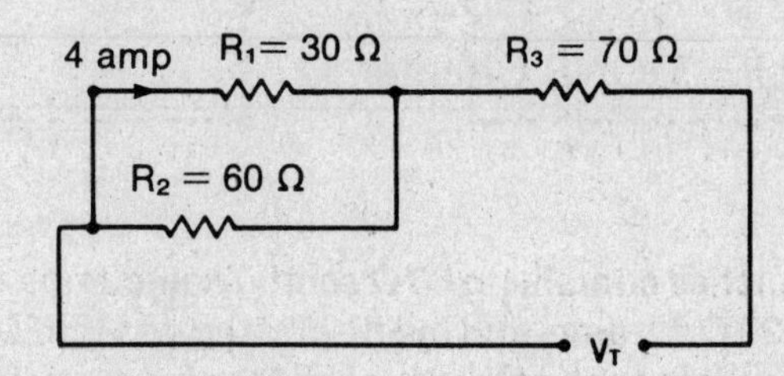

6. The potential difference across R_1 is, in volts, **(A)** 7.5 **(B)** 30 **(C)** 60 **(D)** 120 **(E)** 160

7. V_T is equal to, in volts, **(A)** 90 **(B)** 160 **(C)** 400 **(D)** 500 **(E)** 540

8. The rate at which R_1 uses electrical energy **(A)** is 120 watts **(B)** is 240 watts **(C)** is 360 watts **(D)** is 480 watts **(E)** can't be calculated with the given information

9. The heat developed by R_1 in 5 seconds is **(A)** 120 joules **(B)** 600 joules **(C)** 1800 joules **(D)** 2400 joules **(E)** none of the above

10. One thousand watts of electric power are transmitted to a device by means of two wires, each of which has a resistance of 2 ohms. If the resulting potential difference across the device is 100 volts, the potential difference across the source supplying the power is **(A)** 20 V **(B)** 40 V **(C)** 100 V **(D)** 140 V **(E)** 500 V

For 11-12. A battery has an emf of 6.0 volts and an internal resistance of 0.4 ohm. It is connected to a 2.6-ohm resistor through a SPST (single pole, single throw) switch.

11. When the switch is open, the potential difference between the terminals of the battery is, in volts, **(A)** 0 **(B)** 0.8 **(C)** 2.6 **(D)** 5.2 **(E)** 6.0

12. When the switch is closed, the potential difference between the terminals of the battery is, in volts, **(A)** 0 **(B)** 0.8 **(C)** 2.6 **(D)** 5.2 **(E)** 6.0

13. The current through the 10-ohm resistor is, in amperes, **(A)** 6 **(B)** 2 **(C)** 1 **(D)** $^2/_3$ **(E)** 0

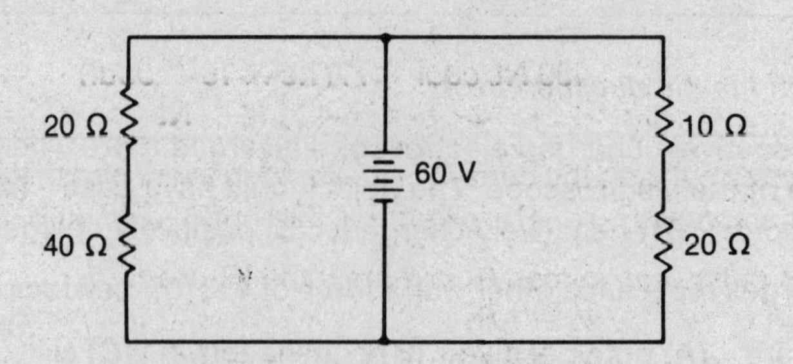

14. In the circuit shown, a good lead-acid storage battery (negligible internal resistance) is used. The battery supplies 6.0 volts. F_1, F_2, and F_3 are fuses with ratings of 10 amp, 10 amp, and 2 amp, respectively. When switches S_1, S_2, and S_3 are closed, the fuses that will blow are: **(A)** F_1 only **(B)** F_2 only **(C)** F_3 only **(D)** F_1 and F_2 **(E)** F_1 and F_3

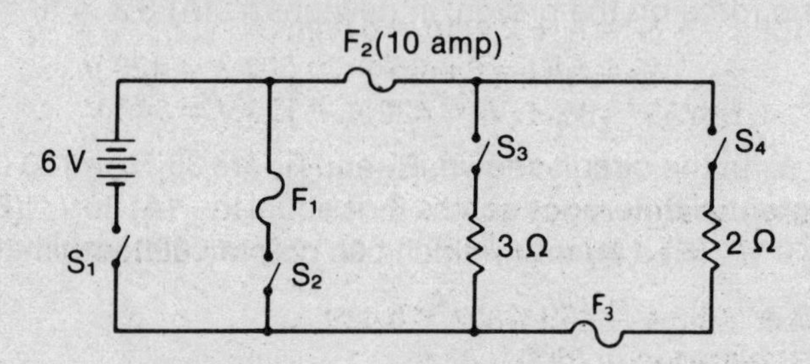

EXPLANATIONS TO QUESTIONS Chapter 12

Answers

1. **(D)**	4. **(B)**	7. **(E)**	10. **(D)**	13. **(B)**
2. **(C)**	5. **(D)**	8. **(D)**	11. **(E)**	14. **(A)**
3. **(A)**	6. **(D)**	9. **(D)**	12. **(D)**	

Explanations

1. (D) The leaves of a negatively charged electroscope have an excess of electrons. Since like charges repel, the negatively charged object will repel some (negatively charged) electrons from the knob of the electroscope to the leaves, increasing the amount of negative charge on the leaves. Since the force of repulsion increases with the charge, the leaves will diverge more. The negatively charged object may be rubber, but it may also be some other insulator or a metal.

2. (C) Two like charges repel each other. Both may be positive or both may be negative. They may be equal or unequal.

3. (A) The force between two point charges varies inversely as the square of the distance between them. Since the distance is doubled, the force becomes ¼ as great.

4. (B) The plates are charged oppositely by the battery and this produces a uniform electric field between the plates. Its intensity is equal to the ratio of the difference of potential between the plates to the distance between the plates:

$$E = V/d$$
$$= (80 \text{ volts}/0.2 \text{ m}) = 400 \text{ Nt/coul}$$

The field intensity is defined as the force on a unit positive charge placed in the field:

$$E = F/Q$$
$$400 \text{ Nt/coul} = F/(1.6 \times 10^{-19} \text{ coul})$$
$$F = 6.4 \times 10^{-17} \text{ Nt}$$

5. (D) This is a series circuit and the current is the same in every part of the circuit: $I_1 = I_2 = 4$ amp. The potential difference across $R_2 = I_2R_2 = 4$ amp $\times$ 60 ohms $= 240$ volts

6. (D) The potential difference across R_1 is merely the IR-drop.

$$V_1 = I_1R_1$$
$$V_1 = 4 \text{ amp} \times 30 \text{ ohms} = 120 \text{ V}$$

7. (E) Since R_1 and R_2 are in parallel, $V_1 = V_2$

$$\therefore I_1R_1 = I_2R_2$$
$$4 \text{ amp} \times 30 \text{ ohms} = I_2 \times 60 \text{ ohms}$$
$$I_2 = 2 \text{ amp}$$
$$I_3 = I_1 + 2 \text{ amp} = 6 \text{ amp}$$
$$V_3 = I_3R_3 = 6 \text{ amp} \times 70 \text{ ohms} = 420 \text{ V}$$
$$\text{But } V_T = V_3 + V_1 = 420 \text{ V} + 120 \text{ V} = 540 \text{ V}$$

8. (D) The rate of using energy is power.

$$P_1 = I_1^2 R_1 = (4 \text{ amp})^2 \times 30 \text{ ohms} = 480 \text{ watts}$$

9. (D) Energy = power $\times$ time = 480 watts $\times$ 5 sec
= 2400 watt-sec = 2400 joules

The energy used by a resistor is converted to heat. Therefore the heat developed by the resistor is 2400 joules. Notice that any unit of energy may be used. If desired, we may convert to calories.

10. (D) The power used by the device, $P = VI$.

$$1000 \text{ watts} = 100 \text{ volts} \times I$$
$$I = 10 \text{ amp}$$

Since this is a series circuit, the current in each of the wires is also 10 amp. The voltage across each of these wires is an IR-drop = 10 amp $\times$ 2 ohms = 20 volts. The voltage across the source = voltage across device + voltage across the wires = 100 V + 20 V + 20 V = 140 V.

11. (E) When the switch is open, the circuit is incomplete and no current flows. When a battery supplies no current, the potential difference across its terminals is the emf.

12. (D) When the switch is closed, current flows. The terminal voltage = emf $- Ir$, where I is the current flowing and r is the internal resistance of the battery.

$$I = \frac{\text{total voltage}}{\text{total resistance}} = \frac{6.0 \text{ V}}{(2.6 + 0.4)\ \Omega}$$
$$I = 2 \text{ amp.}$$

terminal voltage = 6.0 V $-$ (2 amp $\times$ 0.4 Ω) = 5.2 V.

13. (B) This is really a parallel circuit with the two branches drawn on opposite sides of the battery. You may be able to see it better if you draw both branches on the same side of the battery. In any case, the full 60 V is applied across the branch containing the 10-ohm and the 20-ohm resistors in series. For that branch $R_T = 30\ \Omega$, $V_T = 60$ V, and $I = V/R = 60 \text{ V}/30\ \Omega = 2$ amp.

14. (A) The fuse rating gives the maximum current that the fuse can carry continuously. The resistance of the fuses is usually quite small—a fraction of an ohm at these current ratings. When switches S_1 and S_2 are closed, F_1 is directly across the battery; with practically no resistance in the circuit, a current much larger than 10 amp will flow momentarily through F_1, burning out the fuse. The additional closing of S_3 closes the circuit for F_2 and the 3-ohm resistor. The current in this circuit is $V/R = 6\text{V}/3\Omega = 2$ amp. This current is handled safely by fuse F_2.

13

Magnetism; Meters, Motors, Generators

What happens to an ammeter when it is connected across a generator? What happens when a dc generator is connected across a battery? To answer these questions we need to understand the principle of operation of the generator and meters. This requires a knowledge of magnetism.

SOME BASIC TERMS

A *magnet* attracts iron and steel; when freely suspended it assumes a definite position (because of the earth's magnetism). A *magnetic* substance is one that can be attracted by a magnet. The magnetic materials include iron, nickel, cobalt, alloys of iron like alnico. *Nonmagnetic* substances are only feebly affected by a magnet; e.g., glass, wool, brass. *Paramagnetic* substances are attracted feebly; *diamagnetic* ones repelled feebly. The effect of paramagnetic and diamagnetic substances is so feeble that we usually ignore it. *Ferromagnetic* substances like iron are attracted strongly by a magnet. A *magnetized* substance is a magnetic substance which has been made into a magnet. A magnetic *pole* is the region of a magnet where its strength is concentrated; every magnet has at least two poles. The *North pole* (*N-pole,* or *north-seeking pole*) of a suspended magnet points towards the earth's magnetic pole in the northern hemisphere. A *lodestone* is a natural magnet. Usually we use artificial (man-made) magnets.

FACTS AND THEORY OF MAGNETS

The *law of magnets* states that like poles repel; unlike poles attract. If the poles are concentrated at points, this force between two poles is proportional to the product of their strengths and varies inversely as the square of the distance between them. (Note similarity to Coulomb's law for electric charges.) The old *molecular theory of magnetism* states that every molecule of a *magnetic* substance is a magnet with an N- and S-pole. When the substance is fully magnetized, all the like poles face in the same direction. When it is not magnetized, the molecules are arranged in helter-skelter fashion. A more modern theory speaks about *domains* rather than molecules. A domain is about 0.001 in. long and consists of many, many atoms; some of its properties are like those of the above molecular magnets. In general, a moving electric charge produces a magnetic field. In the atoms the spinning of the electrons produces the magnetism. (The explanation of diamagnetism involves the orbital motion of the electrons around the nucleus.)

Magnetic Field and Magnetic Flux

The *magnetic field* is the region around the magnet where its influence can be detected as a force on another magnet. The direction of the field at any point is the direction in which the N-pole of a compass would point. The magnetic field can be represented by fictitious *lines of flux* with the following properties: they are closed curves; they leave the magnet at an N-pole and enter at an S-pole; they never cross each other; they are more crowded at the poles; and, in general, the greater the crowding, the greater is the force on a pole placed at that point. A *permeable* substance is one through which magnetic lines of force go readily; e.g., soft iron. A representation of the magnetic field can be obtained by sprinkling iron filings on a horizontal surface in the field. A compass needle takes a position in the direction of the magnetic field in which it is located and can be used to get a point-by-point representation of the magnetic field.

Typical Magnetic Field Diagram

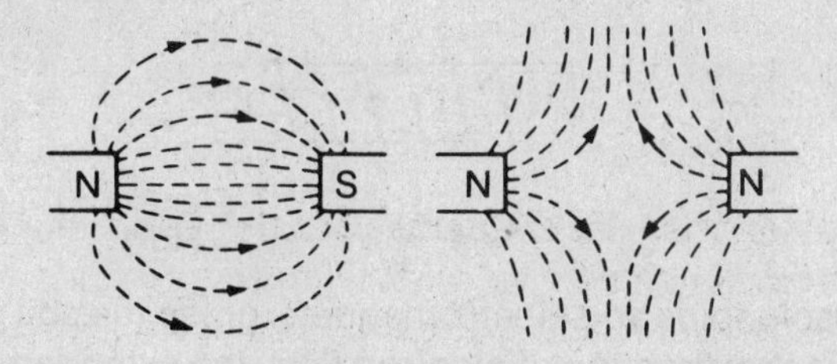

Terrestrial Magnetism

In a rough way the earth acts as if it had a huge bar magnet inside with its S-seeking pole in the northern hemisphere. (However, this pole is called the North Magnetic Pole.) Remember that it attracts the N-pole of the compass needle and that the magnetic poles do not coincide with the geographic poles, which are the points where the earth's axis of rotation intersects the surface of the earth.

The angle between the direction of the compass needle at a particular location and the direction to geographic north (or true north) is the *angle of declination* or compass variation of that location. *Isogonic lines* are the irregular lines drawn through locations having the same angle of declination. The *agonic line* is the isogonic line drawn through points having zero declination. Places east of the agonic line have west declination because the compass needle points west of the direction to true north.

The *dipping needle* is a compass needle mounted on a horizontal axis; it measures the *angle of dip* or angle of *inclination* at a given place. At the magnetic poles the angle of dip is 90° The magnetic equator is the line drawn through points of zero dip. *Isoclinic lines* are lines drawn through points having the same dip.

Electromagnetism

Oersted discovered that a wire carrying current has a magnetic field around it. The magnetic field around a long straight wire carrying current is represented by lines of magnetic flux which are concentric circles (in the plane perpendicular to the wire) with the wire at the center. (The magnetic lines of flux are somewhat analogous to the electric lines of force.) To determine the direction of these magnetic lines of flux we may use the *left-hand rule:* Grasp the wire with the left hand so that the thumb will point in the direction of *electron flow;* the fingers will then circle the wire in the direction of the lines of flux. (If conventional current is used, the right hand will be employed with the thumb pointing in the direction of the conventional current. The direction of the lines of flux is the same, no matter what the convention of current.) The unit of magnetic flux in the MKS system is the *weber.*

In describing the effect of the magnetic field it is convenient to introduce another concept. Magnetic *flux density* (also called magnetic induction) is the number of flux lines perpendicular to a unit area and is proportional to the strength of the magnetic field. The unit for flux density in the MKS system is the weber per square meter. In the CGS system the unit is the gauss.

In a metallic solid the electric current does consist of moving electrons, and the left-hand rule is natural for determining the direction of the magnetic field. In a liquid and gas the electric current usually consists of moving negative and positive ions. What is the direction of the magnetic field produced by the moving ions? For the negative ions we use the left-hand rule just as for electrons. For the positive ions we may still use the left-hand rule, but we must have the thumb point in the direction opposite to the flow of the positive ions; the other fingers will then point in the direction of the magnetic lines of the flux.

A *solenoid* is a long spiral coil carrying electric current. It produces a magnetic field similar to that of a bar magnet, and has an N-pole and S-pole at each end, respectively. Its strength is increased by the use of a permeable core such as soft iron. The location of the poles can be found by the use of

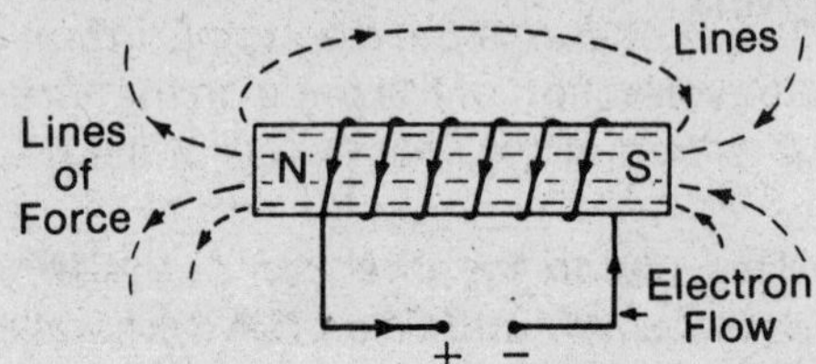

another *left-hand rule* for solenoids and electromagnets: grasp the coil with the left hand so that the *fingers* will circle the coil in the direction of *electron flow;* the extended thumb will then point to the N-pole and in the direction of the lines of flux. An *electromagnet* is a solenoid with a permeable core such as soft iron. The *strength* of an electromagnet depends on the number of turns in the coil, the current through the coil, and the nature of the core. The first two effects are combined in the concept of *ampere-turns:* it is the product of the number of turns and the number of amperes of current through the wire. A solenoid with 50 turns and 2 amperes going through it has 100 ampere-turns. Over a wide range the strength of an electromagnet is proportional to its ampere-turns. This is affected by the nature of the core. Iron is said to have a high permeability; it helps to give a strong electromagnet. Air, wood, paper, etc. have a low permeability; their presence does not affect the strength of the solenoid. (*Strength* here refers to the magnet's ability to attract iron.) The electromagnet can be made very strong; its strength can be changed by changing the current; its strength becomes practically zero when the current is turned off; it is, therefore, a temporary magnet. The electromagnet has many practical applications: electric motor, meter, generator, relay, telegraph sounder, bell, telephone.

Circular Loop

The magnetic field around a long straight wire carrying current does not have any poles. As we saw above, when the wire is wound in the form of a coil, magnetic poles are produced. The magnetic field around a single circular loop carrying current is such that the faces show polarity. We may apply the same left-hand rule as above to see this. Notice that both the left half and the right half of the loop produce lines of flux which have the same direction near the center of the loop. In this region the

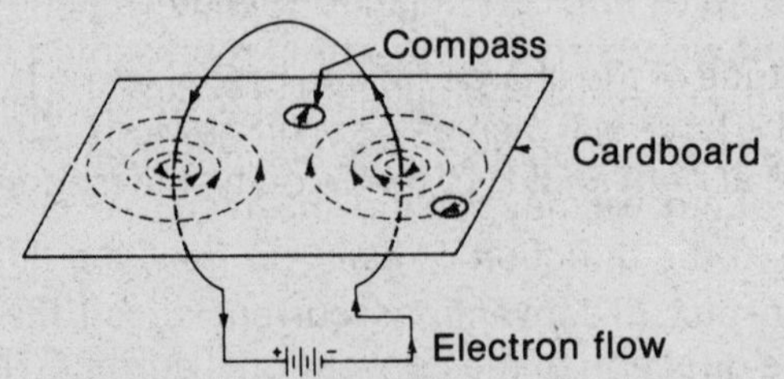

two fields add up to give a relatively strong field. The face of the loop from which the lines are directed is the north pole: in this diagram, it is the face away from us.

Force on a Current-carrying Conductor

When electric charges move across a magnetic field, a force acts on the moving charges which is not present when the charges are stationary. In the diagram imagine the magnetic lines of flux between the north and south poles of some magnet. In this magnetic field is located a wire perpendicular to the paper and to the lines of flux. The current through the wire is shown going into the paper. We can determine the direction of the force on the wire by using the first left-hand rule we mentioned. Grasp the wire with the left hand with the thumb pointing (in this case) into the paper. The fingers then indicate that the lines of flux due to the current are counterclockwise around the wire; they oppose the magnet's lines of flux above the wire, but reinforce them below the wire. The result is that the magnetic field is stronger below the wire than above, and the wire is pushed up. (*If* you *must* learn another rule, here is a three-finger rule for the *right* hand. Hold the *F*orefinger in the direction of the lines of *F*lux of the magnet and the *C*enterfinger in the direction of the electron *C*urrent; then the thu*M*b will show the direction of *M*otion.) Notice that the direction of the force on the wire is at right angles to both the direction of the current and the direction of the lines of flux of the magnet.

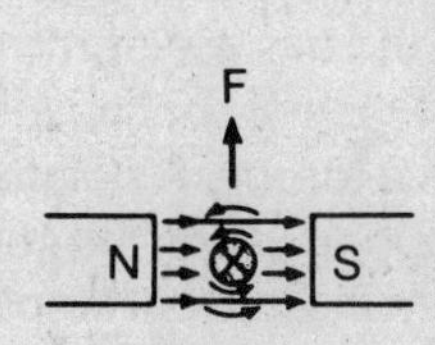

The *magnitude of the force* depends on the direction of the wire in the magnetic field. It is maximum when it is at right angles to the field and zero when it is parallel to it. It is proportional to the current and the strength of the magnetic field.

When the wire is at right angles to the field, the force on the wire is given by the following formula:

$$\boxed{F = BIL}$$

where B is the flux density, L the length of wire in the magnetic field, and I the current in the wire. This formula indicates that in the MKS system we have another unit for B: the *newton per ampere-meter*. This unit is also called the *tesla*.

When two parallel wires carry current, a force acts on both wires. The reason for this is that each wire is in the magnetic field produced by the other. When the two currents are in the same direction, the two wires attract each other. When the currents are in opposite directions, the wires repel each other. The force is proportional to the product of the two currents and inversely proportional to the distance between them:

$$F \propto \frac{I_1 I_2}{d}$$

Force on a Moving Charge

The electric current consists of moving charges. We mentioned before that, when an electric charge moves across a magnetic field, a force acts on the moving charge which is not present when the charge is stationary. The direction of the force is perpendicular to the field and the velocity and can be found with one of the hand rules given above for the electric current. The force is maximum when the charge's velocity is at right angles to the flux. Its magnitude is then given by:

$$F = BQv$$

where Q is the magnitude of the charge (in coulombs), B the flux density (in webers/m^2), and v the velocity (in m/sec). The force will then be given in newtons. The path of this charge will be circular. Since the force will be at right angles to the velocity, the magnitude of the velocity will not change.

DC Meters

A *galvanometer* measures the relative strength of small currents. The common ones are of the moving-coil type; a coil free to rotate is placed in the magnetic field provided by a permanent magnet.

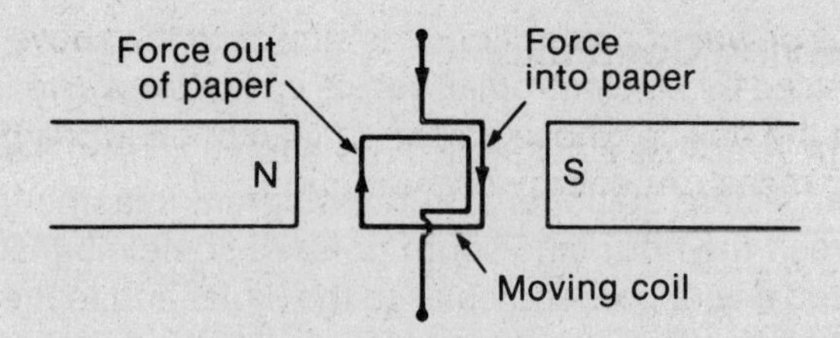

When current goes through the coil, the force on two sides of the coil produces a torque. A pointer attached to the coil rotates with it. In the Weston type of galvanometer a spiral spring keeps the coil from rotating too much; the deflection is usually proportional to the current. When the current stops, the spring restores the coil to its zero position. The resistance of the galvanometer coil is usually small (about 50 ohms); usually the coil can carry safely only small currents (a few milliamperes).

The *ammeter* is an instrument for measuring current; it is calibrated to give the actual magnitude of the current. The ammeter usually consists of a galvanometer with a low resistance in parallel with the moving-coil; this resistance is called a *shunt.* The ammeter is connected in series with the device

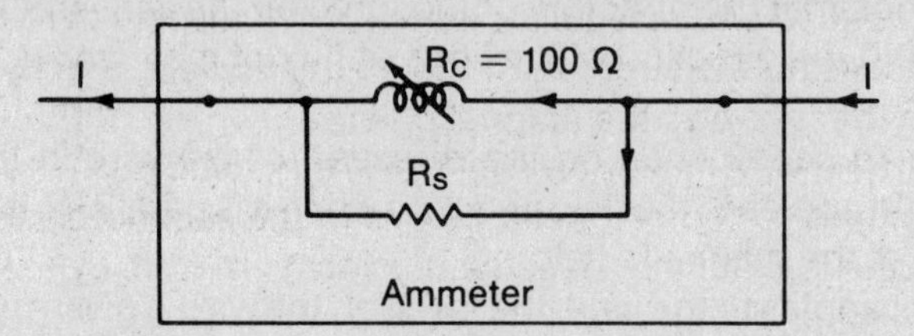

whose current is to be measured. All the current to be measured enters the ammeter. Internally the current divides: a part of the current goes through the coil; the rest goes through the shunt. We can calculate the required resistance of the shunt (R_s) if we apply the rules for a parallel circuit. The shunt is in parallel with the coil (R_c), and therefore the voltage drops are equal: $I_sR_s = I_cR_c$. If the coil has a resistance of 100 ohms and gives full-scale deflection for 10 milliamperes, we can convert it to a 0-1 ampere ammeter: remember that the coil gives full-scale deflection when 0.01 amp goes through it. The other 0.99 amp must then go through the shunt.

$$0.99\ R_s = 0.01 \times 100$$

and R_s is approx. one ohm.

A *voltmeter* is an instrument calibrated to measure the potential difference connected to its terminals. If we want to measure the potential difference across a lamp, we connect a voltmeter in

Rc = 100 Ω

Rm

Voltmeter

parallel with the lamp. A voltmeter usually consists of a galvanometer with a high resistance connected in series with it. This resistance is often called a *multiplier.* The range of a voltmeter can be changed by changing the multiplier. Suppose we want to convert the above galvanometer to a voltmeter with full-scale deflection at 100 V. (Range: 0-100 V.) At full-scale deflection 0.01 amp goes through both the multiplier (R_m) and the coil. The IR-drop across both together is 100 V.

$$I_TR_T = 100$$
$$0.01\ R_T = 100$$

and $R_T = 10{,}000$ ohms. The coil supplies 100 ohms of this total resistance; therefore the required multiplier resistance is 9,900 ohms. In general the resistance of voltmeters is high; that of ammeters is low.

The *wattmeter* measures power consumption. It is similar to the above voltmeter except that the permanent magnet is replaced by an electromagnet; its coil gets a current which is proportional to the current going through the device whose power consumption is being measured.
The *kilowatt-hour meter* measures energy consumption.

ELECTROMAGNETIC INDUCTION

Faraday in England and Henry in the United States independently discovered that a magnetic field can be used to produce an electric current. If a wire moves so that it cuts magnetic lines of flux, an emf is induced in the wire; an *induced emf* is produced in a wire whenever there is relative motion between a wire and a magnetic field. If the wire is part of a complete conducting path, current flows in this circuit in accordance with Ohm's law; this current is referred to as *induced current.*
The *magnitude* of the *induced emf* may be increased by increasing the length of the wire cutting the magnetic field, by increasing the strength of the magnetic field, and by increasing the relative speed of motion between the wire and the magnetic field.
No emf is induced when the wire moves parallel to the magnetic flux. The emf is maximum when the wire moves at right angles to the magnetic flux as in the second diagram below. In the MKS system the magnitude of the maximum induced emf is given in volts by the following formula:

$$E = Blv$$

where B is the flux density in webers per m_2, l is the length of wire cutting the flux, and v is the speed of cutting the flux. It is the *relative* motion between the wire and the magnetic flux that matters. The same result is obtained if the wire is stationary and the magnet moves.
An emf is induced in *a single loop* of wire when the magnetic flux going through the loop changes. The magnitude of the induced emf is proportional to the speed with which the flux changes. The emf is 1 volt when the change in flux is 1 weber per second. If a coil has several turns of wire, the emf is proportional to the number of turns in the coil.
The *direction* of the induced current can be figured out by the use of *Lenz's law:* the direction of the

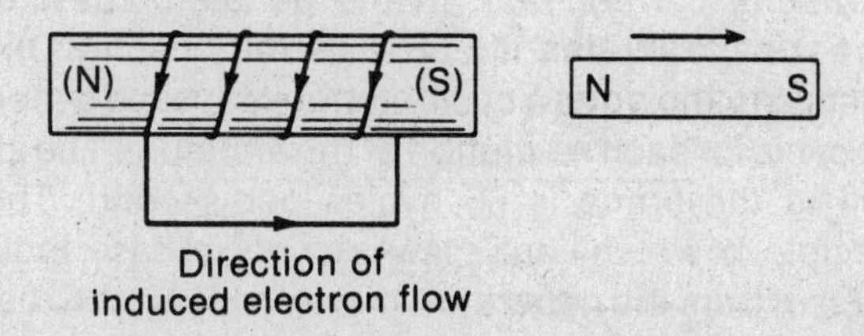

induced current is such as to produce a magnetic field which will hinder the motion that produced the current. For example, if a permanent magnet is moved away from a stationary coil, an emf is induced in the coil. In the diagram shown, the motion of the magnet is opposed if the induced current produces an S-pole to attract the retreating N-pole. This will happen if the induced current has the direction shown. Of course, an N-pole will be produced simultaneously at the left end of the coil. If we have a single wire moving across a magnetic field, we can determine the direction of the current this way: In the diagram assume that the wire is perpendicular to the page and pulled towards the bottom edge of the paper. If the induced current is directed into the paper, the left-hand rule tell us that the magnetic field is weakened above the wire, strengthened below it (closer to you). This tends to push the wire toward the top edge of the paper, opposing the motion which produced the induced current. Therefore we guessed the direction of the induced current correctly: into the paper.

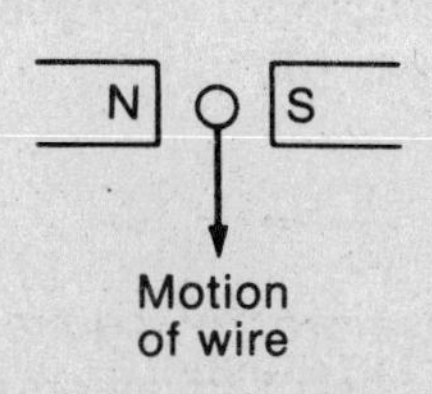

Electric Generators

The term *dynamo* is sometimes used to refer to an electric generator which converts mechanical energy to electrical energy. Usually a coil is made to rotate in a magnetic field, and an emf is induced in the coil. The source of the energy for turning the coil may be a waterfall, steam under pressure, etc. Imagine the coil in the diagram turning clockwise in the diagram. The axis of the coil's rotation is perpendicular to the paper and is represented by the dot. The coil is rotated clockwise. The

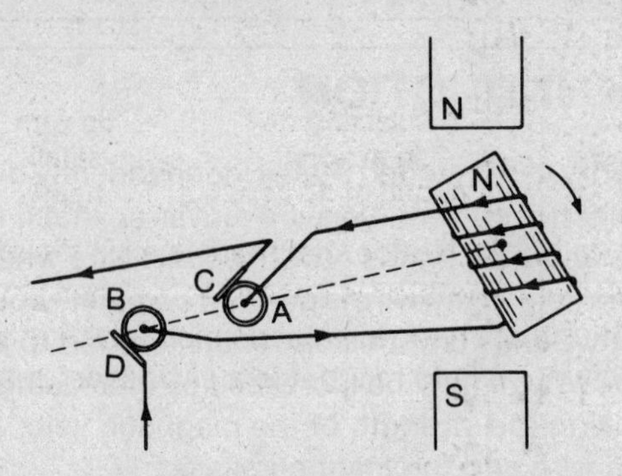

stationary magnet provides a magnetic field which is cut by the rotating coil; this magnet is called a *field magnet.* Because the rotating coil cuts across magnetic lines of flux, an emf is induced in it and electric energy can be taken from it; this coil is called an *armature* coil. We can use Lenz's law to figure out the direction of the current in the armature. At the instant shown the direction of the current in the armature must be such as to produce the poles marked in the diagram; then the motion will be opposed by the repulsion of like poles. If we use our left-hand rule for a solenoid, we will find that the direction of electron flow in the armature is as shown in the diagram. *Slip rings A and B,* insulated from each other, are on the same shaft as the armature and rotate with it. *Brushes C* and *D* are stationary and make wiping electrical contact with the slip rings. Current can therefore flow through an external circuit connected to the brushes.

However, a moment later the armature will be in a new position. In order for the motion to be opposed now, the polarity of the armature, and therefore the direction of the current in the armature, must change. This is repeated as the armature keeps rotating. Current whose direction is constantly reversing is known as *alternating current* (ac). In the generator described, using slip rings, the current in the external circuit is also alternating. One complete back-and-forth variation is known as a *cycle.* In the simple a c generator shown there would be one cycle for each revolution of the armature. The alternating current ordinarly used in the home is 60 cycles per second. The conservation of energy principle, of which Lenz's law is a special case, indicates that, as more current is drawn from the generator, more energy has to be used to turn the armature.

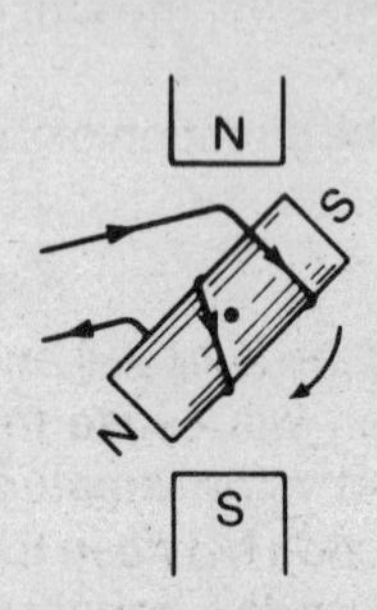

A *dc* (*direct current*) *generator* can be constructed with a slight modification of the above ac generator. We use the same field magnet and armature coil, but instead of slip rings we use a *commutator,* a metallic ring split into two insulated parts of equal size. (In more complicated dc

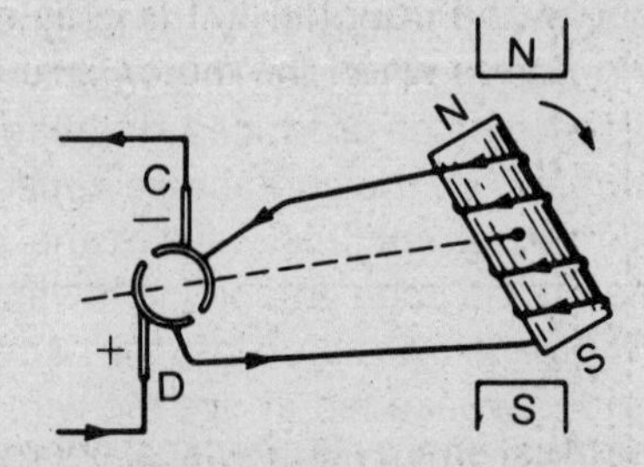

generators the ring is split into several commutator segments.) Each end of the armature coil is connected to a commutator segment. As the armature rotates, the stationary brushes alternately make contact with one segment and then with the other. The brushes are set so that this change in contact occurs just when the direction of current in the armature coil changes. As a result, the current in the external circuit is always in one direction (dc), but its magnitude is constantly changing; such current is known as *pulsating* or *fluctuating dc.* This fluctuation is reduced in more complicated dc generators by the use of more commutator segments and more armature coils. The

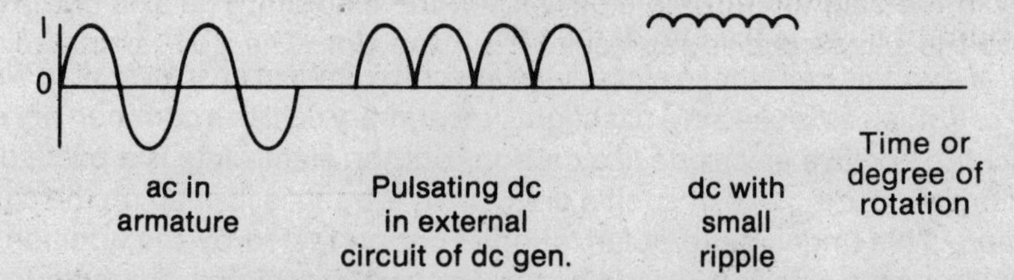

residual fluctuation is known as *ripple.* Notice that there is ac in the armature of both the dc and ac generators. The function of the commutator is described as changing ac to dc. Also note that the polarity of the field magnet does not change. This means that if it is an electromagnet, dc must be used. A *magneto* is a dynamo whose field magnet is a permanent magnet.

Electric Motors—DC

The dc motor is similar to the dc generator except that electrical energy is supplied to the motor to be converted to mechanical energy. The diagram shown is similar to the one for the dc generator. A *field winding* or field coil has been shown for the field magnet. The diagram is for a *shunt-wound motor:*

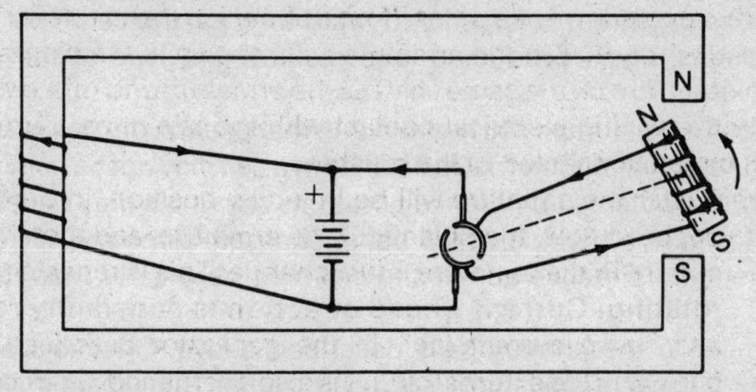

The field winding is in parallel with the armature coil. (In a series-wound motor the field coil and armature coils are in series.) Notice that the connection has been made which will cause the armature to rotate counterclockwise. To keep the coil rotating that way, the current in the armature must reverse at the right instant to change the polarity of the armature. The interaction between the poles of the field magnet and the poles of the armature provides the force to keep the armature rotating. However, because the armature coil rotates in a magnetic field, an emf is induced in the armature. According to Lenz's law, this emf will be in such a direction as to oppose the rotation. Therefore this induced emf is called a *counter-emf* or *back emf.* This counter-emf tends to reduce the current in the armature. Because the counter-emf is greatest when the armature is rotating fastest, the current in the armature is less when the motor is running at full speed than when it is starting.

VOLTAIC CELLS

Another important source of electrical energy is chemical energy. A *voltaic cell* converts chemical energy into electrical energy; it consists of two dissimilar *electrodes* immersed in an electrolyte which acts on at least one of them. An *electrolyte* is a liquid which conducts electricity by the motion of ions, such as a solution of salt in water—also, a substance that becomes such a conductor when dissolved in a suitable solvent. The electrodes are conductors. Examples of voltaic cells are: flashlight cell, automobile battery. A *battery* is a combination of two or more cells. Because of the chemical action of the electrolyte on the electrodes, there will be an excess of electrons on one electrode as compared with the other. This results in a potential difference between the electrodes, the one with the excess of electrons becoming the negative terminal. The potential difference that exists when no current flows is the emf of the cell.

A *primary cell* is a voltaic cell whose electrodes are consumed in an irreversible way when the cell is used. The *dry cell,* such as is used in a flashlight, is a primary cell. In a common dry cell the negative electrode is zinc; the positive electrode is a carbon rod; the electrolyte is a paste of sal ammoniac (ammonium chloride). During use of such a dry cell, hydrogen is formed on the carbon rod; this is called *polarization.* This undesirable polarization is counteracted by the addition to the paste of manganese dioxide, which acts as the depolarizer. Another undesirable characteristic of this kind of dry cell is *local action*-a wasting away of the zinc because of impurities in it. This is counteracted by coating the zinc with mercury in a process known as amalgamating the zinc. (In recent years another kind of dry cell has been introduced—the mercury cell.)

A *secondary cell* is a voltaic cell whose electrodes can be used over and over again. The automobile battery uses storage cells. In the U.S. the most commonly used storage cell is the *lead storage cell.* In this cell the positive electrode contains lead dioxide; the negative electrode contains spongy lead; the electrolyte is sulfuric acid in water. When the cell is *discharging,* it is supplying electrical energy to an external circuit. During the discharging process both the lead and the lead dioxide are converted to lead sulfate, which adheres to the electrodes. At the same time the specific gravity of the electrolyte goes down. Therefore a hydrometer can be used to measure the specific gravity of the electrolyte in the automobile battery to determine the condition of the battery. When the cell is *being charged,* current is forced backward through the cell. This reverses the chemical action at the electrodes, again producing lead dioxide and lead at the positive and negative electrodes, respectively.

The emf of each lead acid storage cell is about 2 V; that of each dry cell is about 1.5 V. Remember from the previous chapter that the internal resistance of a cell may not be negligible. Therefore, when a cell is discharging, the potential difference across its terminals may be less than its emf (by the internal IR-drop). However, when the cell is being charged, since the current through the cell is backwards, the potential difference across the cell may be greater than its emf (by the internal IR-drop).

Fuel cells are similar to the primary cells described above. Chemical energy is converted to electrical energy, but conventional fuels are used, rather than metals like zinc. For example, hydrogen may be continuously fed to the anode, and oxygen to the cathode.

Solar cells convert the sun's energy to electrical energy. They have been used extensively in space vehicles. The potential of solar cells is great since sunlight is free and available everywhere. There should be no harmful by-products, but progress must be made to make the cells less expensive and more efficient. (Also see *photovoltaic cell.*)

ELECTROLYTIC CELLS

In *electrolytic cells* the energy conversion is the opposite of that in voltaic cells: electrical energy is converted to chemical energy. An example is the electrolysis of water. Water is dissociated by the electric current into hydrogen and oxygen. (The hydrogen and oxygen are chemicals; when a spark is applied to their mixture they combine to give water, and at the same time energy is released.)

ANSWERING THE QUESTIONS Now let us return to the questions at the beginning of the chapter. What happens when an ammeter is connected across a generator? We saw that the resistance of an ammeter is small. The one we described had a resistance of about one ohm and a full-scale deflection for one ampere. If we connected the ammeter to a generator of 30 V, we could expect a current of 30 amp (Ohm's law) to flow momentarily through the ammeter; momentarily, because the moving coil of the meter would burn out unless a protective fuse would burn out first. In general, don't connect an ammeter directly to a source of emf.

What happens when a dc generator is connected across a battery? If the emf of the generator is less than that of the battery (for example, when the armature is not rotating), the battery will turn the generator like a motor. In practice this is usually not desirable. If the emf of the generator is greater than the emf of a storage battery connected across it, the battery will be recharged. However, unless precautions are taken, this current may be excessive and damage the battery.

QUESTIONS Chapter 13

Select the choice which fits the question best.

1. Heating a magnet will **(A)** weaken it **(B)** strengthen it **(C)** reverse its polarity **(D)** produce new poles **(E)** have no effect

2. The emf produced by a generator operating at constant speed depends mainly on **(A)** the thickness of the wire on the armature **(B)** the thickness of the wire on the field magnet **(C)** the strength of the magnetic field **(D)** the length of time the generator operates **(E)** the size of the brushes

3. A factory gets its electrical power from a generator two miles away. The two wires connecting the generator to the factory terminals have a resistance of 0.04 ohm/mile. When the generator supplies 50 amp to the factory, the terminal voltage at the generator is 120 volts. A voltmeter connected to the factory terminals should then read **(A)** 100 V **(B)** 110 V **(C)** 112 V **(D)** 118 V **(E)** 120 V

For 4-6. In the circuit each of the three resistors has a resistance of 30 ohms.

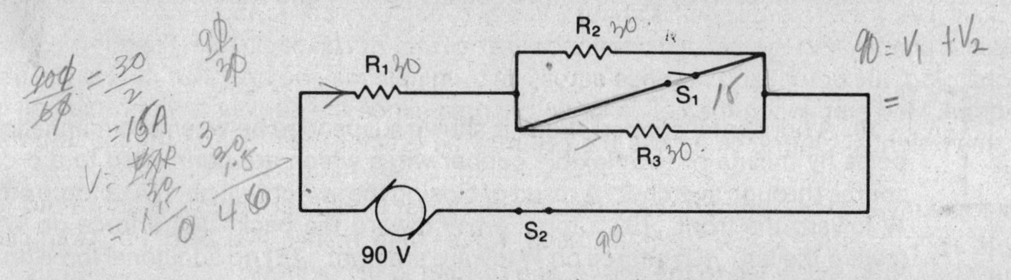

4. When switch S_1 is open and S_2 is closed as shown, the potential difference across R_1 is, in volts, **(A)** 0 **(B)** 30 **(C)** 45 **(D)** 60 **(E)** 90

5. When switch S_1 and S_2 are closed, the potential difference across R_1 is, in volts, **(A)** 0 **(B)** 30 **(C)** 45 **(D)** 60 **(E)** 90

6. When switches S_1 and S_2 are open, a voltmeter connected across S_2 read **(A)** 0 **(B)** 30V **(C)** 45 V **(D)** 60 V **(E)** 90 V

7. X is a coil of wire with a hollow core. The permanent magnet is pushed at constant speed from the right into the core and out again at the left. During the motion **(A)** there will be no current in wire YZ **(B)** electron flow in wire YZ will be from Y to Z **(C)** electron flow in wire YZ will be from Z to Y **(D)** electron flow in wire YZ will be from Z to Y and then from Y to Z **(E)** electron flow in wire YZ will be from Y to Z and then from Z to Y

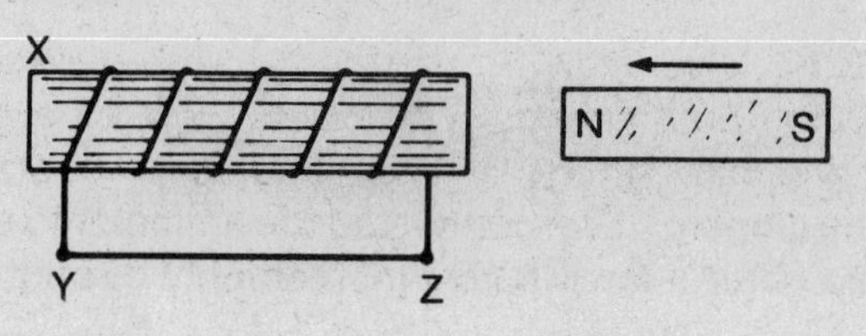

8. The soft-iron armature *X* is allowed to fall into position on top of the poles of the horseshoe magnet. While it falls,

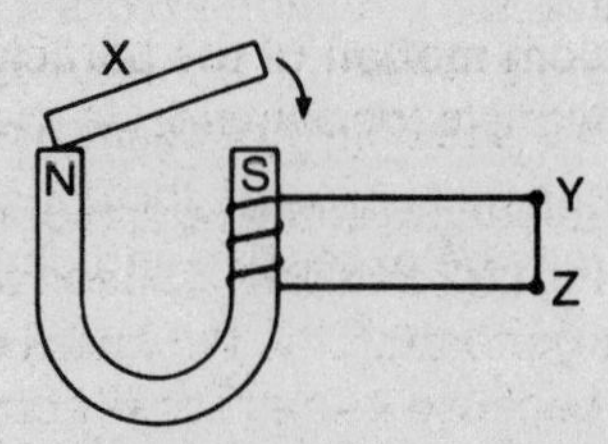

(A) there will be no current in wire *YZ* **(B)** electron flow in wire *YZ* will be from *Y* to *Z* **(C)** electron flow in wire *YZ* will be from *Z* to *Y* **(D)** electron flow in wire *YZ* will be from *Z* to *Y* and then from *Y* to *Z* **(E)** electron flow in wire *YZ* will be from *Y* to *Z* and then from *Z* to *Y*

9. *G* is a sensitive galvanometer connected to coil 2. Coil 1 is insulated from coil 2. Both coils are wound on the same iron core and are insulated from it. When the variable contact on the rheostat is

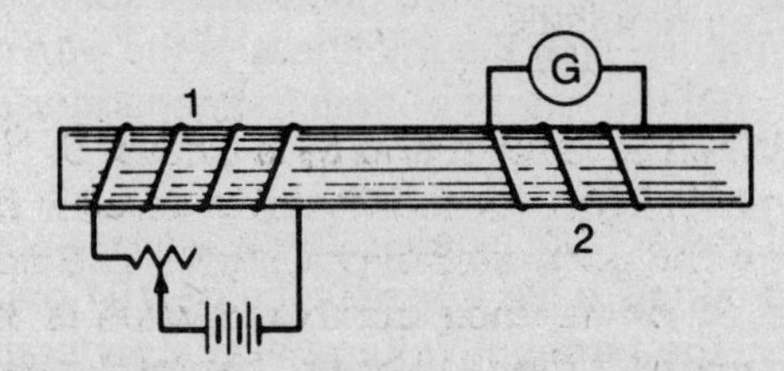

moved half way to the left, the needle on the galvanometer **(A)** moves to the left **(B)** moves to the right **(C)** moves momentarily and returns to its starting position **(D)** doesn't move because coils 1 and 2 are insulated from each other **(E)** doesn't move because coils 1 and 2 are insulated from the iron

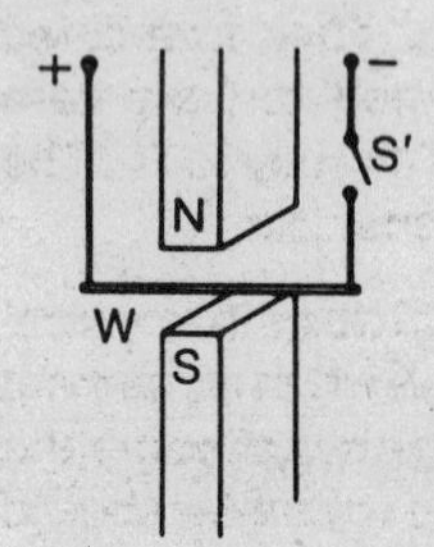

10. A horizontal copper wire *W* is shown suspended between two magnetic poles by means of two flexible copper wires which are connected to a d c source through switch *S′*. A result of closing the switch will be **(A)** a force on *W* toward the front **(B)** a force on *W* toward the back **(C)** a force on *W* toward the left **(D)** a force on *W* toward the right **(E)** no additional force on *W*

11. A 3-cm length of wire is moved at right angles across a uniform magnetic field with a speed of 2.0 m/sec. If the flux density is 5.0 webers/m^2, what is the magnitude of the induced emf? **(A)** 0.03 V **(B)** 0.3 V **(C)** 0.6 V **(D)** 10 V **(E)** 20 V

EXPLANATIONS TO QUESTIONS Chapter 13

Answers

1. **(A)**	4. **(D)**	7. **(D)**	10. **(B)**
2. **(C)**	5. **(E)**	8. **(C)**	11. **(B)**
3. **(C)**	6. **(E)**	9. **(C)**	

Explanations

1. **(A)** Heating increases the random motion of the particles in the magnet. This leads to a less orderly arrangement of molecular magnets (or domains) in the magnet, making the magnet weaker.

2. **(C)** The emf induced in a wire depends on the length of the wire cutting the magnetic field, on the strength of the magnetic field being cut, and on the speed with which the field is being cut at right angles. In the practical design of the generator, the thickness of the wire used is important but does not affect directly the emf produced because it does not affect the three factors mentioned. Once the generator reaches constant speed, the length of time the generator operates does not affect the above three factors.

3. **(C)** Four miles of wire are used to connect the generator to the factory terminals. Since the wire has a resistance of 0.04 ohm/mile, 4 miles will have a resistance of 0.16 ohm. This resistance is in series with the factory load, which gets 50 amp. The IR-drop in the connecting wires = (50 amp) $\times$ (0.16 ohm) = 8 volts. Since the generator supplies 120 V to the circuit and the drop in the connecting wires is 8 V, 112 V will be left for the factory load (In a series circuit, $V_T = V_1 + V_2$). A voltmeter should measure this voltage at the factory terminals.

R_1 15 Ω
30 Ω
90 V

4. **(D)** When S_1 is open and S_2 is closed, S_1 has no effect. We then have a series-parallel circuit: R_2 and R_3 are in parallel; their combination is in series with R_1 and the generator supplying 90 V. Since R_2 and R_3 have the same resistance, 30 ohms, their combined value is 30/2 ohms, or 15 ohms. The equivalent circuit is then as shown. The total resistance of the circuit then is 45 ohms. (In a series circuit, $R_T = R_1 + R_2$.)

$$I_T = V_T/R_T = 90\ \text{V}/45\ \Omega = 2\ \text{amp.}$$
$$V_1 = I_1R_1 = 2\ \text{amp} \times 30\ \Omega = 60\ \text{volts}; \quad I_T = I_1$$

5. **(E)** When switch S_1 is closed, we have a short-circuit across R_2 and R_3. Their combined resistance is zero ohms, because the switch, in parallel with them, is assumed to have zero resistance. The circuit is then the equivalent of having the generator connected directly to R_1. The potential difference across R_1 is then the same as the voltage supplied by the generator.

6. **(E)** When switch S_2 is open, there is a break in the whole circuit and no current flows in any part. We can find the answer to the problem in different ways. One way is to realize that, when no current flows through a resistor, there is no voltage drop across it. Therefore the voltage supplied by the generator will be available at the break, and this is what the voltmeter should measure. Another way is to think of the voltmeter as completing the circuit; the voltmeter then acts as a resistor as well as a meter. The resistance of the voltmeter is high: several thousand ohms. The current through it and the other resistors in the circuit will be small. The IR-drop across the other resistors in the circuit (having small resistance) will be negligibly small. Practically the full generator voltage will be across the voltmeter, and the voltmeter will record that voltage.

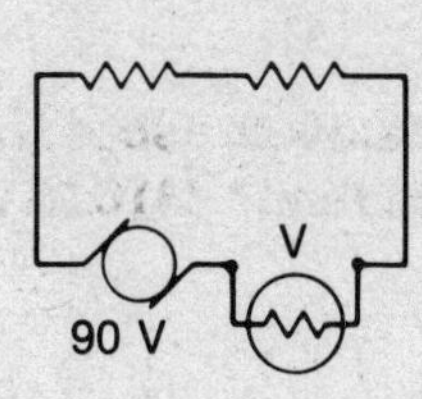

7. **(D)** When the magnet approaches the coil, electrons will flow through the coil and from Z to Y. This follows from Lenz's law. An emf is induced in the coil (because the magnetic field moves with respect to the coil); the direction of the induced current must be such as to oppose the approach of the magnet by producing poles in the coil, as shown. The left-hand rule applied to the coil (thumb pointing in direction of N-pole) then gives the direction of electron flow indicated. When the magnet has moved through the coil and is moving away from it, the direction of the induced current must be reversed: In that case the poles of the coil will be reversed, which is needed to oppose the motion.

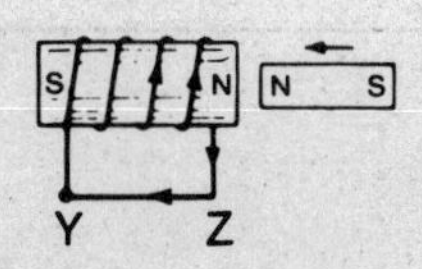

8. **(C)** The soft-iron armature *X* is magnetized by induction: the part on the left becomes an S-pole; the part on the right an N-pole. The falling of *X* is then equivalent to a magnet moving near a coil and a current will be induced in the coil. According to Lenz's law, the direction of the current should be such as to oppose the falling of *X*. This will be so if the top of the coil becomes a North pole, opposing the falling N-pole. The left-hand rule for the coil indicates that electron flow is from *Z* to *Y*.

9. **(C)** There will be a change of current in coil 1; the resulting change in the magnetic field in the iron core affects coil 2 in spite of any insulation. The current is momentary, because an emf is induced only while there is a change in the magnetic field. We can't tell whether the needle moves to the left or right because we don't have enough information about the connection to the galvanometer.

10. **(B)** The closing of the switch results in electron flow through *W* from right to left. The left-hand rule for a wire tells us the direction of the resulting magnetic field around the wire: in front of the wire downward, behind the wire upward. The magnetic field due to the two poles is thus reinforced in front, weakened in back of the wire. The resulting force is toward the back.

11. **(B)** $E = Blv$
$= (5 \text{ webers/m}^2)\ (0.03 \text{ m})\ (2 \text{ m/sec})$
$= 0.3$ volt

14

Alternating Current Circuits

Does a coil with an iron core use more electrical power on 120-V dc than on 120-V ac? Does a capacitor use more electrical power on 120-V dc than on 120-V ac?

AC can do some things better than dc, and some other things just as well. AC can be transmitted more efficiently over long distances and its voltage can be changed more readily. AC as well as dc can be used conveniently for heating and lighting. Only dc can be used for electroplating, charging batteries, and operating some electronic circuits. Some efficient motors run on ac only, but the best variable-speed motors operate on dc only. *Rectifiers* can be used to change ac to dc (see next chapter).

In the last chapter we showed that the potential difference at the terminals of an ac generator constantly varies in magnitude and direction. The graph indicated that this voltage varies with time like a sine wave; such an AC voltage is known as a *sinusoidal voltage*. AC voltages can be obtained which vary like a square wave (flat-topped) or like a sawtooth wave (pointed-top), but we shall restrict our description to sinusoidal voltages.

When a resistor is connected across an ac generator, the current through the resistor will vary in step with the voltage change: the current through the resistor is always *in phase* with the voltage across it, even when there are other devices in the circuit; i.e., when the voltage is zero the current is zero, and when the voltage has a peak value the current has a peak value. Ohm's law applies directly to the resistor: the current through the resistor at any instant is equal to the voltage across the resistor at that instant divided by the resistance.

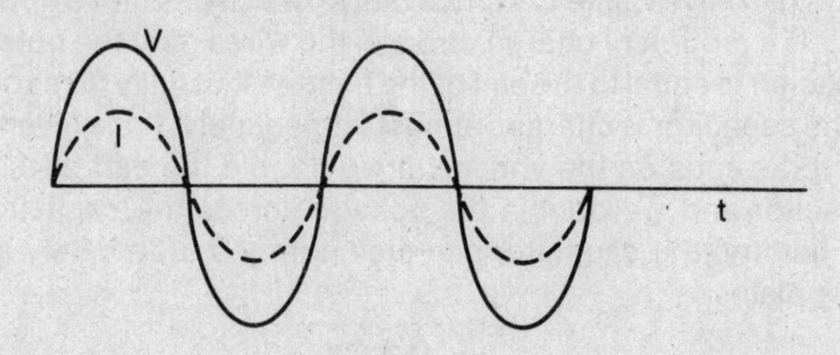

Usually we describe the magnitude of the alternating current and voltage in terms of their *effective values.* An alternating current has an effective value of one ampere if it produces the same heat in one second in a given resistor as a direct current of one ampere. The effective value of a sinusoidal current is approximately 0.707 of the peak current; the effective value of the voltage is approximately 0.707 of the peak voltage. The effective value is sometimes referred to as the *root mean square* value.

Usually in describing the magnitude of an alternating current or voltage the term "effective" will be omitted.

$$I = I_{eff} = I_{rms} = 0.707\ I_p$$
$$V = V_{eff} = V_{rms} = 0.707\ V_p$$

Example: The ac used in the U.S. is often referred to as 115 V, 60 cps. This means that the frequency of the alternating current is 60 cycles per second, and that the effective value of the voltage is 115 V. The peak value of this voltage, therefore, is (115/0.707) or approximately 162 V.

Ohm's law can also be applied directly to resistors using effective values. For example, if the above 115 volts are applied to a device having a resistance of 5 ohms, the current ($I = V/R$) is 115/5, or 23 amp. This, of course, is the effective value of the current.

AC meters are usually calibrated to give effective values. The *hot-wire meter* depends on the fact that alternating current through a wire produces heat in the wire and that the resulting rise in temperature increases the length of the wire. Some meters use this heat to produce a dc voltage which can then be measured on the dc meters described in the previous chapter. This is done with a *thermocouple:* two different wires (e.g., iron and lead) are twisted together at one end; when this junction is heated an emf appears between the other ends. In some meters the ac is changed directly to dc by means of rectifiers (see next chapter); this dc is then measured on the dc meters.

CAPACITOR AND CAPACITANCE

A *capacitor* consists of two conductors separated by an insulator or *dielectric.* The capacitor was formerly called a condenser. Common dielectrics are paper, mica, air, ceramic. Capacitors are often named by the dielectric used; e.g., paper capacitors. In some capacitors the relative position of the two conductors can be changed; these capacitors are known as *variable capacitors;* the others are known as *fixed capacitors.* The *Leyden jar* is an old type of capacitor in which the insulator is a glass jar. The *function* of a capacitor is to store an electric charge. Capacitors are used in tuning a radio, in reducing hum in a radio by reducing the ripple in the dc, and in reducing sparks at electrical contacts.

How does a capacitor store a charge? Let us first imagine a capacitor in a dc circuit—connect a capacitor to a battery. On the negative terminal of the battery is an excess of electrons. These electrons repel each other and we have provided a path for these electrons. They start moving along the connecting wire *A* to *B,* the right plate of the capacitor. At the same time, electrons are attracted from wire *D* and plate *C* of the capacitor to the positive terminal of the battery, where there is a deficiency of electrons. As a result there is a difference of potential between the plates of the capacitor. For every electron that gets to plate *B* there is another electron that leaves plate *C.* Notice that no electrons move through the dielectric. There is a motion of charge through the wires until the potential difference between the plates of the capacitor is equal to the emf of the battery. It usually takes only a fraction of a second for this to happen; the capacitor is charged almost immediately. For a given capacitor the amount of charge that is stored depends on the voltage; if we double the emf of the battery, we double the charge on the capacitor and we double the voltage across the capacitor. If we disconnect the capacitor from the battery, the capacitor can stay charged indefinitely and have a difference of potential between its plates.

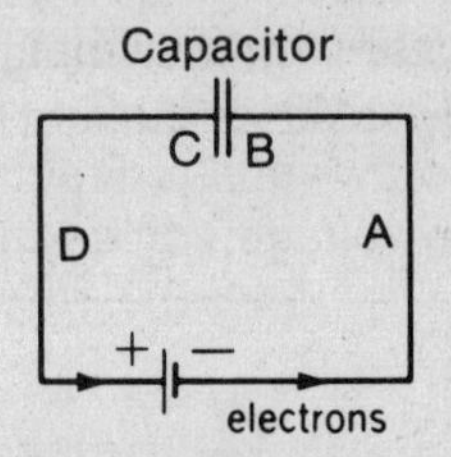

$$Q = CV$$

where Q is the charge on either plate of the capacitor, C is the capacitance of the capacitor, and V is the voltage or potential difference of the capacitor. If the voltage is expressed in volts and the charge in coulombs, then the capacitance is given in *farads.* The farad is a rather large unit. Capacitance is usually expressed in *microfarads* (μfd).

$$1 \text{ farad} = 10^6 \text{ microfarads}$$

The capacitance of a capacitor is proportional to the area of its plates. If we double the area of the plates, we provide twice as much space for storing electrons. The capacitance of a capacitor is inversely proportional to the distance between the plates; if we double the distance between the plates, we reduce the capacitance to one-half of its original value. The capacitance also depends on the nature of the dielectric.

Capacitors in Series AC Circuits *

We saw before that, if a capacitor is connected in a dc circuit, there is a momentary current. After the initial current we think of the capacitor as blocking dc. If we connect a capacitor in an ac circuit, the capacitor charges and discharges; the voltage across it goes up and down and changes direction at the same frequency as the emf of the generator to which it is connected. However, the current in the circuit is not in phase with the voltage across the capacitor. If we think of one complete cycle as representing 360°, we find that the current "through the capacitor" is 90° ahead of the voltage across it.

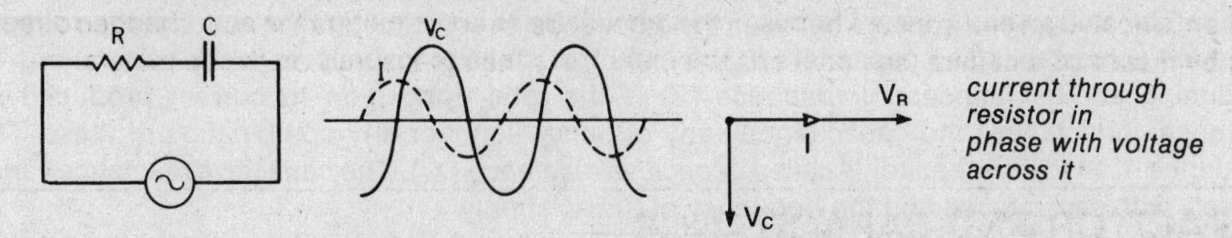

This means that, when the current reaches its peak value, the voltage across the capacitor is just reaching zero; when the current reaches zero, the voltage is just becoming maximum. We speak about the current through the capacitor, but, as in dc, no current actually goes through the dielectric. Notice that, since we have a series circuit, the current through the resistor is the same as that through the capacitor. Also, the voltage across the resistor is still in phase with the current through it. Therefore, the voltage across the resistor is 90° ahead of the voltage across the capacitor. We can represent this phase relation by means of sine curves; this tends to be a little confusing. We usually use a vector notation to represent the effective values of the current and voltages. We can use different scales for the current and voltages, but, of course, must use the same scale for all the voltages.

When a capacitor is charged, we think of it as storing an electric charge. This also means that it stores electrical energy. When the capacitor is discharged it returns this energy to the circuit.

COILS IN SERIES—AC CIRCUITS

We saw in the previous chapter that an emf is induced in a coil if there is relative motion between the coil and a magnetic field. If alternating current goes through a coil, the magnetic field which is produced moves away from and back to the coil and also keeps changing direction. As a result an emf is induced in the coil which is in a direction such as to oppose the change in the current. *Self-induction* is the production of an emf in a coil because of the current in the coil. (A coil is sometimes called an *inductor.*) The induced voltage is proportional to the rate at which the current changes and to a property of the coil called *inductance.* The inductance of a coil depends on its shape, on the number of turns of wire in the coil, and on the nature of the core. If we increase the number of turns, we increase the inductance. If we replace the air core by iron, we also increase the inductance. The unit of inductance is the *henry.* If we connect a coil in an ac circuit, the self-induced emf goes up and down and changes direction at the same frequency as the emf of the generator to which it is connected. However, the current through the coil is not in phase with the voltage across it. As in the case of the capacitor, there is a 90° phase difference (if the resistance of the coil is negligible). But this time the voltage (V_L) is ahead of the current.

*Before studying the rest of this chapter, re-read the Introduction.

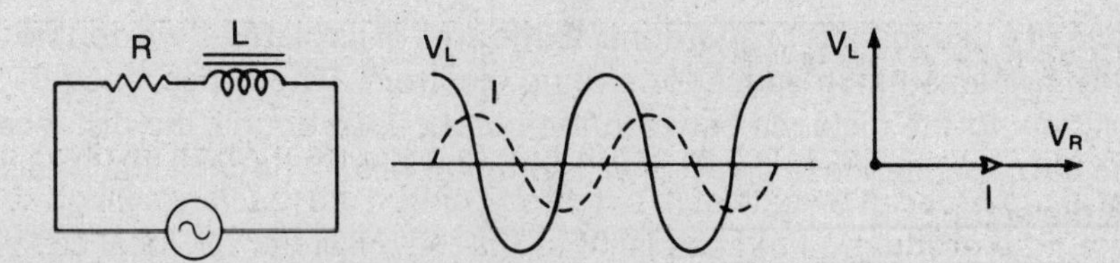

Again the current through the circuit is in phase with the voltage across the resistor.

Notice that in an ac circuit the magnitude of the current depends not only on the resistance in the circuit and the applied emf, but also on the inductance of the coil and the capacitance of any capacitor in the circuit. If the inductance is increased, the current *decreases*, but if the capacitance is increased, the current *increases*.

REACTANCE AND IMPEDANCE

We can try to figure out what the current in an ac circuit is by taking into account the voltages developed across coils and capacitors in the circuit. It is easier, usually, to make calculations by introducing another concept. *Impedance* (Z) is the total opposition to current produced by resistance, inductance, and capacitance—any of these alone, or any combination of these. The impedance of an ideal capacitor is called *capacitive reactance* (X_c). The capacitive reactance varies inversely with capacitance and the frequency of the ac supply.

$$\mathbf{X_c = \frac{1}{2\pi fc}}$$

When the frequency is in cycles per second and the capacitance is in farads, the reactance is in ohms. Most capacitors are close enough to the ideal so that we can ignore any other effect of a capacitor.

The impedance of an ideal coil is called its *inductive reactance* (X_L). This is the opposition of a coil due only to its inductance and is proportional to the inductance of the coil and the frequency of the ac supply.

$$\mathbf{X_L = 2\pi fL}$$

When the frequency is in cycles per second and the inductance is in henries, the reactance is in ohms. Actual coils are often far from ideal; that is, coils usually have appreciable resistance in addition to inductance. For the purpose of most calculations we can think of such a coil as having its resistance in series with its inductance.

The total impedance is not obtained by ordinary addition. Because of the phase differences between voltages and current which we saw earlier, the impedance is obtained by application of the Pythagorean theorem. In a series ac circuit:

V

R L C

$$Z = \sqrt{R^2 + (X_L - X_c)^2}$$

$$X_L = 2\pi fL$$

$$X_c = \frac{1}{2\pi fc}$$

$$I = \frac{V}{Z}$$

$$V_L = IX_L;\ V_c = IX_c;\ V_R = IR$$

Note several things about the use of these relationships. If the circuit contains only resistance and capacitance, the formula for impedance does not contain the term for inductive reactance: $Z = \sqrt{R^2 + X_c^2}$. We use a generalization of Ohm's law: $I = V/Z$. To calculate the current in the circuit we use total impedance rather than just resistance. This generalization also applies to any part of the circuit. Therefore the voltage across any part of the circuit is equal to the current times the impedance of that part of the circuit; e.g., $V_c = IX_c$.

POWER IN AN AC CIRCUIT

Remember that we defined resistance as opposition to the current which involved the production of heat. This heat is produced at the rate of I^2R. In the ac circuit the reactance involves opposition to the current without the production of heat. Neither the ideal coil nor the ideal capacitor uses any power. For example, a capacitor in an ac circuit gets charged during one-half of the cycle, thus storing electrical energy; then it discharges, sending its charge through the generator and gets recharged in the opposite direction during the second half of the cycle, again storing the same amount of energy. Therefore, as the cycles get repeated, the generator does not have to supply any energy to the capacitor. A similar thing is true for the ideal coil, except that the energy is stored in the magnetic field of the coil. Therefore the power used by an ac circuit is calculated by calculating the power used by the resistance in the circuit. The best formula is:

$$\mathbf{P = I^2R}$$

Remember that the magnitude of the current is affected by the presence of reactance. This is the expression for *true power* and gives the power in watts when the current is in amperes and the resistance is in ohms. The other expressions for electric power usually do not give the correct value of true power. For example, the product VI is called *apparent power* (unit is volt-ampere) and is usually greater than true power, I^2R. If we multiply the apparent power by a certain number called the *power factor* of the circuit, we get the true power consumption of the circuit.

$$\textbf{power factor} = \frac{\textbf{true power}}{\textbf{apparent power}}$$

The value of the power factor depends on the phase relation between the current and the voltage.

It may be easier to visualize some of these things if we again represent some of the quantities by vectors. Recalling that the voltage across a capacitor is 90° behind the current, while the voltage across an ideal coil is 90° ahead of the current, we note that there is a phase difference of 180° between the voltage across the coil and the voltage across the capacitor. The voltage across the resistor is in phase with the current. Let us combine the three voltages:

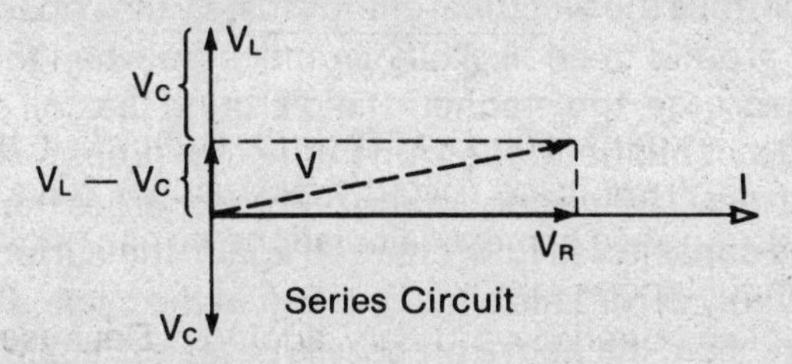

Series Circuit

Take the difference between the voltage across the coil and the voltage across the capacitor; then combine this difference with the voltage across the resistor by completing the parallelogram (rectangle) and drawing the concurrent diagonal, V. This represents the voltage across the whole circuit. Notice that usually V is not in phase with I, the current through the circuit. (The angle between these two is the phase angle of the circuit; the power factor of the circuit is the cosine of this angle.) Notice that the component of this total voltage which is in phase with the current is the voltage across the resistor, $V_R = IR$. Because the voltages are proportional to the current, we may use the same representation for reactances and impedance. Notice that the diagonal of this parallelogram gives the same expression for impedance that we had before.

$$Z = \sqrt{R^2 + (X_L - X_C)^2}$$

RESONANT CIRCUITS

By looking at the expression for impedance [$Z = \sqrt{R^2 + (X_L - X_c)^2}$], we notice that the impedance of a series ac circuit is least if

$$X_L = X_c$$

This equality may be used as a definition of a resonant circuit. Other conditions follow. Since the current is the same throughout the series circuit

$$IX_L = IX_c$$

that is, the voltage across the coil is equal to the voltage across the capacitor. These voltages may be very large, in fact larger than the applied voltage from the generator. Since these voltages are opposite in phase, they cancel each other, and the applied voltage is equal to the IR-drop across the resistor. This also indicates, as can be seen from the vector diagram, that at resonance the *current and voltage* across the whole circuit *are in phase.* Furthermore, since the impedance is least, for a given applied voltage, the *current is maximum* at resonance. One more but very important item—since $X_L = X_c$, $2\pi f L = \frac{1}{2\pi f C}$; clearing of fractions and solving for *f* gives the expression for the *resonant frequency* of the circuit:

$$f_r = \frac{1}{2\pi\sqrt{LC}}$$

The resonant frequency is also known as the *natural frequency.*

Why is it that the voltage across the coil or capacitor may be larger than the voltage applied to the circuit by the generator? Imagine a capacitor which has been charged in some way and is then connected across a coil. Nothing else is in the circuit—only a coil and a capacitor. As soon as the terminals of the coil are connected to the terminals of the capacitor, the voltage across the two has the same magnitude and stays equal. However, the capacitor starts to discharge and the common voltage starts to decrease. As the capacitor discharges, the current going through the coil builds up a magnetic field around the coil; thus the electrical energy lost by the capacitor is converted to energy stored in the magnetic field around the coil. Things don't stop when the capacitor is discharged. When the current starts to decrease, the magnetic field around the coil starts to collapse, and as a result the current is forced to continue in the same direction (Lenz's law: the magnetic field will oppose the change in the current ...). The continuing current charges the capacitor again, but in the opposite direction. When the capacitor is fully charged again, the current is momentarily zero and the magnetic field is momentarily zero. This ends one-half of the cycle. Then the capacitor starts to discharge in the opposite direction; the above process is repeated, again leading to a conversion of electrical energy to magnetic energy. There is a voltage across the coil and capacitor, but there is no generator in the circuit. Ideally this could keep up indefinitely. Why doesn't it? There is always resistance in the wires of the circuit, and when current flows heat is produced; (power = I^2R). If we want to keep the oscillating current from dying down, we can insert into the circuit a generator which will supply just enough energy to compensate for the heat produced by the resistance. The voltage of this generator can be quite small if its frequency is the same as the natural frequency of this coil-capacitor combination: the frequency at which this charging and discharging tends to take place. We saw from the above formula that this resonant frequency is determined by the inductance and capacitance. If either is increased, the resonant frequency is decreased. Resonant circuits are *used in tuning a radio.* Usually when we tune a radio, we vary a capacitor so that the natural frequency of the radio circuit is the same as the frequency of the signal sent out by the desired radio station. Then a small signal will produce a relatively large voltage across the capacitor in the radio's resonant circuit.

MUTUAL INDUCTION AND TRANSFORMERS

We saw before that an emf is induced in a coil if the current in that coil is changing. This is self-induction. We have *mutual induction* if an emf is induced in a coil because of current changes in a second coil. The current may be ac or dc. As long as there is a change in the current there will be a change in the magnetic field surrounding the coil. If this changing magnetic field cuts the second coil, an emf is induced in the second coil. The *primary* coil (*P*) is the coil which is directly connected to the source of electrical energy, such as a dynamo. The *secondary* coil (*S*) is the coil from which electrical energy may be taken as a result of the emf induced in it. In the *induction coil* the source of energy is dc, usually a battery.

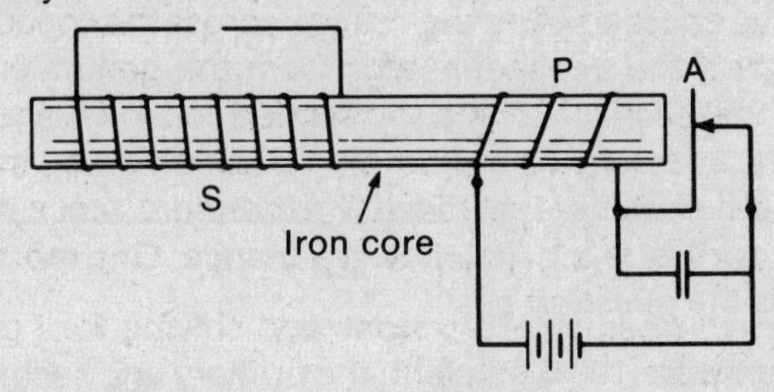

The purpose is to obtain a high voltage. In an automobile using a gasoline engine the induction coil is used as the *ignition coil:* it provides the high voltage which, at the spark plug, results in a spark that ignites the fuel mixture. A steady dc would produce no emf in the secondary. An interrupter is used to interrupt the current in the primary automatically. The changing magnetic field then cuts across the many turns of the secondary. The greater the number of turns, the greater the induced emf. In the diagram, *A* is springy, magnetic material. When the iron core is sufficiently magnetized, it will pull the armature (A) to the left, thus breaking the primary circuit. When, as a result of this, the current decreases, the magnetism becomes less and the spring *A* returns to its original position. This completes the primary circuit again; primary current and the magnetic field increase, inducing an emf in the secondary. Notice that the increase and decrease of the current result in an alternating emf in the secondary. Usually this emf is greater in one direction than in the other, and the gap in the secondary circuit can be arranged so that there will be current in only one direction in the secondary. A spark at the gap is evidence of current in the secondary. (In the automobile the armature of *A* is opened by cams or projections on a rotating shaft.)

In the *transformer* we also depend on mutual induction; usually ac is applied to one coil and ac is obtained from a second coil. In the transformers used on low-frequency ac (up to few thousand cycles per sec), the following formulas apply:

$$\frac{\text{secondary emf}}{\text{primary emf}} = \frac{\text{number of turns on secondary}}{\text{number of turns on primary}}$$

$$\text{power supplied by secondary} = \text{efficiency} \times \text{power supplied to primary}$$

$$V_s I_s = V_p I_p \times \text{efficiency}$$

When the efficiency is 100%, $V_s I_s = V_p I_p$

The efficiency of practical transformers is high and constant over a wide range of power. When more power is used in the secondary circuit, more power is supplied to the primary circuit automatically. If we want a higher voltage than the generator supplies, we use a *step-up transformer:* more turns on the secondary than on the primary. A *step-down* transformer has fewer turns on the secondary than on the primary and therefore produces a lower voltage than the generator supplies.

Eddy Currents and Hysteresis

Some transformers are made with a silicon steel core and two coils of wire wound on the core. The coils are insulated from each other. Ac power is supplied to the primary winding. As more and more power is supplied by the secondary winding to devices connected to it, the current in the primary winding also increases. This can be explained by considering the magnetizing effect of the currents on the core and the emf induced in the coils. However, the changing magnetic field in the core also has two undesirable effects, eddy currents and hysteresis losses. Eddy currents result from the emf induced in the core. This current is reduced by making the core out of thin laminations insulated from each other. Hysteresis losses result from the fact that the magnetic domains tend to be twisted back and forth as the direction of the applied emf changes. This requires energy.

ANSWERING THE QUESTIONS: Does a coil with an iron core use more electrical power on 120-V dc than on 120-V ac? Yes. In both cases the power used is given by the expression I^2R. The resistance is practically the same on dc as on ac. However, the current is considerably less on ac than on dc because the impedance of the coil will be high: $Z = \sqrt{R^2 + X_L^2}$; the reactance of the coil will be high because its core is made of iron. (If we move the iron core back and forth we will vary the reactance of the coil.) The reactance of the coil is zero on dc.

QUESTIONS Chapter 14

Select the choice which fits the question best.

1. If a step-up transformer were 100% efficient, the primary and secondary windings would have the same **(A)** current **(B)** power **(C)** number of turns **(D)** voltage **(E)** direction of winding

For 2-4. A certain capacitor consists of two metal plates in air placed parallel and close to each other but not touching. The capacitor is connected briefly to a dc generator so that the difference in potential between the two plates becomes 12,000 V. The capacitance of the capacitor is 1×10^{-6} microfarads.

2. The charge on either plate is, in coulombs, **(A)** 1.2×10^{-8} **(B)** 1.2×10^{-6} **(C)** 1.2×10^{-4} **(D)** 1.2×10^{-2} **(E)** 1.2

3. If a thin uncharged glass plate is then slipped between the two metal plates without touching them, the charge on each plate **(A)** will increase by an amount depending on the thickness of the glass **(B)** will decrease by an amount depending on the thickness of the glass **(C)** will increase by an amount depending on the distance the glass plate was inserted **(D)** will decrease by an amount depending on the distance the glass plate was inserted **(E)** will not change

4. As a result of the insertion of the glass plate, the potential difference between the plates **(A)** increases by an amount depending on the thickness of the glass **(B)** increases by an amount depending on the distance the glass is inserted **(C)** decreases somewhat **(D)** decreases to zero **(E)** does not change

5. A 100-volt battery is connected in series with a 100-volt, 60 cps, generator of the type used to supply power to homes. The peak voltage reached by the combination is, in volts, approximately, **(A)** 160 **(B)** 171 **(C)** 200 **(D)** 241 **(E)** 260

For 6-10. A series ac circuit consists of a 30-ohm resistor and an inductor (coil) with a reactance of 40 ohms connected to a source of 350 V whose frequency is $200/\pi$ cycles/sec.

6. The current through the inductor is, in amperes, **(A)** 5 **(B)** 7 **(C)** 9 **(D)** 10 **(E)** 12

7. The potential difference across the resistor is **(A)** 150 V **(B)** 210 V **(C)** 270 V **(D)** 300 V **(E)** 350 V

8. The potential difference across the coil is **(A)** 200 V **(B)** 210 V **(C)** 280 V **(D)** 300 V **(E)** 350 V

9. The power actually used by the circuit is, in watts, **(A)** 100 **(B)** 500 **(C)** 1000 **(D)** 1500 **(E)** 1800

10. The induction of the coil is, in henries, **(A)** 0.1 **(B)** 1.0 **(C)** π **(D)** 2π **(E)** 10

11. At 100 cycles/sec the reactance of a capacitor is 4000 ohms and the reactance of a coil is 1000 ohms. The frequency at which they will be in resonance is, in cycles/sec, **(A)** 60 **(B)** 200 **(C)** 100π **(D)** 300 **(E)** 400

For 12-13. When a transformer is connected to 120 volts ac, it supplies 3000 V to a device. The current through the secondary winding then is 0.06 amp and the current through the primary is 2 amp. The number of turns in the primary winding is 400.

12. The number of turns in the secondary winding is **(A)** 16 **(B)** 30 **(C)** 1000 **(D)** 2000 **(E)** 10,000

13. The efficiency of the transformer is **(A)** 75% **(B)** 80% **(C)** 85% **(D)** 90% **(E)** 95%

EXPLANATIONS TO QUESTIONS Chapter 14

Answers

1. **(B)**	4. **(C)**	7. **(B)**	10. **(A)**	13. **(A)**
2. **(A)**	5. **(D)**	8. **(C)**	11. **(B)**	
3. **(E)**	6. **(B)**	9. **(D)**	12. **(E)**	

Explanations

1. **(B)** Efficiency = (work or power output)/(work or power input). In order for this ratio to be equal to 1 (or 100%), the numerator and denominator must be equal; i.e., the power supplied to the primary of the transformer equals the power supplied by the secondary to the device connected to it.

2. **(A)** The capacitance equals the charge on either plate divided by the resulting difference of potential. $C = \frac{Q}{V}$; $Q = CV = (1 \times 10^{-6} \times 10^{-6}$ farads$) \times (1.2 \times 10^{4}$ V$)$ $Q = 1.2 \times 10^{-8}$ coulomb.

3. **(E)** Since the glass is an insulator, and the air is an insulator, no path is provided for electrons to go to or from the metal plates. Even accidental touching of the metal by the glass will not result in a significant transfer of electrons.

4. **(C)** Since $C = Q/V$, $V = Q/C$. In question 3 we saw that the charge Q did not change. What happened to the capacitance C? By inserting the glass we replaced some of the air by a solid dielectric. This increased the capacitance. Since the denominator increased (C), while the numerator (Q) remained the same, the value of the fraction (V) decreased.

5. **(D)** The two voltage sources are in series. Therefore, their combined voltage is the algebraic sum of the two. The dc voltage remains constant at 100 volts in one direction. The ac voltage varies from a peak value of approx. 141 V (100/0.707) in the same direction as the dc to 141 V in the opposite direction. The combined voltage is maximum when the dc and ac are in the same direction, and when the ac reaches its peak value. The maximum value then is 100 V + 141 V or 241 V.

6. **(B)** The impedance $Z = \sqrt{R^2 + X_L^2} = \sqrt{30^2 + 40^2} = 50\ \Omega$.

(Recognize the 3-4-5 right triangle?) The current through the inductor is the same as the current in the whole circuit.

$$I = V_T/Z_T = 350\ \text{V}/50\ \Omega = 7\ \text{amp}$$

7. **(B)** $V_R = IR = (7\ \text{amp}) \times (30\ \Omega) = 210$ V.

8. **(C)** $V_L = IX_L = (7\ \text{amp}) \times (40\ \Omega) = 280$ V.

9. **(D)** The actual power used is the power used by the resistor.

$$P = I^2R = (7\text{ amp})^2 \times (30\ \Omega) = 1470\text{ watts}$$

10. **(A)** $X_L = 2\pi fL$;

$$L = \frac{X_L}{2\pi f} = \frac{40}{2\pi \times \frac{200}{\pi}} = \frac{40}{400} = 0.1\text{ h}$$

11. **(B)** At resonance the two reactances are equal. The reactance of the coil is proportional to the frequency of the current; the reactance of the capacitor is inversely proportional to the frequency. If we increase the frequency, the reactance of the coil will go up; that of the capacitor will go down. This changes the reactance of the two in the necessary direction. (This eliminates the first choice, 60 cycles/sec.) With an eye on the second choice, if we double the frequency, the reactance of the coil will be doubled to 2000 ohms; at the same time the reactance of the capacitor will be reduced to one-half of its original value, to 2000 ohms. Doubling the frequency to 200 cycles/sec is then the proper choice.

12. **(E)** In such a transformer,

$$\frac{\text{number of turns on the secondary}}{\text{number of turns on the primary}} = \frac{\text{secondary emf}}{\text{primary emf}}$$

$$\frac{N_s}{400} = \frac{3000\text{ V}}{120\text{ V}};\ N_s = 10{,}000$$

13. **(A)**

$$\text{efficiency} = \frac{V_sI_s}{V_pI_p} = \frac{3000\text{ V} \times 0.06\text{ amp}}{120\text{ V} \times 2\text{ amp}} = 0.75 = 75\%$$

15

Basic Electronics *

DIODES AND TRIODES

Edison Effect

Thomas A. Edison discovered that electric current will flow in a vacuum between a heated filament and a cold metal plate if the positive terminal of a battery is connected to the plate and the negative terminal to the filament.

Thermionic emission

Thermionic emission is the giving off of electrons by a heated metal. The higher the temperature of the metal, the greater is the electron emission. The TV picture tube is an example of vacuum tubes which depend on this principle. They are sometimes called *thermionic* tubes. In these tubes, the part that is heated so that it will give off electrons is called the *cathode.* The *directly heated* cathode is heated because it is connected directly to a source of emf. The *indirectly heated* cathode is a sleeve slipped over a filament from which it is insulated. The filament is connected to a source of emf; the hot filament heats the cathode by radiation.

Diode

A vacuum tube containing only two elements, the cathode and the plate, is a *diode.* One of the chief uses of a diode is to *rectify,* that is, to change ac to dc. The diagram shows the circuit for a *half-wave rectifier* using an indirectly heated diode.

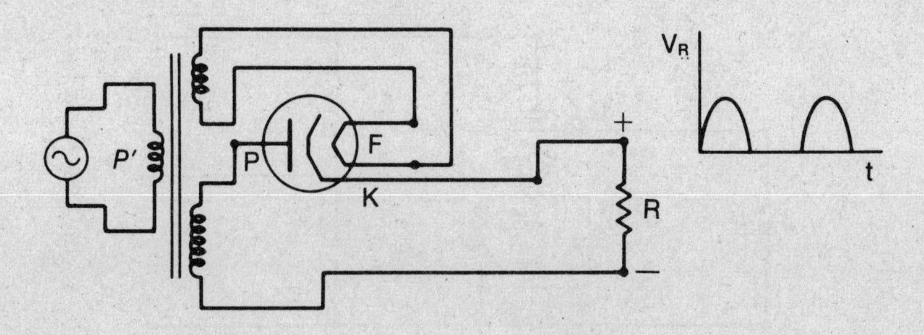

*Re-read the Introduction.

Standard ac voltage is applied to the primary *P'* of a transformer. This voltage may be 115 V. The secondary coil at the top provides the proper voltage to the filament or heater of the diode; this is usually a low voltage, about 5 or 6 volts. The other secondary winding, sometimes referred to as a tertiary winding, provides the ac voltage to be rectified. One end of this coil is connected to the plate, *P*, of the diode; the other end is connected, in this circuit, to the bottom of resistor *R*. When the cathode *K* gets hot it emits electrons. These electrons form a cloud near the cathode. The ac voltage to be rectified makes the plate alternately positive and negative with respect to the cathode. The cold plate itself does not emit electrons. When the plate gets negative with respect to the cathode, it repels the electrons emitted by the cathode. When the plate gets positive with respect to the cathode it attracts the electrons emitted by the cathode. These attracted electrons then move through the bottom secondary winding, through resistor *R* towards the top, and back to the cathode. The electron current goes through the resistor always from the bottom towards the top. Therefore the bottom is negative, and the top is positive. We have direct current, but only during one-half of each cycle. This pulsating current produces a corresponding voltage across the resistor.

For some purposes such fluctuating dc is not desirable. The fluctation can be reduced by using a *filter*. The diagram is the same as before, except that an inductance *L* and two capacitors *C* have

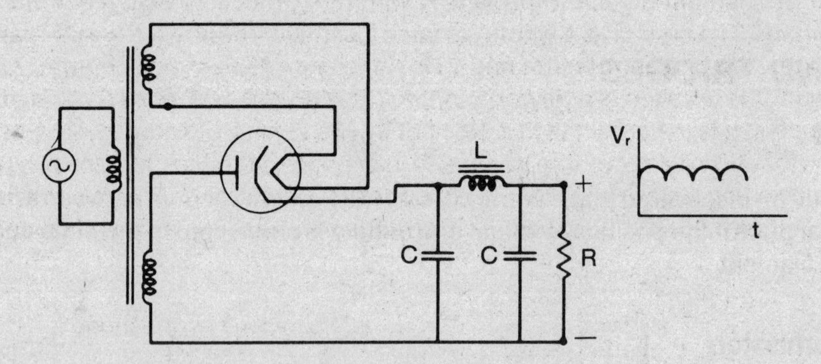

been added. This combination is one type of filter. We still have fluctuating dc, but the ripple has been reduced.

Triode

A vacuum tube containing three elements (the cathode, plate, and grid) is known as a *triode*. The *grid* is a wire mesh placed between the cathode and plate which serves to regulate the flow of electrons from cathode to plate. Because of its location and structure the grid can control relatively large currents by means of small changes in grid voltage (voltage between grid and cathode). When the grid becomes more negative, it tends to reduce the electron flow from the cathode to the plate. This can be compensated for by a relatively large increase in the plate voltage (voltage between the plate and cathode). The ratio of this compensating plate voltage to the corresponding grid voltage is the amplification factor of the triode. The amplification factor tells us to what degree a triode can amplify a signal. If this factor is 20, then, with the proper circuit, the triode can change a signal of 0.1 V to as much as 2 V.

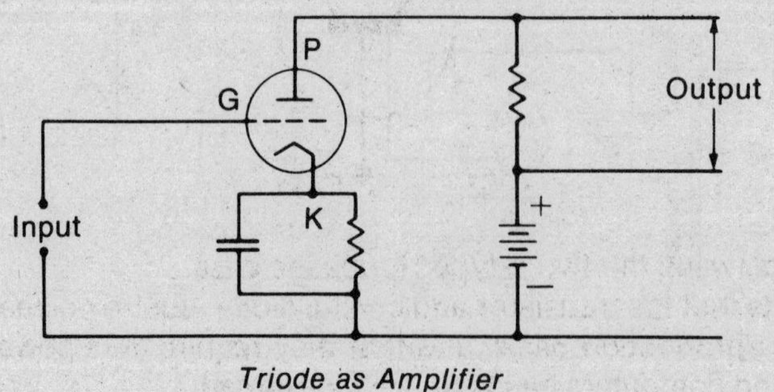

Triode as Amplifier

The diagram shows a circuit for using the triode to amplify signals. (The indirectly heated filament is not shown.) Notice that the signal to be amplified (INPUT) is connected between the grid *G* and the cathode *K* (through a resistor and capacitor connected in parallel). The output voltage is obtained across a resistor connected between the plate and the positive terminal of a battery or other supply of a dc voltage (sometimes referred to as the B-battery or B-supply).

A triode can also be used as an oscillator: in that case it converts dc power into ac power. It can produce an ac voltage of a few cycles per second up to a few million cycles per sec.

Semiconductors

In many applications such as radios and computers semiconductor devices have replaced vacuum tubes. A *semiconductor* is a material whose conductivity is very small by comparison with conductors like copper, but greater than that of insulators like glass. Common semiconductors are germanium and silicon. In practice a small precise amount of a selected impurity is added to the pure semiconductor to give it the desired characteristics. For example, a few parts per million of arsenic are added to a germanium crystal to provide it with free or loosely held electrons. This is called *n-type* germanium. (The crystal as a whole remains electrically neutral.) If we use gallium instead of arsenic as the impurity, the germanium has an insufficiency of electrons, referred to as holes, in its crystal structure. This is called *p-type* germanium. (This crystal, too, as a whole is neutral.) When a piece of n-type crystal is in contact with a piece of p-type crystal, the contact surface is called a *p-n junction.* At this junction electrons can go readily in only one direction, from the n-type crystal to the p-type crystal. The conductivity is poor in the opposite direction. Notice that this is similar to the characteristic of the vacuum tube diode. This arrangement is called a *crystal diode.* You probably

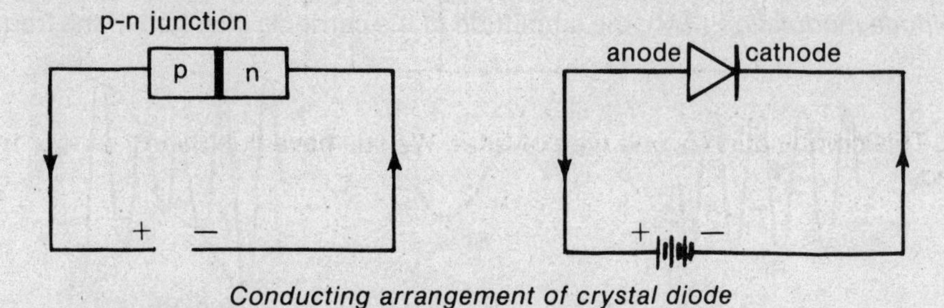

Conducting arrangement of crystal diode

noticed that the cathode (the n-type crystal) is not heated.

Just as the crystal diode can replace the vacuum tube diode, the transistor can replace the vacuum tube triode. Such a *transistor* has two junctions, which may be formed by a p-type wafer between two sections of n-type, giving a n-p-n transistor. We can also have a p-n-p transistor, in which an n-type wafer is between two p-types. The symbol for a p-n-p transistor is:

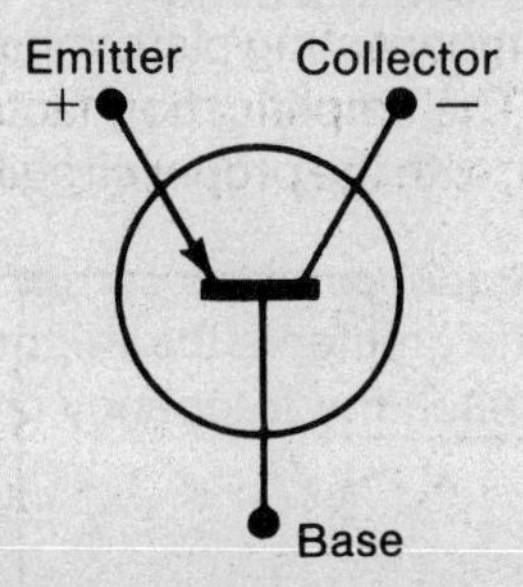

The n-type wafer in between the two p-types forms the base.

Because of the facts that the transistor and crystal diode require no heat, that they can be made very small on a mass-production basis, and that they require little power, the miniaturization of radios, calculators, and computers has become widespread.

TRANSMITTER—RECEIVER

A radio system consists of a radio *transmitter,* which sends out the information, such as speech or music, which is to be communicated, and of a radio *receiver,* which translates the message into longitudinal waves of the same frequencies as were originally used at the transmitter. In ordinary speech we refer to the receiver as the radio. The information to be communicated is sometimes referred to as the intelligence. It has been found that, in order to radiate waves over significant distances through space, we must use electromagnetic waves of frequencies higher than the audio range. Those that are suitable for radio transmission are called *radio* frequencies (RF). These radio frequencies are used as the carrier for the audio frequencies (AF) which composed the speech or music. The longitudinal sound waves of the speech or music are converted to electrical currents of the same frequencies with the aid of a microphone. These audio frequencies may then be amplified with the aid of a triode. These voltages are then impressed on the carrier; the AF is said to *modulate* the carrier. The carrier is produced by the oscillator. The oscillator circuit usually includes a quartz crystal. This crystal is known for its *piezoelectric effect:* a mechanical pressure on the crystal produces an emf across the crystal. The mechanical energy is changed to electrical energy, and then back again. This is similar to the behavior of the coil and capacitor combination we described before, in which magnetic energy is changed to electrical energy and back again. The crystal has a natural frequency of oscillation which depends on its dimensions. As a result of this effect the crystal oscillator produces an extremely constant frequency.

Amplitude Modulation

There are different ways in which the audio frequency voltages can modulate the carrier. One of these is *amplitude modulation* (*AM*): the amplitude of the carrier is changed at the frequency of the

RF + AF = Mod. RF [AM]

AF. The greater the amplitude change of the AF, the greater is the amplitude change of the RF. Notice that both the top and the bottom of the RF are changed in amplitude to the same degree, and that the outline of both top and bottom of the modulated RF is the shape of the AF. The carrier frequencies used for commercial AM broadcasting are 550 kilocycles per sec (kc) to 1650 kilocycles per sec. This range of frequencies is known as the broadcast band.

Frequency Modulation

In *frequency modulation* (*FM*) the frequency of the carrier is changed at the frequency of the AF. The greater the amplitude of the AF, the greater is the frequency change of the carrier; that is, the amplitude of the AF determines how much the frequency changes, while the frequency of the AF

RF + AF = FM

determines the rate at which this change takes place. The FM band of frequencies is about 88 Megacycles per sec (Mc)—108 Mc. (One Mega- is equal to one million.) An FM system can be designed readily to be relatively static-free as compared with AM.

The *antenna* at the transmitter is used to radiate electromagnetic waves whose frequency and variation is the same as the modulated RF described before. These electromagnetic radio waves travel through space with the same speed as other electromagnetic waves such as light: about 3×10^{10} cm/sec, or 3×10^{8} meters/sec, or 186,000 miles/sec. The length of the antenna at the transmitter is often made about one-half of a wavelength long. This is easily calculated. For example, if we want to transmit a frequency of 100 Mc,

$$\textbf{wavelength (L)} = \frac{\textbf{speed of wave}}{\textbf{frequency}} = \frac{3 \times 10^{8}\ \textbf{meters/sec}}{100 \times 10^{6}\ \textbf{cy/sec}}.$$

$$\textbf{L} = \textbf{3 meters}$$
$$\tfrac{1}{2}\textbf{L} = \textbf{1.5 meter}$$

The antenna would be about 1.5 meter long.

At the *receiver* the antenna length is usually not critical; electromagnetic waves of many frequencies pass the antenna and induce electric currents in it of the same frequency as the carrier. When we tune the receiver, we usually vary a capacitor to establish resonance for the desired carrier frequency. In the receiver we need a *detector* circuit in which a crystal or vacuum tube diode may be used in a circuit similar to the half-wave rectifier described before. In the detector, however, after rectifying the modulated RF, we use a filter circuit to recover the AF which was used to modulate the RF. The detector is said to *demodulate* or detect. The AF electrical currents are used to operate a *reproducer* which converts the electrical currents to longitudinal sound waves in the surrounding medium. The reproducer may be a loudspeaker or earphones.

Television transmission consists of the sending of two messages: sound and picture. FM is used for sound transmission, AM for picture transmission. At the transmitter a camera is used to translate information about the picture into electrical signals whose frequency range is up to about 4 Mc (video frequencies). This is done by a scanning process which is repeated in a somewhat different way in the picture tube of the TV receiver. In the cathode ray tube or picture tube at the receiver the screen is coated with chemicals (phosphors) which glow when hit by electrons. The greater the energy of these electrons, the greater the intensity of the glow. In the scanning process an electron beam (cathode ray) is deflected horizontally and vertically across the screen. The energy of the electron beam is controlled by a grid in the tube which determines the speed and quantity of electrons reaching the screen. The action of the grid is kept in step (synchronized) with the brightness of the scene being televised by the TV camera.

In the television camera, and elsewhere, an important principle is the photoelectric effect: when light shines on some metals at room temperatures electrons are emitted. A *photoelectric cell* is a vacuum tube in which the cathode is made of such metal; another metal is used for the plate or collector to which the electrons go when an external voltage is used to keep the collector positive with respect to the cathode. In a photovoltaic cell a somewhat similar process takes place when light falls on two specially selected materials which are in contact with each other; no external voltage is required.

The photoelectric effect will be discussed further in the next chapter.

QUESTIONS Chapter 15

Select the choice which fits the question best.

1. When a metal is heated sufficiently, electrons are given off by the metal. This phenomenon is known as **(A)** thermionic emission **(B)** photoelectric effect **(C)** piezoelectric effect **(D)** secondary emission **(E)** canal ray emission

2. When a radio station is broadcasting a musical program, the antenna of its transmitter **(A)** radiates RF electromagnetic waves **(B)** radiates AF electromagnetic waves **(C)** radiates RF longitudinal waves **(D)** radiates AF longitudinal waves **(E)** none of the above

3. Cathode rays are **(A)** sharp projections on the cathodes of diodes **(B)** sharp projections on the cathodes of triodes **(C)** striations in a gas discharge tube **(D)** proton beams **(E)** electron beams

4. A cathode ray oscilloscope has two pairs of deflection plates. If no voltage is applied to one pair, while the voltage applied to the other pair is a-c of the type used for power in most homes, the pattern on the face of the oscilloscope will be most like **(A)** a square wave **(B)** a sine wave **(C)** a saw-tooth wave **(D)** an ellipse **(E)** a straight line

5. Two good students, *X* and *Y*, connect the cathode and plate of a diode in a series with a battery and a milliammeter. The cathode is indirectly heated with a transformer. When *X* performs the experiment, he gets a zero reading on his milliammeter. *Y* gets the desired half-scale deflection on his meter. If both use identical equipment, the most probable reason for the difference is **(A)** *X* has reversed connections to the terminals of the transformer secondary **(B)** *Y* has improper connections to the terminals of the transformer secondary **(C)** *Y* has improper connections to the terminals of the battery **(D)** *X* has reversed connections to the terminals of the battery **(E)** *X* has reversed connections to the transformer primary.

6. When vacuum tube diodes are used as a full-wave rectifier, **(A)** electrons go from cathode to plate and then from plate to cathode **(B)** electrons go from cathode to plate only **(C)** electrons go from plate to cathode only **(D)** electrons go from cathode to plate and protons go from plate to cathode **(E)** protons go from cathode to plate and electrons go from plate to cathode

7. In a vacuum tube triode, electrons for the electron current come chiefly from **(A)** the grid **(B)** the plate **(C)** the hot filament **(D)** the getter **(E)** ionization of residual gases in the tube.

8. A vacuum tube triode is connected with the proper voltages applied to all the elements. A milliammeter is inserted between the plate and the positive terminal of the B-supply. The grid voltage (voltage between grid and cathode) is —3 V and the milliammeter reads 10 ma. When the grid voltage is changed to —6 V, the milliammeter reading will **(A)** not change **(B)** become 20 ma. **(C)** become 5 ma. **(D)** decrease to some value between 5 and 10 ma. **(E)** decrease, possibly even to zero.

9. The operation of a transistor requires **(A)** that the emitter be heated **(B)** that the base be heated **(C)** that the collector be heated **(D)** that it be enclosed in a vacuum **(E)** none of the above

EXPLANATIONS TO QUESTIONS Chapter 15

Answers

1. **(A)**	3. **(E)**	5. **(D)**	7. **(C)**
2. **(A)**	4. **(E)**	6. **(B)**	8. **(E)**
			9. **(E)**

Explanations

1. **(A)** Thermionic emission is the giving off of electrons by a metal when it is heated. Secondary emission may occur in vacuum tubes: this is the emission of electrons from a cold electrode, such as the plate, when it is bombarded by fast electrons.

2. **(A)** All the waves radiated by an antenna are electromagnetic waves; these are transverse. It is not practical to radiate and transmit AF electromagnetic waves over long distances. When a radio

station is On The Air, it radiates an RF electromagnetic wave. This wave may be modulated or not, but it is RF.

3. **(E)** The term *cathode ray* came into use before its actual nature was known. Further study showed that they were fast-moving electrons.

4. **(E)** In the cathode ray oscilloscope a fine beam of fast-moving electrons is accelerated towards a fluorescent screen, as in a TV picture tube. Where the electrons hit the screen there is a bright spot of light. One pair of deflection plates deflects the electrons horizontally, the other pair vertically, when a difference of potential is applied to the plates. The amount of deflection depends on the magnitude of the applied voltage. When an ac voltage is applied to one pair of plates, the fast-moving electrons will be deflected varying amounts depending on the instantaneous voltage. Therefore, on the screen there will be a succession of bright spots of light which will appear as a line. If no voltage is applied to the other pair of plates, the deflection is along one straight line, and the pattern on the fluorescent screen will be a straight line.

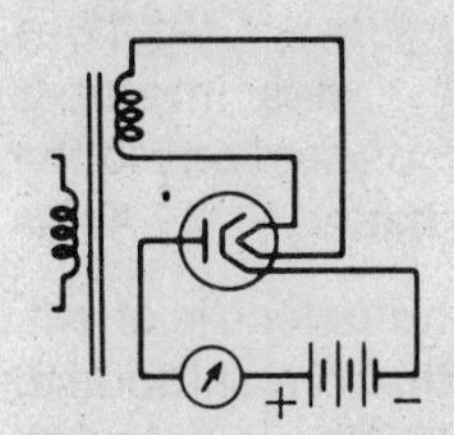

5. **(D)** Study the diagram of the half-wave rectifier. You will see that reversing the connections to the transformer windings, primary or secondary, cannot affect the operation. However, if, like *X* and *Y*, we use a battery instead of the bottom secondary winding, no current can go from cathode to plate unless the plate is connected to the positive terminal of the battery (through the milliammeter). If the battery connections were reversed, the plate would be negative with respect to the cathode and repel electrons.

6. **(B)** A vacuum tube diode is like a valve: electrons can go only one way through it, from cathode to plate. Ordinarily there are no protons in the common vacuum tube.

7. **(C)** In the common vacuum tube we depend on thermionic emission to give us the electrons to move through the space in the tube. Of the choices, the hot filament is the only one that can act as the cathode, the emitter of electrons in the vacuum tube. The *getter* is a piece of metal used inside the vacuum tube during manufacture to help get a good vacuum.

8. **(E)** Ohm's law does not apply in the usual way to a vacuum tube. When the grid voltage is made more negative, there will be less current to the plate; how much less depends on the particular tube. If the grid is made negative enough, there will be no plate current.

9. **(E)** One of the advantages of a transistor is that no heated cathode is used to supply electrons. Electrons are made available by using semiconductors, which have free electrons. The transistor is a solid-state device; the motion of electrons in it takes place within a solid and no vacuum is needed.

16

Atoms, Photons, Nuclei

At the beginning of Chapter 10 we mentioned that some phenomena, like the photoelectric effect, are not adequately explained by the wave theory of light and that the quantum theory can do this. Historically, the notion of the quantum of energy was introduced by Planck in 1900 to explain the continuous spectrum produced by a hot blackbody radiator. This ideal object is also a perfect absorber of light (page 85). The quantum theory also helps to explain the production of the bright-line spectrum and the structure of the atom. We shall first look at the photoelectric effect.

PHOTOELECTRIC EFFECT

When light falls on certain materials, electrons come off the material. If careful measurements are made when monochromatic light is used, the following observations can be made:

1. If the intensity of the light is increased, the *number* of emitted electrons increases.
2. If the intensity of the light is increased, the *speed* of the emitted electrons does not increase.
3. If we decrease the frequency of the light gradually, we reach a frequency below which no electrons come off. This is known as the *threshold frequency.* No matter how much we increase the intensity of the radiation, if it is below the threshold frequency, no electrons will be emitted by the substance.
4. For each substance, the speed of the emitted electrons is not the same for all electrons. The maximum speed and kinetic energy increase with the frequency of the incident radiation.
5. Different substances have different threshold frequencies.
6. If the frequency which is used is above the threshold frequency, even the weakest radiation will result in the emission of electrons, and their maximum speed will be the same as when very intense radiation of that frequency is used.

All of these observations except the first one can not be explained well by the wave theory. Albert Einstein explained this simply by using an expanded concept of the quantum. He suggested that we think of electromagnetic energy as being granular, traveling through space as little "grains", and being emitted and absorbed this way. The amount of energy, or quantum, in each grain is given by Planck's formula, $E = hf$, where f is the frequency of the radiation. (The term *photon* is often used to refer to the "grain," and *quantum* to the amount of energy in it; sometimes the terms are used interchangeably.) All photons in a monochromatic radiation of a given frequency are the same, but they are different from the photons in radiation of a different frequency. The higher the frequency, the greater the energy of the photon. All photons travel with the same speed, the speed of light.

Energy has to be used to remove an electron from a material. Some electrons are removed more readily than others. The *work function* (W) of a material is the minimum amount of energy needed to remove an electron from the material. Different materials have different work functions. If light shines on a material and a photon has energy greater than the work function, the photon may be absorbed by an electron, and the electron comes off with a kinetic energy equal to the difference between the quantum of energy and the work function of the material:

$$E_K = hf - W$$

For a given material, if we increase the frequency of the monochromatic radiation, the kinetic energy of the fastest photoelectron will increase. If we plot a graph of the kinetic energy of the fastest electrons against frequency of radiation used, we get a straight line:

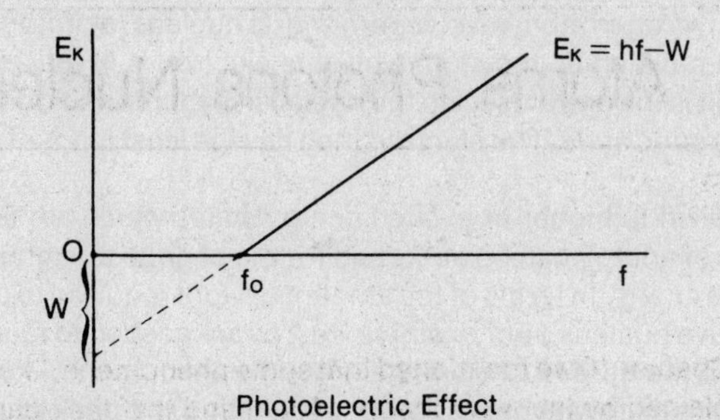

Photoelectric Effect

The plots for other substances will be straight lines parallel to this one. The slope of the straight line is Planck's constant, $h = 6.63 \times 10^{-34}$ joule-sec. The magnitude of the y-intercept is the work function of the material. The photon which has that amount of energy is just barely able to remove an electron from this material and has a frequency equal to the threshold frequency:

$$W = hf_o$$

For example, suppose that we have a material whose work function is 3.0×10^{-19} joule, and light with a wavelength of 4000 angstrom units is shining on it. Let's calculate a few relevant quantities. What is the energy of each photon? $E = hf = hc/\lambda$

$$E = (6.63 \times 10^{-34}\text{ joule})\ (3 \times 10^{8}\text{ m/sec})/(4000 \times 10^{-10}\text{ m})$$
$$= 5.0 \times 10^{-19}\text{ joule}$$

What is the speed of the fastest photoelectron?

$$E_K = hf - W$$
$$= 5.0 \times 10^{-19}\text{ joule} - 3.0 \times 10^{-19}\text{ joule}$$
$$= 2.0 \times 10^{-19}\text{ joule}$$
$$E_K = \tfrac{1}{2} mv^2$$
$$2.0 \times 10^{-19}\text{ joule} = \tfrac{1}{2}\ (9.1 \times 10^{-31}\text{ kg})\ v^2$$
$$v = 6.6 \times 10^{5}\text{ m/sec}$$

When dealing with small energies such as those possessed by electrons and photons, we often find it convenient to introduce another unit of energy. The *electron volt* (*ev*) is the energy one electron acquires when it is accelerated through a difference of potential of one volt.

$$1\text{ ev} = 1.6 \times 10^{-19}\text{ joule}$$

THE BRIGHT-LINE SPECTRUM

Ernest Rutherford performed a number of experiments in which alpha particles given off by radioactive materials were shot at thin gold foil. On the basis of observations and theoretical calculations he decided (1910) that (1) atoms are mostly empty space; (2) practically all of the atom's mass is concentrated in a very small nucleus; (3) the nucleus is positively charged because of its protons; and (4) the neutral atom has as many electrons outside the nucleus as there are excess protons in the nucleus. (Neutrons were not discovered until 1932. Rutherford assumed that there were enough other protons and an equal number of electrons in the nucleus to provide the rest of the mass which the atom has.) The electrons are relatively far from the nucleus. Alpha particles shot into the atom are scattered at various angles because of the coulomb force of repulsion between its positive charge and the positive charge of the nucleus. The closer its path is to the nucleus, the greater is the angle of scattering.

Niels Bohr (1913) used the Rutherford model of the atom and the quantum theory to develop an *explanation of the hydrogen spectrum.* The known spectrum of hydrogen at that time consisted of a few lines in the visible spectrum whose wavelengths were well known.

1. The electron of the hydrogen atom revolves around its nucleus (a proton) in only definite, allowed orbits. No energy is radiated by the atom while the electron is in these orbits.
2. The most stable orbit is the one closest to the nucleus. In this condition the atom (or the electron) is said to be in the *ground state.* The electron then has the least amount of energy with respect to the nucleus.
3. The atom can be raised to higher or excited energy states which are fairly stable. Only definite quantized amounts of energy can be absorbed for the change to higher energy states or levels. (Bohr thought of the energy in terms of the amount of work required to pull the negative electron away from the positive nucleus.) For example, 10.2 ev are needed to raise it to the second energy level, and 12.08 ev to the third level from the ground state.
4. When the atom returns to a lower energy state, it can only jump back to one of the allowed energy states, and the difference in energy is given off as a photon. The jump may be to the ground state or an intermediate one.
5. Only specified orbits are permitted. For all orbits of the electron: $mvr = \frac{nh}{2\pi}$ where m is the mass of the electron, v its orbital speed, r is the radius of the orbit, h Planck's constant. The product mvr is *known as the angular momentum* of the electron due to its orbit around the nucleus.

The diagram shows some of the energy levels of hydrogen. On the left, under *n,* the allowed levels are numbered, starting with the ground state as number 1. They are known as the *principal quantum*

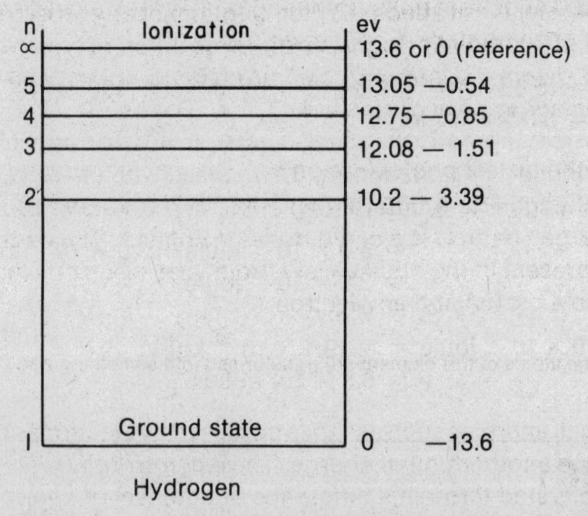

Hydrogen

numbers. On the right, under *ev,* are two columns of numbers. Both give the energy of each level in electron volts. One uses the ground state as the reference level and labels it the *zero level.* Since it is the lowest energy level, the other levels are positive with respect to it. Frequently the ionization state is taken as the reference and is assigned the zero value. The other levels are lower in energy than the ionization state, and therefore the numbers are negative. The difference between any two energy levels is, of course, the same (allowing for significant figures) in both columns. Notice, for example, that both columns tell us that 13.6 ev are needed to ionize the hydrogen atom from the ground state. If the atom is in the $n=3$ state, only 1.5 ev is needed to ionize the atom, but if a photon whose energy is 1.5 ev hits a hydrogen atom in the ground state, it will not be absorbed. The collision is elastic.

ATOMS AND RADIOACTIVITY

There are over 100 different chemical elements. One way in which they differ from each other is in the charge of the nucleus of the individual atom. The *atomic number* (Z) of an element is the number of protons in the nucleus of the atom; since the neutral atom is electrically uncharged, the atomic number is also the number of electrons surrounding the nucleus. (Hydrogen has an atomic number of 1; oxygen an atomic number of 8; uranium an atomic number of 92.) All elements whose atomic number is greater than 92 are man-made; they are known as the *transuranic* elements. Many of these were first made in the U.S. One of these is nobelium ($Z = 102$). Lawrencium has an atomic number of 103. All atoms, except those of common hydrogen, contain neutrons in the nucleus in addition to protons. The *mass number* is the total number of protons and neutrons in the nucleus. Most chemically pure elements contain atoms that are identical in atomic number but different in mass number because they differ in the number of neutrons; these are called *isotopes* of the element. The actual mass of the proton is extremely small; the mass of the neutron is slightly greater. Each of these is about 1840 times as massive as the electron.

The mass of the nucleus is less than the sum of the masses of the *nucleons* (protons and neutrons) which compose it. Einstein had predicted that mass and energy might be converted into one another.

$$\mathbf{E = mc^2}$$

The equivalent amounts are given by the relation in which m is the mass converted, c the speed of light, and E the equivalent amount of energy. If m is in kg, and c in m/sec, E is in joules. The *binding energy* of a nucleus is the energy equivalence of the difference in mass between the nucleus and the sum of the masses of the nucleons; it is the energy which would have to be supplied to break the nucleus into its component protons and neutrons. The nature of the nuclear force which holds the nucleus together is not fully understood. It is greater than electromagnetic and gravitational forces, but acts over short distances only.

Some elements were found whose nuclei break up spontaneously and in this process were found to give off some "radiation" which affects a photographic plate; these elements are said to be *radioactive.* This radioactivity is not affected by temperature or any other treatment we can give it. At first it was not known what this radiation consists of. Three different types were found, and they were called alpha, beta, and gamma. *Alpha particles* are helium nuclei; *beta particles* are electrons; *gamma rays* are electromagnetic waves similar to X rays. The *half-life* of a radioactive element is the length of time it takes for one-half of a given mass of this element to disintegrate.* Artificial radioactivity can be produced by bombarding the nucleus with charged particles, neutrons, or X rays. An element can change into a different element; this is known as *transmutation,* a process taking place when a charged particle is ejected from the nucleus. The particle may be a proton, but, although not normally present in the nucleus, electrons may also come out of the nucleus, since a neutron can change into a proton and an electron.

*In other words, one-half of the atoms of this element are transformed into something else.

COMPTON EFFECT

In the description of the photoelectric effect we stressed a very important characteristic. In order to raise the hydrogen atom to a higher energy state, such as from the ground state to $n = 2$, the photon which arrives must have the exact quantum of energy, 10.2 ev. If it has 11.5 ev it will not raise the electron to a higher level. In 1923, A.H. Compton discovered that if very high energy photons hit a material like carbon, a different effect is observed. When he used high-energy X rays, he found that some of the scattered X rays had a lower frequency than the original X rays. He was able to explain this effect quantitatively by reasoning that each photon has a momentum of hf/c $(=h/\lambda)$; the photon collides with an electron in the carbon atom and loses *some* of its energy and momentum to the electron; energy and momentum are conserved. The photon behaves like a particle.

momentum of the photon = Planck's constant / wavelength

$$p = \frac{h}{\lambda}$$

NUCLEAR CHANGES

Atomic Mass Unit

When dealing with nuclear changes we often find it convenient to use a small mass unit. The *atomic mass unit* (*amu*) is defined as one-twelfth of the mass of an atom of carbon-12, that is, of the isotope of carbon which has a mass number of 12. One amu then turns out to be approximately equal to 1.66×10^{-27} kg. Often conventional energy units are used instead of mass units:

$$\begin{aligned} E &= mc^2 \\ &= (1.66 \times 10^{-27}\text{ kg})\,(3 \times 10^{8}\text{ m/sec})^2 \\ &= 1.49 \times 10^{-10}\text{ joule} = \frac{1.49 \times 10^{-10}}{1.6 \times 10^{-19}}\text{ ev} \\ &= 0.931 \times 10^{9}\text{ ev} \end{aligned}$$

In other words, 1 amu is approximately equal to 931 Mev.

SOME NUCLEAR REACTIONS

We mentioned above that there are naturally radioactive elements and that their nuclei break up. Changes in nuclei also occur in the laboratory as a result of bombardment with fast-moving particles. Rutherford did this with alpha particles obtained from radioactive substances. Now it is frequently done by using particles speeded up with particle accelerators. The latter will be described later. Now let us look at some equations that represent the reactions.

In all of these nuclear reactions, atomic number and mass number are conserved; that is, the algebraic sum of the atomic numbers on the left side of the equation must equal the sum on the right; and the sum of the mass numbers on the left side must equal the sum on the right. Atomic number is written as a subscript of the chemical symbol of the element and mass number as the superscript.

Decay of Radium

$$^{226}_{88}\text{Ra} \rightarrow {}^{222}_{86}\text{Rn} + {}^{4}_{2}\text{He}$$

Note that the atomic numbers on each side add up to 88 and the mass numbers to 226. You will recall that the alpha particle is the nucleus of the common helium atom, which escapes this reaction with great speed. This is an example of *alpha decay.*

Beta Decay

$$^{24}_{11}\text{Na} \rightarrow {}^{24}_{12}\text{Mg} + {}^{\;0}_{-1}e$$

In beta decay an electron (the negative beta particle) is given off. Note that its mass number is zero and its charge is −1. For the electron and neutron, atomic number is replaced by charge. Algebraically the subscripts on the right add up to 11, which is what we started with on the left side. In beta decay neutrinos are also given off, a fact not always shown. The charge and mass of the neutrino are zero. (Recently some scientists expressed the belief that the rest mass is not exactly zero.)

Emission of Neutron

$$^{9}_{4}Be + ^{4}_{2}He \rightarrow ^{12}_{6}C + ^{1}_{0}n$$

Beryllium is bombarded with alpha particles. An isotope of carbon is produced and a neutron comes off. Note that the mass number of the neutron (like that of the proton) is one, but its charge is zero.

FISSION AND FUSION

Fission

Fission is the splitting of a massive nucleus into two large fragments with the simultaneous release of energy and some particles which can also produce fission. For example, the massive uranium nucleus (mass number 235) fissions on capture of a neutron and releases a large amount of energy and two or more neutrons. The total mass of all the particles produced is less than the mass of the original uranium atom plus neutron. The difference in mass accounts for the energy produced in the form of kinetic energy (fast-moving particles) and gamma rays. If the uranium is arranged so that the released neutrons will, in turn, fission other uranium nuclei, we have a *chain reaction*. This is done in a *nuclear reactor* or pile, in which the fissionable material is arranged so that the release of energy can be controlled. Boron rods are used as *control rods*: boron absorbs neutrons readily, and if the pile gets too active, the rods are inserted to cut down on the rate. Cadmium may also be used in the control rod. Fission of Uranium-235 is most likely to proceed with slow-moving neutrons; *moderators* are used to slow down the neutrons. Moderator materials are graphite and deuterium (used in heavy water). In the so-called A-bomb or fission bomb relatively large amounts of fissionable material are split in a short time with the consequent release of tremendous amounts of energy and radioactive materials. Radiation hazards in case of a nuclear reactor breakdown and in disposal of nuclear wastes are a serious concern.

Fusion

In *fusion* the nuclei of some light elements like lithium and hydrogen combine; again there is a loss of some mass with the consequent release of energy. The fusion reaction requires a temperature of millions of degrees. This is therefore called a *thermonuclear reaction* and has been achieved in an uncontrollable way in the H-bomb; here an A-bomb is first used to produce the required high temperature. Scientists are working towards achieving a controllable fusion reaction. This will make available tremendous amounts of energy everywhere by the use of the hydrogen in water.

PARTICLES AND PARTICLE ACCELERATORS

As indicated above, it is believed that the nuclei of all atoms except common hydrogen contain neutrons and protons (the *nucleons*). The nucleus is surrounded by electrons. In the neutral atom the number of these electrons is equal to the number of protons in the nucleus (the atomic number). Other particles have been discovered, some after their existence was predicted by theory. Well over 100 of these particles, referred to as *elementary particles,* are now known. There is no theory available yet which completely accounts for the existence, characteristics, and behavior of these particles. Most of these are now believed to be composed of a small number of *quarks,* whose charge is only a fraction of that of the electron. Particle accelerators have been built to obtain more facts and check some theories. Practical applications of particle accelerators will be mentioned later.

In 1934 Yukawa predicted the existence of a particle with mass intermediate between that of the electron and that of the proton. Later several different ones were actually found and are called *mesons.* Neutral mesons as well as positive and negative mesons have been found. The negative pi-meson fits Yukawa's description. The pi-mesons (also known as *pions*) have a mass about 270 times that of the electron. The mesons are unstable. The breakdown of the pi-meson results in the production of a lighter particle, the mu-meson or *muon.* Yukawa's pi-meson is sometimes referred to as the *cosmic glue,* because it seems to hold the nucleons together.

Particles heavier than the neutron have been found. An example is the upsilon particle, which has a mass about 10 times that of the proton.

Antiparticles have been found. The antiparticles of the charged particles have the same mass as the corresponding particle but an opposite charge. The *positron* is the antiparticle of the electron.

These various particles are often referred to as elementary particles, but this no longer implies that they are indivisible. For example, the neutron can last indefinitely inside the nucleus, but outside the nucleus it is unstable and decays into a proton and electron. In this and similar reactions conservation of energy and conservation of momentum suggested the existence of the *neutrino*–a neutral particle with practically zero rest mass. The concept of such a particle was introduced by Pauli in 1931 and carried further by Fermi, who suggested the name. The neutrino was detected experimentally in 1956. (See the last two sentences on p. 146.)

The list of elementary particles usually includes the *photon.* It always moves with the speed of light, has zero rest mass, and has an amount of energy depending on the frequency of the electromagnetic wave it represents.

Particle accelerators

Particle accelerators are helpful for examination of the nucleus. The greater the energy of the particles used to bombard or "smash" the nucleus, the more detail the physicist expects to find about the structure of the nucleus and its particles. This bombardment has led to the discovery of some of the particles mentioned above.

The energy possessed by the bombarding particles is usually described in terms of million electron volts (*Mev*) or billion electron volts (*Gev: G* is for *giga-;* it equals 10^9). One *electron volt* is the energy one electron acquires on being accelerated through a difference of potential of one volt. Electrons or protons are frequently used as the particles or bullets of these atom smashers. These, like other charged particles, can be accelerated by being placed in an electric field. In the *Van de Graaff* machine and in the *linear accelerator* the particle gains speed along a straight path. In the linear accelerator a series of cylindrical electrodes is used and the particles acquire more and more energy as they pass between successive pairs of electrodes. To achieve high energies a very long evacuated pipe is required.

In order to avoid the need for such long pipes, circular particle accelerators have been constructed. The equivalent length of path is achieved by having the particles go around the circular path many times. The charged particles are forced to go in a circular path by a magnetic field directed at right angles to the path of the particles. The first of these was the *cyclotron* invented by E. O. Lawrence in 1932. In the cyclotron a flat, evacuated cylindrical box is placed between the poles of a strong electromagnet. Inside the box are two hollow, **D**-shaped electrodes called *dees.* The protons or deuterons to be accelerated are fed into the space between the electrodes. An a-c voltage of high frequency (about 10^7 cycles per sec) is applied to the dees, and the particles are accelerated

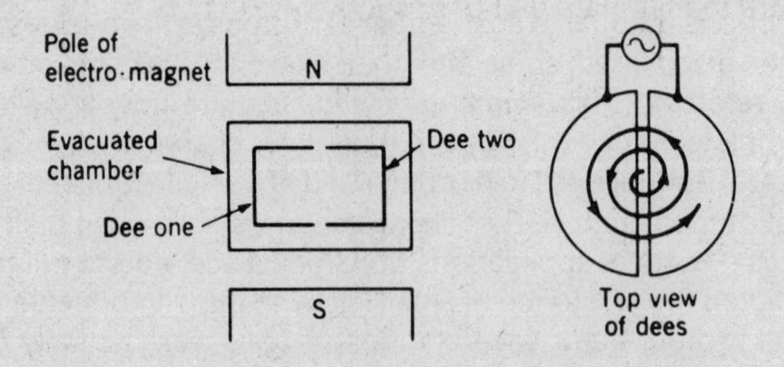

across the gap. The magnetic field bends the moving particles into a circular path. The voltage reverses in time to accelerate the particles emerging from the dee. After a 100 trips around the cyclotron the particles may acquire an energy of several Mev. Successive semicircular paths in the dees are of larger radius, but each of these is covered in equal lengths of time because the greater speed of the particles just compensates for the increased length of path.

When a proton has acquired a speed corresponding to an energy of about 10 Mev a relativity effect sets in: its mass starts to increase noticeably with further increases in energy. It therefore takes the particle longer to cover each successive semicircle. The frequency of the ac voltage applied to the dees must therefore change to compensate for this increase in mass. This is done in the *synchro-cyclotron*, which uses a constant magnetic field but an electric field of changing frequency. (It uses a frequency modulated oscillator.) In the *synchrotron* both the magnetic field and the frequency of the electric field are changed. This results in a circular path of almost constant radius for the particles and also in a saving of material for the magnet. The *cosmotron* at Brookhaven National Lab., Upton, N.Y., and the *bevatron* in California are of this type and allow production of protons with energies of a few Gev. Other accelerators are now available in the U.S. and Europe which produce particles with energies several times as great as the older ones.

Slow Collisions

When a stream of electrons passes through a gas, the electrons may lose energy as they collide with the atoms of the gas. In some collisions practically no energy is lost; in some collisions practically all the energy may be lost by the electron. For example, an electron having energy of 7 ev may lose nearly all of it when hitting the same atom. Careful experiments show that during such collisions atoms can absorb only definite bundles of energy—quanta.

QUESTIONS Chapter 16

Select the choice which fits the question best.

1. Of the following, the particle whose mass is closest to that of the electron is the **(A)** positron **(B)** proton **(C)** neutron **(D)** neutrino **(E)** deuteron

2. When a beta particle is emitted from the nucleus of an atom, the effect is to **(A)** decrease the atomic number by 1 **(B)** decrease the mass number by 1 **(C)** increase the atomic number by 1 **(D)** increase the mass number by 1 **(E)** decrease the atomic number by 2

3. Gamma rays consists of **(A)** helium nuclei **(B)** hydrogen nuclei **(C)** neutrons **(D)** high-speed neutrinos **(E)** radiation similar to X rays

4. Neutrons penetrate matter readily chiefly because they **(A)** occupy no more than one-tenth of the volume of electrons **(B)** occupy no more than one-tenth of the volume of protons **(C)** have a smaller mass than protons **(D)** are electrically neutral **(E)** are needlelike in shape

5. It is characteristic of alpha particles emitted from radioactive nuclei that they **(A)** are sometimes negatively charged **(B)** usually consist of electrons **(C)** are helium nuclei **(D)** are hydrogen nuclei **(E)** are the ultimate unit of positive electricity

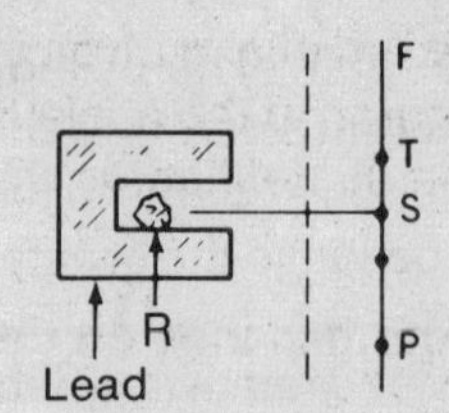

For 6-7. *R* is a natural radioactive material. *F* is a fluorescent screen. A magnetic field is to be imagined perpendicular to the paper. Bright spots appear on the screen at *T*, *S*, and *P*. The space between *R* and the screen is evacuated.

6. The emanation from *R* which produced spot *P* is **(A)** electrons **(B)** protons **(C)** alpha particles **(D)** N-poles **(E)** S-poles

7. If paper is inserted at *X*, the one of the following that is most likely to be blocked by the paper is **(A)** N-pole **(B)** S-pole **(C)** electrons **(D)** alpha particles **(E)** gamma rays

8. When describing isotopes of the same element, the most accurate statement is that they have **(A)** the same spin as a top **(B)** the same atomic mass but different atomic number **(C)** the same atomic number but different atomic mass **(D)** the same chemical properties and therefore cannot be separated **(E)** a coexistence limit; that is, no element can have more than three isotopes.

9. A metal plate is illuminated with monochromatic light, and photoelectrons are observed to come off. If the intensity of this light is reduced to one-fourth of its original value, then the maximum kinetic energy of the photoelectrons will be **(A)** zero **(B)** unchanged **(C)** reduced to one-fourth of the original value **(D)** reduced to one-half of the original value **(E)** reduced to one-sixteenth of the original value.

10. A photon whose energy is E_p joules strikes a photosensitive surface whose work function is W joules. The maximum energy of the ejected photoelectron is equal to **(A)** E_p **(B)** W **(C)** $W + E_p$ **(D)** $W - E_p$ **(E)** $E_p - W$

11. In the nuclear reaction $^{2}_{1}H + ^{3}_{1}H \rightarrow ^{4}_{2}He + ^{1}_{0}n + Q$
Q represents the energy released. The masses of the nuclei are: $^{2}_{1}H = 2.01472$ amu; $^{3}_{1}H = 3.01697$ amu; $^{4}_{2}He = 4.00391$ amu; $^{1}_{0}n = 1.00897$ amu. This reaction is primarily an example of **(A)** fission **(B)** fusion **(C)** ionization **(D)** alpha decay **(E)** neutralization

12. In the reaction shown in question 11, the value of Q in amu, is closest to **(A)** 5.03169 **(B)** 5.01288 **(C)** 0.01881 **(D)** 5.01288 **(E)** 2.01472

13. In the nuclear reaction shown below, what is the value of the coefficient y?
$^{235}_{92}U + ^{1}_{0}n \rightarrow ^{144}_{56}Ba + ^{89}_{36}Kr + y\,^{1}_{0}n$
(A) 0 **(B)** 1 **(C)** 2 **(D)** 3 **(E)** 4

EXPLANATIONS TO QUESTIONS Chapter 16

Answers

1. **(A)**	4. **(D)**	7. **(D)**	10. **(E)**
2. **(C)**	5. **(C)**	8. **(C)**	11. **(B)**
3. **(E)**	6. **(A)**	9. **(B)**	12. **(C)**
			13. **(D)**

Explanations

1. **(A)** The positron is sometimes referred to as a positive electron to suggest that it is similar to the negative electron in all respects except electric charge. The mass of the neutrino is practically zero; that of the proton and neutron is more than 1800 times as great as that of the electron. The deuteron is a combination of a proton and neutron and forms the nucleus of an isotope of hydrogen. The triton is the nucleus of the heavy hygrogen isotope containing one proton and two neutrons.

2. **(C)** The emission of an electron from the nucleus is the result of the break-up of a neutron in the nucleus into an electron and a proton. The electron is emitted, leaving the proton in the nucleus and thus raising the positive charge in the nucleus by 1. Since atomic number is defined as the number of protons in the nucleus, the atomic number went up by 1.

3. **(E)** Gamma rays and X rays are electromagnetic waves. Their wavelength depends on the method of production, but is roughly about 1 Angstrom unit or 10^{-8} cm.

4. **(D)** Neutrons are electrically neutral. Their volume is about the same as that of electrons and protons. The mass of a neutron is slightly greater than that of a proton. There is a great deal of empty space in all atoms; because neutrons are electrically neutral, they can go through this space near electrons and nuclei without experiencing electric forces.

5. **(C)** Alpha particles emitted from radioactive sources are helium nuclei; i.e., they have a positive charge and consist of two protons and two neutrons. Hydrogen nuclei are protons. Both protons and positrons are smaller units of positive charge, each having ½ of the charge of the alpha particle.

6. **(A)** The magnetic field at right angles to the path of the emanation from the radioactive material deflects positive charges in one direction, negative charges in the opposite direction. Gamma rays are not deflected at all and produce a spot at S. Electrons are usually deflected more than alpha particles because they have a much smaller mass than the alpha particles. The speed of the emitted particles is quite high.

7. **(D)** There are no N-poles or S-poles emitted. (Furthermore, they only occur in pairs.) In natural radio-activity gamma rays are most penetrating. Beta particles are next, and then come alpha particles.

8. **(C)** Isotopes of the same element have the same atomic number but differ in atomic mass because their nuclei differ in the number of neutrons. They are similar in chemical properties, but many methods have been devised for their separation. One of these is roughly similar to the set-up in question 6. Ions of the element are produced. They are projected into a magnetic field. The ions of the isotopes have the same charge and are deflected in the same direction, but because they differ in mass, the amount of deflection will be different for the different isotopes.

9. **(B)** Reducing the intensity of monochromatic light merely reduces the number of photons in the light; its frequency is not changed. Therefore each photon which interacts with an electron in the metal has the same amount of energy as before. Therefore each electron that is emitted can have the same maximum kinetic energy as before, $hf-W$. (However, the number of photoelectrons will be reduced.)

10. **(E)** In the photoelectric effect, the energy which the photon has is used for two things when it interacts with an electron in the metal: part is used to do the necessary work to overcome the force which holds the electron in the metal; the rest is used to give the electron kinetic energy: $E_p = W + E_k$. This, of course, is equivalent to $E_k = E_p - W$.

11. **(B)** Fusion is the combining of two nuclei to produce a more massive nucleus. Here the nuclei of two isotopes of hydrogen combine to produce helium. It is true that the alpha particle is a helium nucleus, but in alpha decay the alpha particle is emitted from a single unstable heavy nucleus.

12. **(C)** You can save time if you recognize that this is a mental problem. The value of Q is the difference between the sum of the amu's on the left side of the equation and the sum of the amu's on the right side. The sum on the left side is a little over 5, and the sum on the right side is also a little over 5. The difference, therefore, has to be less than one. The only possible choice is (C).

13. **(D)** The sum of the mass numbers (superscripts) on the left side is 236. The sum on the right side has to be the same. For the first two products, Ba and Kr, the sum is only 233. We need 3 neutrons to provide the additional 3.

17

Graphs, Mathematical and Others

Formulas describe relationship among variables. In elementary physics the relationships described are simple. Graphs are drawn to give a visual representation of these relationships. Many formulas are represented by similar graphs.

ONE QUANTITY IS PROPORTIONAL TO ANOTHER QUANTITY

When one quantity is doubled, the other is doubled; when one is tripled, the other is tripled. This can be represented by $v = kt$, where k is a constant, or by a proportion: $\frac{v_1}{v_2} = \frac{t_1}{t_2}$. For plotting the graph, the form $v = kt$, is more convenient. If we know the value of k, we can make a table of values of v and t, and then draw the graph. We get a straight line passing through the origin. The greater the value of k, the steeper is the straight line. The steepness is called *slope.* What is the physical significance of such a graph? For example, it may represent the speed of an object starting from rest and moving with constant accleration. We gave this relationship as $v = at$. Graph 2 represents greater acceleration than graph 1. Thus v would represent the speed of the object; t the time elapsed since the start of the motion. The graph can also represent Ohm's law, $V = IR$, showing the relationship between voltage and current for a given resistor. Then v would represent the voltage across the resistor; t the current through the resistor. Graph 2 would be for a greater resistance than graph 1.

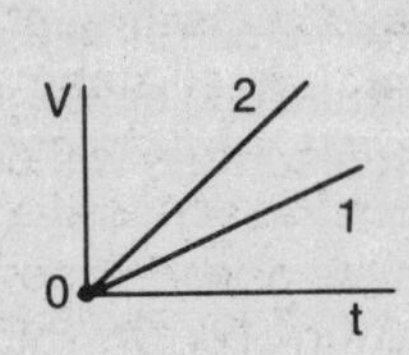

ONE QUANTITY VARIES INVERSELY WITH ANOTHER QUANTITY

When one quantity is doubled, the other is divided by two; when one quantity is tripled, the other is divided by three. This can be represented by $pV = k$, or by $p = k/V$, where the variable quantities are p and V, and k is a constant. The graph for this is a hyperbola.

Boyle's law is represented by such a graph. The p would represent the pressure of the gas and V its volume. At point X would be a condition of large pressure but small volume; at Y the pressure is small but the volume is large.

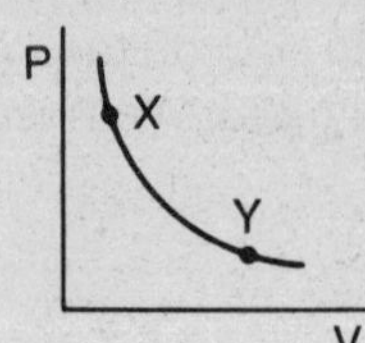

ONE QUANTITY VARIES AS THE SQUARE OF ANOTHER QUANTITY

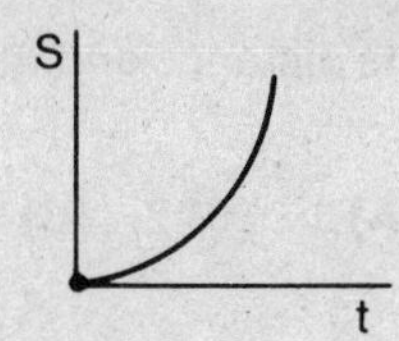

When one quantity is doubled, the other is quadrupled. The formula for such a relationship is $S = kt^2$, where the variable quantities are S and t; k is constant. The graph for such a variation is a parabola. This could represent the distance covered by a freely falling object, or any object with constant acceleration, in time t.

ONE QUANTITY VARIES LINEARLY WITH ANOTHER QUANTITY

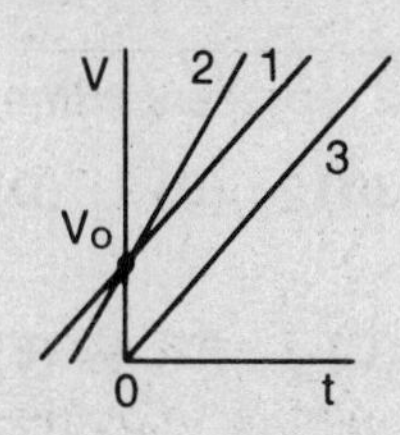

The graph for such a variation is a straight line. The proportional relationship given at the beginning of this chapter is a special case of this variation. A formula for this linear variation is $v = v_o + at$. It describes the relationship between v and t, with v_0 and a as constants. Then v_0 is the value which v has when $t = 0$. The equation could represent the speed v of an object moving with constant acceleration a. The speed of the object when the timing started is v_0, and t represents the elapsed time. Graphs 1 and 2 represent motions with the same initial speed; graph 2 represents the one with the greater acceleration. Graph 3 is the special case mentioned in which the initial velocity is zero. The acceleration for the motion described by graph 3 is the same as for graph 1 since the slopes of the two graphs are the same.

EMPIRICAL GRAPHS

These are graphs based on experimental data. All measurements in the laboratory except counting involve some error. Some errors result from the fact that no perfect instruments can be made; no one can avoid these errors. Some errors result from carelessness; everyone can try to avoid these errors. Suppose we measure the voltage across a resistor as we send various currents through the resistor. Let us plot the voltage against the current. The experimental points will not fall along a straight line. What graph shall we draw? When we had the mathematical formula first, we obtained the points for the graph by substitution in the formula. The construction of the graph then consisted of drawing a smooth curve through all the plotted points. In the experimental graph we know that each plotted point includes a representation of error. This error may make some measured quantities too large, some too small. We try to take this into account when drawing the empirical graph. We usually draw a smooth curve that has approximately as many plotted points on one side of the curve as on the other. In this case, a straight line seems to do this well. If the straight line goes through the origin, the graph suggests a proportional relationship between voltage and current. This doesn't surprise us if we recall Ohm's law. Suppose the graph looks more like a parabola? We may have introduced a systematic error; we may have overlooked something; or, rarely, we may have discovered something new. In this case, perhaps, the resistance did not remain constant as the current increased. Unexpected results require further investigation.

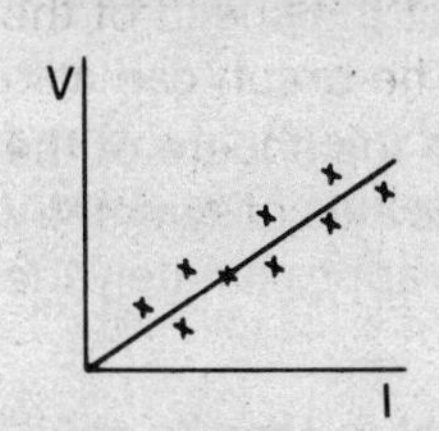

QUESTIONS Chapter 17

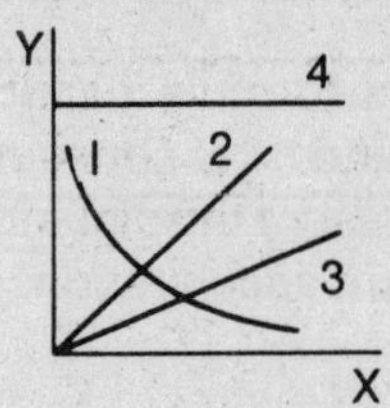

For 1-3. If in the set of graphs shown, the ordinate represents distance and the abscissa represents time,

1. the graph which could well represent no motion is **(A)** 1 **(B)** 2 **(C)** 3 **(D)** 4 **(E)** none of these

2. the graph which could well represent uniformly accelerated motion starting from rest is **(A)** 1 **(B)** 2 **(C)** 3 **(D)** 4 **(E)** none of these

3. the graph which represents motion with the smallest constant speed is **(A)** 1 **(B)** 2 **(C)** 3 **(D)** 4 **(E)** none of these

For 4-5. If, in the graphs shown, the ordinate represents velocity and the abscissa represents time,

4. the graph which represents motion with the greatest acceleration is **(A)** 1 **(B)** 2 **(C)** 3 **(D)** 4 **(E)** none of these

5. the graph which could refer to an object in equilibrium is **(A)** 1 **(B)** 2 **(C)** 3 **(D)** 4 **(E)** none of these

EXPLANATIONS TO QUESTIONS Chapter 17

Answers

1. **(D)**
2. **(E)**
3. **(C)**
4. **(B)**
5. **(D)**

Explanations

1. **(D)** The horizontal line of graph 4 indicates that the distance remains the same as time goes on. This could represent the state of rest with the distance, measured from some arbitrary point to the object, remaining constant. (Graph 4 could also represent circular motion, the unchanging distance being measured from the object to the center of its circular orbit.)

2. **(E)** For uniformly accelerated motion, distance covered is proportional to the square of the elapsed time ($s = \frac{1}{2}at^2$). The graph of distance against time is a parabola with the axis of symmetry parallel to the *Y*-axis. If graph 1 were rotated 90° counterclockwise, it would have approximately the right shape.

3. **(C)** For motion with constant speed, the distance covered is proportional to time ($s = vt$). Graphs 2 and 3 represent such motion. The slope of graph 3 is smaller than that of graph 2: the object whose motion is described by graph 3 covers less distance in unit time than the object described by graph 2; its speed is less. Graph 1 does not describe motion with constant speed.

4. **(B)** When acceleration is positive, velocity increases with time. This is represented by graphs 2 and 3. For the same change in time, the velocity increases more rapidly in graph 2 than in graph 3. This means that the acceleration in graph 2 is greater.

5. **(D)** If an object is in equilibrium, the unbalanced force acting on it is zero; therefore, according to Newton's second law, its acceleration is zero. If an object's acceleration is zero, its speed and direction of motion do not change. Graph 4 represents motion with constant speed; it could describe the motion of an object in equilibrium.

PRACTICE TEST 1

Answer Sheet

1. (A) (B) (C) (D) (E)
2. (A) (B) (C) (D) (E)
3. (A) (B) (C) (D) (E)
4. (A) (B) (C) (D) (E)
5. (A) (B) (C) (D) (E)
6. (A) (B) (C) (D) (E)
7. (A) (B) (C) (D) (E)
8. (A) (B) (C) (D) (E)
9. (A) (B) (C) (D) (E)
10. (A) (B) (C) (D) (E)
11. (A) (B) (C) (D) (E)
12. (A) (B) (C) (D) (E)
13. (A) (B) (C) (D) (E)
14. (A) (B) (C) (D) (E)
15. (A) (B) (C) (D) (E)
16. (A) (B) (C) (D) (E)
17. (A) (B) (C) (D) (E)
18. (A) (B) (C) (D) (E)
19. (A) (B) (C) (D) (E)
20. (A) (B) (C) (D) (E)
21. (A) (B) (C) (D) (E)
22. (A) (B) (C) (D) (E)
23. (A) (B) (C) (D) (E)
24. (A) (B) (C) (D) (E)
25. (A) (B) (C) (D) (E)
26. (A) (B) (C) (D) (E)
27. (A) (B) (C) (D) (E)
28. (A) (B) (C) (D) (E)
29. (A) (B) (C) (D) (E)
30. (A) (B) (C) (D) (E)
31. (A) (B) (C) (D) (E)
32. (A) (B) (C) (D) (E)
33. (A) (B) (C) (D) (E)
34. (A) (B) (C) (D) (E)
35. (A) (B) (C) (D) (E)
36. (A) (B) (C) (D) (E)
37. (A) (B) (C) (D) (E)
38. (A) (B) (C) (D) (E)
39. (A) (B) (C) (D) (E)
40. (A) (B) (C) (D) (E)
41. (A) (B) (C) (D) (E)
42. (A) (B) (C) (D) (E)
43. (A) (B) (C) (D) (E)
44. (A) (B) (C) (D) (E)
45. (A) (B) (C) (D) (E)
46. (A) (B) (C) (D) (E)
47. (A) (B) (C) (D) (E)
48. (A) (B) (C) (D) (E)
49. (A) (B) (C) (D) (E)
50. (A) (B) (C) (D) (E)
51. (A) (B) (C) (D) (E)
52. (A) (B) (C) (D) (E)
53. (A) (B) (C) (D) (E)
54. (A) (B) (C) (D) (E)
55. (A) (B) (C) (D) (E)
56. (A) (B) (C) (D) (E)
57. (A) (B) (C) (D) (E)
58. (A) (B) (C) (D) (E)
59. (A) (B) (C) (D) (E)
60. (A) (B) (C) (D) (E)
61. (A) (B) (C) (D) (E)
62. (A) (B) (C) (D) (E)
63. (A) (B) (C) (D) (E)
64. (A) (B) (C) (D) (E)
65. (A) (B) (C) (D) (E)
66. (A) (B) (C) (D) (E)
67. (A) (B) (C) (D) (E)
68. (A) (B) (C) (D) (E)
69. (A) (B) (C) (D) (E)
70. (A) (B) (C) (D) (E)
71. (A) (B) (C) (D) (E)
72. (A) (B) (C) (D) (E)
73. (A) (B) (C) (D) (E)
74. (A) (B) (C) (D) (E)
75. (A) (B) (C) (D) (E)

PRACTICE TEST 2

Answer Sheet

1. (A) (B) (C) (D) (E)
2. (A) (B) (C) (D) (E)
3. (A) (B) (C) (D) (E)
4. (A) (B) (C) (D) (E)
5. (A) (B) (C) (D) (E)
6. (A) (B) (C) (D) (E)
7. (A) (B) (C) (D) (E)
8. (A) (B) (C) (D) (E)
9. (A) (B) (C) (D) (E)
10. (A) (B) (C) (D) (E)
11. (A) (B) (C) (D) (E)
12. (A) (B) (C) (D) (E)
13. (A) (B) (C) (D) (E)
14. (A) (B) (C) (D) (E)
15. (A) (B) (C) (D) (E)
16. (A) (B) (C) (D) (E)
17. (A) (B) (C) (D) (E)
18. (A) (B) (C) (D) (E)
19. (A) (B) (C) (D) (E)
20. (A) (B) (C) (D) (E)
21. (A) (B) (C) (D) (E)
22. (A) (B) (C) (D) (E)
23. (A) (B) (C) (D) (E)
24. (A) (B) (C) (D) (E)
25. (A) (B) (C) (D) (E)
26. (A) (B) (C) (D) (E)
27. (A) (B) (C) (D) (E)
28. (A) (B) (C) (D) (E)
29. (A) (B) (C) (D) (E)
30. (A) (B) (C) (D) (E)
31. (A) (B) (C) (D) (E)
32. (A) (B) (C) (D) (E)
33. (A) (B) (C) (D) (E)
34. (A) (B) (C) (D) (E)
35. (A) (B) (C) (D) (E)
36. (A) (B) (C) (D) (E)
37. (A) (B) (C) (D) (E)
38. (A) (B) (C) (D) (E)
39. (A) (B) (C) (D) (E)
40. (A) (B) (C) (D) (E)
41. (A) (B) (C) (D) (E)
42. (A) (B) (C) (D) (E)
43. (A) (B) (C) (D) (E)
44. (A) (B) (C) (D) (E)
45. (A) (B) (C) (D) (E)
46. (A) (B) (C) (D) (E)
47. (A) (B) (C) (D) (E)
48. (A) (B) (C) (D) (E)
49. (A) (B) (C) (D) (E)
50. (A) (B) (C) (D) (E)
51. (A) (B) (C) (D) (E)
52. (A) (B) (C) (D) (E)
53. (A) (B) (C) (D) (E)
54. (A) (B) (C) (D) (E)
55. (A) (B) (C) (D) (E)
56. (A) (B) (C) (D) (E)
57. (A) (B) (C) (D) (E)
58. (A) (B) (C) (D) (E)
59. (A) (B) (C) (D) (E)
60. (A) (B) (C) (D) (E)
61. (A) (B) (C) (D) (E)
62. (A) (B) (C) (D) (E)
63. (A) (B) (C) (D) (E)
64. (A) (B) (C) (D) (E)
65. (A) (B) (C) (D) (E)
66. (A) (B) (C) (D) (E)
67. (A) (B) (C) (D) (E)
68. (A) (B) (C) (D) (E)
69. (A) (B) (C) (D) (E)
70. (A) (B) (C) (D) (E)
71. (A) (B) (C) (D) (E)
72. (A) (B) (C) (D) (E)
73. (A) (B) (C) (D) (E)
74. (A) (B) (C) (D) (E)
75. (A) (B) (C) (D) (E)

PRACTICE TEST 3

Answer Sheet

1. (A) (B) (C) (D) (E)	26. (A) (B) (C) (D) (E)	51. (A) (B) (C) (D) (E)
2. (A) (B) (C) (D) (E)	27. (A) (B) (C) (D) (E)	52. (A) (B) (C) (D) (E)
3. (A) (B) (C) (D) (E)	28. (A) (B) (C) (D) (E)	53. (A) (B) (C) (D) (E)
4. (A) (B) (C) (D) (E)	29. (A) (B) (C) (D) (E)	54. (A) (B) (C) (D) (E)
5. (A) (B) (C) (D) (E)	30. (A) (B) (C) (D) (E)	55. (A) (B) (C) (D) (E)
6. (A) (B) (C) (D) (E)	31. (A) (B) (C) (D) (E)	56. (A) (B) (C) (D) (E)
7. (A) (B) (C) (D) (E)	32. (A) (B) (C) (D) (E)	57. (A) (B) (C) (D) (E)
8. (A) (B) (C) (D) (E)	33. (A) (B) (C) (D) (E)	58. (A) (B) (C) (D) (E)
9. (A) (B) (C) (D) (E)	34. (A) (B) (C) (D) (E)	59. (A) (B) (C) (D) (E)
10. (A) (B) (C) (D) (E)	35. (A) (B) (C) (D) (E)	60. (A) (B) (C) (D) (E)
11. (A) (B) (C) (D) (E)	36. (A) (B) (C) (D) (E)	61. (A) (B) (C) (D) (E)
12. (A) (B) (C) (D) (E)	37. (A) (B) (C) (D) (E)	62. (A) (B) (C) (D) (E)
13. (A) (B) (C) (D) (E)	38. (A) (B) (C) (D) (E)	63. (A) (B) (C) (D) (E)
14. (A) (B) (C) (D) (E)	39. (A) (B) (C) (D) (E)	64. (A) (B) (C) (D) (E)
15. (A) (B) (C) (D) (E)	40. (A) (B) (C) (D) (E)	65. (A) (B) (C) (D) (E)
16. (A) (B) (C) (D) (E)	41. (A) (B) (C) (D) (E)	66. (A) (B) (C) (D) (E)
17. (A) (B) (C) (D) (E)	42. (A) (B) (C) (D) (E)	67. (A) (B) (C) (D) (E)
18. (A) (B) (C) (D) (E)	43. (A) (B) (C) (D) (E)	68. (A) (B) (C) (D) (E)
19. (A) (B) (C) (D) (E)	44. (A) (B) (C) (D) (E)	69. (A) (B) (C) (D) (E)
20. (A) (B) (C) (D) (E)	45. (A) (B) (C) (D) (E)	70. (A) (B) (C) (D) (E)
21. (A) (B) (C) (D) (E)	46. (A) (B) (C) (D) (E)	71. (A) (B) (C) (D) (E)
22. (A) (B) (C) (D) (E)	47. (A) (B) (C) (D) (E)	72. (A) (B) (C) (D) (E)
23. (A) (B) (C) (D) (E)	48. (A) (B) (C) (D) (E)	73. (A) (B) (C) (D) (E)
24. (A) (B) (C) (D) (E)	49. (A) (B) (C) (D) (E)	74. (A) (B) (C) (D) (E)
25. (A) (B) (C) (D) (E)	50. (A) (B) (C) (D) (E)	75. (A) (B) (C) (D) (E)

18

Practice Tests

PRACTICE TEST 1

Part A

INSTRUCTIONS–*Read the statement in the Introduction on* HOW TO TAKE THE TEST.

DIRECTIONS: Each set of lettered choices below refers to the numbered questions immediately following it. Select the one lettered choice that best answers each question. A choice may be used once, more than once, or not at all in each set.

QUESTIONS 1-3 refer to the graph which represents the speed of an object moving along a straight line. The time of observation is represented by *t*.

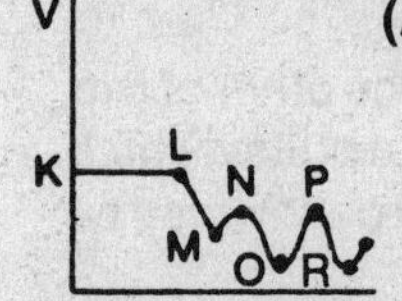

(A) KL **(B)** LM **(C)** NO **(D)** OP **(E)** PR

1. The interval during which the object moves with constant speed is represented by

2. The interval during which the object's speed is increasing is represented by

3. The interval during which the acceleration is constant but not zero is represented by

QUESTIONS 4-8 refer to the following concepts:

(A) Energy **(B)** Power **(C)** Momentum **(D)** Acceleration **(E)** Torque

For each of the following formulas select the choice which is most closely related to it. Use the following key.

4. mgh
5. ms/t
6. $\frac{Q^2R}{t^2}$
7. $\frac{1}{2} ms^2/t^2$
8. $\frac{Q^2R}{t}$

m = mass
g = acceleration due to gravity
h = vertical height
s = displacement
t = time
Q = electric charge
R = electric resistance

QUESTIONS 9-13 may express a relationship to speed or velocity given in these choices:

(A) is proportional to its velocity **(B)** is proportional to the square of its speed **(C)** is proportional to the square root of its speed **(D)** is inversely proportional to its velocity **(E)** is not described by any of the above

For each question select the choice which best completes it:

9. The kinetic energy of a given body

10. The acceleration toward the center of an object moving with constant speed around a given circle

11. The momentum of a given body

12. The displacement per second of an object in equilibrium

13. The displacement of an object starting from rest and moving with constant acceleration

QUESTIONS 14 AND 15 deal with the characteristics of a parallel circuit in the following situation:

The following five lengths of thin wire, all of which have the same diameter, are connected in parallel to a battery:

(A) 3 ft of nichrome wire **(B)** 3 ft of copper wire **(C)** 3 ft of silver wire **(D)** 2 ft of nichrome wire **(E)** 2 ft of silver wire

14. The current is greatest in

15. The greatest power is dissipated by

Part B

DIRECTIONS: Each of the questions or incomplete statements is followed by five suggested answers or completions. Select the one that is best in each case.

16. The resultant of a 3-N and a 4-N force which act on an object in opposite direction to each other is, in Nt, **(A)** 0 **(B)** 1 **(C)** 5 **(D)** 7 **(E)** 12

17. Two forces, one of 6 lb and the other of 8 lb, act on a point at right angles to each other. The resultant of these forces is, in pounds, **(A)** 0 **(B)** 2 **(C)** 5 **(D)** 10 **(E)** 14

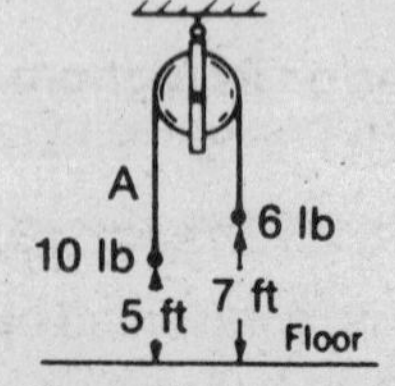

For 18-20. As shown in the diagram, two weights, one of 10 lb and the other of 6 lb, are tied to the ends of a flexible string. The string is placed over a pulley which is attached to the ceiling. Frictional losses and the weight of the pulley may be neglected as the weights and the string are allowed to move.

18. At the instant shown in the diagram, the potential energy of the 10-lb object with respect to the floor is, in ft-lb, **(A)** 0 **(B)** 2 **(C)** 20 **(D)** 50 **(E)** 70

19. At the instant shown, the acceleration of the moving 10-lb object is **(A)** 0 **(B)** less than g **(C)** g **(D)** $5g$ **(E)** $10g$

20. At the instant shown, the tension in rope *A* is **(A)** less than 3 lb **(B)** 3 lb **(C)** 6 lb **(D)** more than 6 but less than 10 lb **(E)** 10 lb

21. How many meters will a 2.00-kg ball starting from rest fall freely in 1.00 second? **(A)** 4.90 **(B)** 2.00 **(C)** 9.81 **(D)** 19.6 **(E)** 32

22. The concept or principle which is most important in the explanation of the fact that a steel wire can float on the top of water is **(A)** surface tension **(B)** adhesion **(C)** kinetic theory of matter **(D)** Archimedes' principle **(E)** Pascal's principle

23. A beam of parallel rays is reflected from a smooth plane surface. After reflection the rays will be **(A)** converging **(B)** diverging **(C)** parallel **(D)** diffused **(E)** focused

For 24-27. In the following table you may find information you need for this set of questions.

24. The number of calories required to melt 10 g of aluminum without change of temperature is **(A)** 2.20 **(B)** 7.7 **(C)** 10 **(D)** 77 **(E)** 770

25. Ten grams of lead are to be vaporized without change of temperature. The number of calories required is **(A)** 0.31 **(B)** 2.2 **(C)** 60 **(D)** 1900 **(E)** not to be calculated with the above information

26. A 100-cm rod of aluminum is heated from 40° F to 50° F. The change in the length of the rod, in cm, is **(A)** 0.0025 **(B)** 0.014 **(C)** 0.025 **(D)** 0.045 **(E)** 0.125

27. A block of aluminum is transferred from boiling water to a beaker containing 200 g of water at 10° C. The final temperature of the water is 30° C. The mass of the block is, in g, **(A)** 200 **(B)** 230 **(C)** 260 **(D)** 290 **(E)** 320

	Aluminum	Lead
melting point	660°C	327°C
heat of fusion (cal/g)	77	6.0
specific heat	0.220	0.031
coeff. of linear expansion (per C deg)	0.000025	0.000029

28. Of the following, the one that cannot be plane polarized is **(A)** sound **(B)** FM signals **(C)** RF signals **(D)** infra red light **(E)** yellow light from sodium

29. It is said that the illumination on a small surface is inversely proportional to the square of its distance from the source. This is a true statement if **(A)** the light consists of parallel rays **(B)** the size of the source is small **(C)** the light has been filtered **(D)** the light has not been filtered **(E)** monochromatic light is used

30. A tuning fork vibrating gently produces the note C. Another tuning fork produces the note C′, an octave higher in pitch than C, and slightly louder than C. The speed of the wave produced by the second fork, as compared with that produced by the first fork, is **(A)** 8 times as great **(B)** 4 times as great **(C)** 2 times as great **(D)** the same **(E)** ½

For 31-34. A ship 4000 ft distant from a large cliff sounds a short note of 222 vib/sec on its foghorn. The speed of the sound is 1110 ft/sec. (Assume a speed of 1090 ft/sec at 0° C.)

31. The time required for the sound to reach the cliff is, approximately, **(A)** less than 0.1 sec **(B)** 0.3 sec **(C)** 1.8 sec **(D)** 3.6 sec **(E)** 7.2 sec

32. The wavelength of the sound in air is, approximately, **(A)** 0.5 ft **(B)** 1 ft **(C)** 2 ft **(D)** 5 ft **(E)** 4×10^6 ft

33. The time between the sounding of the foghorn and the time the echo is heard on the ship is, approximately, **(A)** less than 0.1 sec **(B)** 0.3 sec **(C)** 1.8 sec **(D)** 3.6 sec **(E)** 7.2 sec

34. In question 33, it was necessary to assume **(A)** nothing **(B)** that the amplitude of the sound did not change **(C)** that the frequency of the sound produced by the foghorn remained constant **(D)** that all the sound was beamed towards the cliff **(E)** that the ship was stationary

35. A tank has a volume of 10 ft^3 and contains air at atmospheric pressure. In order to raise the gauge pressure to 135 $lb/in.^2$, the number of cubic feet of air at atmospheric pressure that must be pumped into the tank (assuming temp. to remain constant) is **(A)** 70 **(B)** 80 **(C)** 90 **(D)** 100 **(E)** 110

36. In the circuit represented, the voltage across R_1 and R_2 is 20 V each. R_1 has a resistance of 40 ohms. S_1 is a switch which can be moved to positions 1, 2, 3, and 4. It is shown in position 1.

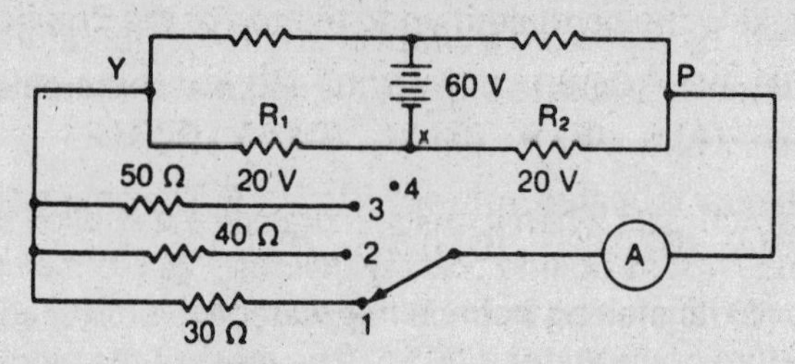

With the switch in position 1, the current in R_1 is **(A)** 0 **(B)** $^1/_3$ A **(C)** ½ A **(D)** 2 A **(E)** 1.5 A

For 37-38 A 5-ohm coil of tungsten wire (M.P. is 3370° C) is immersed in a non-conducting liquid (B.P. is 78° C). The coil is connected to a battery which supplies it with a constant 40 V. When the experiment starts, the liquid is at room temperature; during the experiment the liquid covers the coil.

37. During the experiment the temperature of the liquid **(A)** drops somewhat below room temperature because evaporation is a cooling process **(B)** drops somewhat below room temp. because the wire has to be heated **(C)** rises until a temperature of 78° C is reached **(D)** rises until a temperature of 3370° C is reached **(E)** remains constant because the liquid short-circuits the wire

38. After the circuit has been completed, the current **(A)** remains constant **(B)** decreases somewhat until a temperature of 78° C is reached **(C)** decreases somewhat until a temperature of 3370° C is reached **(D)** increases sharply at 78° C **(E)** increases gradually until a temp. of 78° C is reached

For 39-40 *X* is a coil of copper wire with many turns wound on a soft iron core. Another coil wound on an iron core is near it, as shown.

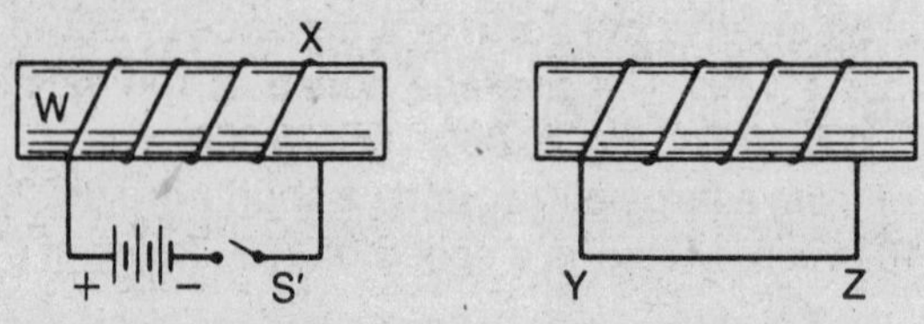

39. If switch *S'* is closed and kept closed, the end of the core which is marked *W* **(A)** becomes an N-pole momentarily **(B)** becomes and stays an N-pole **(C)** becomes an S-pole momentarily **(D)** becomes and stays an S-pole **(E)** exhibits no effect because of the current

40. The instant after switch *S'* is closed **(A)** there will be no current in wire *YZ* **(B)** electron flow in wire *YZ* will be from *Y* to *Z* **(C)** electron flow in wire *YZ* will be from *Z* to *Y* **(D)** electron flow in wire *YZ* will be from *Z* to *Y* and then from *Y* to *Z* **(E)** electron flow in wire *YZ* will be from *Y* to *Z* and then from *Z* to *Y*

41. The diagram represents two equal negative point charges, *Y* and *Z*,

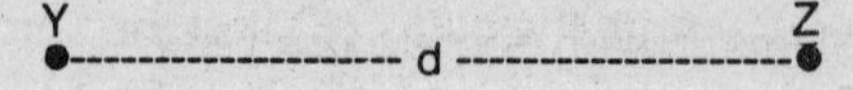

that are a distance *d* apart. Where would the electric field intensity due to these two charges be zero? **(A)** on *Y* **(B)** on *Z* **(C)** on both *Y* and *Z* **(D)** midway between *Y* and *Z* **(E)** none of the above

42. A positive charge is moving with constant speed at right angles to a uniform magnetic field. If the speed of the particle were doubled, the force exerted on the particle by the magnetic field would be **(A)** unaffected **(B)** quadrupled **(C)** doubled **(D)** halved **(E)** reduced to one-fourth of the original value

43. Of the following, the particle whose mass is closest to that of the neutron is **(A)** meson **(B)** deuteron **(C)** neutrino **(D)** proton **(E)** positron

44. The time of one vibration of a simple pendulum may be decreased by **(A)** increasing the length of the pendulum **(B)** decreasing the length of the pendulum **(C)** using a heavier bob **(D)** using a lighter bob **(E)** taking the pendulum up to the top of the Empire State Building

45. In rising from the bottom of a lake to the top, the volume of a bubble triples. The approximate depth of the lake, in feet, is **(A)** 6 **(B)** 9 **(C)** 34 **(D)** 68 **(E)** 102

46. A certain substance contracts when it melts. If pressure is applied to this substance, **(A)** its boiling point is not changed **(B)** its boiling point is lowered **(C)** its melting point is raised **(D)** its melting point is lowered **(E)** its melting point is not changed

47. One can be sure that a rod is electrically charged if it **(A)** repels a pitch ball **(B)** attracts a pitch ball **(C)** attracts the N-pole of a compass needle **(D)** repels the N-pole of a compass needle **(E)** points north

48. The rate of heat production of a wire immersed in ice water and carrying an electric current is proportional to **(A)** the current **(B)** the reciprocal of the current **(C)** the reciprocal of the square of the current **(D)** the square of the current **(E)** the square root of the current

49. Fraunhofer lines are chiefly due to **(A)** absorption in the earth's atmosphere **(B)** absorption in the sun's atmosphere **(C)** reflection by meteors **(D)** absence of certain elements in the sun **(E)** radioactive francium and hafnium

50. When water cools from 7° C to 1° C, **(A)** it contracts only **(B)** it expands only **(C)** it first contracts and then expands **(D)** it first expands and then contracts **(E)** it first expands, then contracts, and then expands again

51. Boyle's law describes the behavior of a gas when **(A)** its pressure is kept constant **(B)** its volume is kept constant **(C)** its density is kept constant **(D)** its mass is kept constant **(E)** nothing is kept constant

52. A satellite in circular orbit around the earth at a distance of about 300 miles from the earth completes each orbit in about 1½ hr. The average speed of the satellite is, in mi/hr, approximately **(A)** 200 **(B)** 400 π **(C)** 3000 **(D)** 16,000 **(E)** 25,000

53. The force acting on a satellite in circular orbit around the earth is chiefly **(A)** the satellite's inertia **(B)** the satellite's mass **(C)** the earth's mass **(D)** the earth's gravitational pull **(E)** the sun's gravitational pull

54. An object which is black **(A)** absorbs black light **(B)** reflects black light **(C)** absorbs all light **(D)** reflects all light **(E)** refracts all light

55. A string's lowest natural frequency is 400 vibrations/sec. Its fundamental frequency is, in vib/sec **(A)** 100 **(B)** 200 **(C)** 400 **(D)** 800 **(E)** 1200

QUESTIONS 56-70 Each of these questions or incomplete statements is followed by four numbered (I–IV) statements. Select all the numbered statements which are correct and indicate those choices by the appropriate lettered answer.

56. A 60-watt, 110-volt tungsten filament lamp operated on 120V

I will consume more than 60 watts while operating
II will have a lower resistance than at 110V
III will be brighter than at 110V
IV will burn out after operation of ½ hour or less

(A) if only I, II, and III are correct
(B) if only I and III are correct
(C) if only II and IV are correct
(D) if only IV is correct
(E) if you do not select any of the above as correct. (On these practice tests, if you select E, jot down your choices so that you can compare them with the Explanations.)

57. A person standing in an elevator is taken up by the elevator at constant speed. The push which he exerts on the floor of the elevator

I is equal to his weight
II is equal to less than his weight
III is equal to more than his weight
IV is dependent on the value of this constant speed

(A) if only I, II, and III are correct
(B) if only I and III are correct
(C) if only II and IV are correct
(D) if only IV is correct
(E) if you do not select any of the above as correct. (On these practice tests, if you select E, jot down your choices so that you can compare them with the Explanations.)

58. An object with a constant mass rests on a perfectly horizontal table. If a horizontal force *F* is applied, acceleration *a* results. If *F* is doubled without changing the direction,

I the acceleration will remain the same
II the acceleration will be doubled
III the acceleration will decrease
IV the acceleration will increase

(A) if only I, II, and III are correct
(B) if only I and III are correct
(C) if only II and IV are correct
(D) if only IV is correct
(E) if you do not select any of the above as correct. (On these practice tests, if you select E, jot down your choices so that you can compare them with the Explanations.)

59. An object with constant mass rests on a horizontal surface whose coefficient of friction is 0.2. If a horizontal force *F* is applied to the object, the object

I may move with constant speed in the direction of *F* once it has been set in motion
II may remain at rest
III may accelerate
IV may move with constant speed in a direction opposite to *F*

(A) if only I, II, and III are correct
(B) if only I and III are correct
(C) if only II and IV are correct
(D) if only IV is correct
(E) if you do not select any of the above as correct. (On these practice tests, if you select E, jot down your choices to that you can compare them with the Explanations.)

60. A man pulls an object up an inclined plane with a force *F* and notes that the object's acceleration is $5 ft/s^2$. When he doubles the force without changing its direction, the acceleration will

I decrease
II increase
III remain the same
IV be doubled

(A) if only I, II, and III are correct
(B) if only I and III are correct
(C) if only II and IV are correct
(D) if only IV is correct
(E) if you do not select any of the above as correct. (On these practice tests, if you select E, jot down your choices so that you can compare them with the Explanations.)

61. On the centigrade (Celsius) scale.

I 0° occurs where it does because it was intended to coincide with the melting point of ice
II 4° occurs where it does because it was intended to coincide with the temperature of the greatest density of water
III 100° occurs where it does because it was intended to coincide with the boiling point of water at standard pressure
IV 37° occurs where it does because it was intended to coincide with normal body temperature

(A) if only I, II, and III are correct
(B) if only I and III are correct
(C) if only II and IV are correct
(D) if only IV is correct
(E) if you do not select any of the above as correct. (On these practice tests, if you select E, jot down your choices so that you can compare them with the Explanations.)

62. Assume that you have two balls of identical volume, one weighing 2 N and the other 10 N. Both are falling freely after being released from the same point simultaneously. It will then be true that

I the 10-N ball falling freely from rest will be accelerated at a greater rate than the 2-N ball
II at the end of 4 s of free fall, the 10-N ball will have 5 times the momentum of the 2-N ball
III at the end of 4 s of free fall, the 10-N ball will have the same kinetic energy as the 2-N ball
IV the 10-N ball possesses greater intertia than the 2-N ball

(A) if only I, II, and III are correct
(B) if only I and III are correct
(C) if only II and IV are correct
(D) if only IV is correct
(E) if you do not select any of the above as correct. (On these practice tests, if you select E, jot down your choices so that you can compare them with the Explanations.)

63. A lighted candle *X* is placed 20 cm from *Y*. An observer places his eye 45 cm on the other side of *Y* and, looking toward *X*, sees an image of *X*. Object *Y* may be

I a new convex mirror of 10 cm focal length of the type often used in the laboratory
II a new convex lens of 10 cm focal length of the type often used in the laboratory
III a new concave mirror of 10 cm focal length of the type often used in the laboratory
IV a new concave lens of 10 cm focal length of the type often used in the laboratory

(A) if only I, II, and III are correct
(B) if only I and III are correct
(C) if only II and IV are correct
(D) if only IV is correct

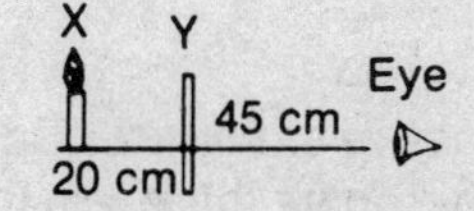

(E) if you do not select any of the above as correct. (On these practice tests, if you select E, jot down your choices so that you can compare them with the Explanations.)

64. A lens is used to produce a sharp image on a screen. When the right half of the lens is covered with an opaque material

I the right half of the image will disappear
II the left half of the image will disappear
III the image size will become approximately ½ of the original size
IV the image brightness will become approximately ½ of the original brightness

(A) if only I, II, and III are correct
(B) if only I and III are correct
(C) if only II and IV are correct
(D) if only IV is correct
(E) if you do not select any of the above as correct. (On these practice tests, if you select E, jot down your choices so that you can compare them with the Explanations.)

65. When a beam of light goes from a rarer to a denser medium like glass and has an angle of incidence equal to zero, the beam of light does not change

I amplitude
II speed
III wavelength
IV direction

(A) if only I, II, and III are correct
(B) if only I and III are correct
(C) if only II and IV are correct
(D) if only IV is correct
(E) if you do not select any of the above as correct. (On these practice tests, if you select E, jot down your choices so that you can compare them with the Explanations.)

66. Capacitor *P* is connected to a battery through switch *S* and wires *Y* and *Z*. The capacitor's dielectric is marked *X*. For a short time after the switch is closed, electrons will move through

I *Y*
II *X*
III *Z*
IV *S*

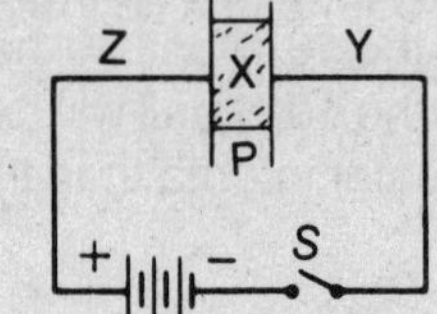

(A) if only I, II, and III are correct
(B) if only I and III are correct
(C) if only II and IV are correct
(D) if only IV is correct
(E) if you do not select any of the above as correct. (On these practice tests, if you select E, jot down your choices so that you can compare them with the Explanations.)

67. A point source of light is placed at the principal focus of a concave lens. The refracted light will

I diverge
II be parallel to the principal axis
III seem to come from a point ½ of the radius of curvature from the lens
IV converge

(A) if only I, II, and III are crrect
(B) if only I and III are correct
(C) if only II and IV are correct
(D) if only IV is correct
(E) if you do not select any of the above as correct. (On these practice tests, if you select E, jot down your choices so that you can compare them with the Explanations.)

68. In an experiment, ice is wedged in at the bottom of a test tube by means of a bright wire mesh. Water is poured in and then heated near the top. Although the water boils, the ice does not melt. The explanation for this is that

I glass is a good insulator
II heated water is less dense than cold water
III water is a good insulator
IV the wire is a good radiator

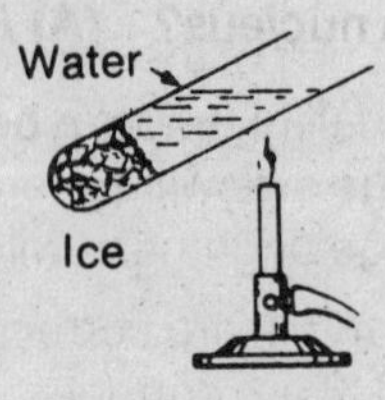

(A) if only I, II, and III are correct
(B) if only I and III are correct
(C) if only II and IV are correct
(D) if only IV is correct
(E) if you do not select any of the above as correct. (On these practice tests, if you select E, jot down your choices so that you can compare them with the Explanations.)

69. As an object starting from rest accelerates uniformly

I its kinetic energy is proportional to its displacement
II its displacement is proportional to the square root of its velocity
III its kinetic energy is proportional to the square root of its speed
IV its velocity is proportional to the square of elapsed time

(A) if only I, II, and III are correct
(B) if only I and III are correct
(C) if only II and IV are correct
(D) if only IV is correct
(E) if you do not select any of the above as correct.(On these practice tests, if you select E, jot down your choices so that you can compare them with the Explanations.)

70. All air is removed from the device shown after it is partly filled with a volatile liquid such as ether. The device is then sealed and kept in the position shown. As ice water is allowed to drip on top of bulb *Y* but not on *X*,

I the level of the liquid in bulb *Y* will rise
II the liquid will reach the same level in *X* as in *Y*
III the level of the liquid in bulb *X* will drop
IV the level of the liquid in bulb *X* will rise

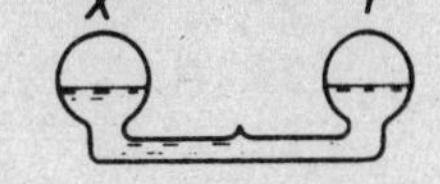

(A) if only I, II, and III are correct
(B) if only I and III are correct
(C) if only II and IV are correct
(D) if only IV is correct
(E) if you do not select any of the above as correct. (On these practice tests, if you select E, jot down your choices so that you can compare them with the Explanations.)

For 71-75. DIRECTIONS: ***Each of the following questions or incomplete statements is followed by five choices. Select the choice which fits best.***

For 71-73. Monochromatic light falls on a metal surface that has a work function of 6.7×10^{-19} joule. Each photon has an energy of 8.0×10^{-19} joule. (Planck's constant = 6.63×10^{-34} joule-sec. One electron-volt = 1.60×10^{-19} joule.)

71. What is the maximum kinetic energy of the photoelectrons emitted by the substance, in joules? (A) 1.3×10^{-19} **(B)** 1.6×10^{-19} **(C)** 2.6×10^{-19} **(D)** 6.7×10^{-19} **(E)** 8.0×10^{-19}

72. What is the energy of each photon, expressed in electron volts? (A) 1.6 **(B)** 1.6×10^{-19} **(C)** 5.0 **(D)** 6.7 **(E)** 8.0

73. What is the frequency of the photon, in cycles per second? (A) 3.7×10^{14} **(B)** 4.2×10^{14} **(C)** 1.2×10^{15} **(D)** 3.7×10^{15} **(E)** 7.0×10^{15}

74. What is the relationship between the atomic number, *R*, the mass number, *M*, and the number of neutrons, *N*, in a nucleus? (A) $R = MN$ **(B)** $R = \frac{M}{N}$ **(C)** $R = \frac{N}{M}$ **(D)** $R = M - N$ **(E)** $R = N - M$

75. When lead, $^{214}_{82}$Pb, emits a beta particle, the resultant nucleus will be (A) $^{214}_{83}$Bi **(B)** $^{214}_{84}$Po **(C)** $^{213}_{82}$Pb **(D)** $^{214}_{81}$Tl **(E)** $^{213}_{81}$Tl

ANSWERS FOR PRACTICE TEST 1

Answer Key for Test 1

1. A	16. B	31. D	46. D	61. B
2. D	17. D	32. D	47. A	62. C
3. B	18. D	33. E	48. D	63. C
4. A	19. B	34. E	49. B	64. D
5. C	20. D	35. C	50. C	65. D
6. B	21. A	36. C	51. D	66. E (I, III, IV)
7. A	22. A	37. C	52. D	67. E (I)
8. A	23. C	38. B	53. D	68. A
9. B	24. E	39. B	54. C	69. B
10. B	25. E	40. C	55. C	70. B
11. A	26. B	41. D	56. B	71. A
12. A	27. C	42. C	57. E (I)	72. C
13. B	28. A	43. D	58. C	73. C
14. E	29. B	44. B	59. A	74. D
15. E	30. D	45. D	60. E (II)	75. A

Explained Answers for Test 1

1. (A) If the graph is drawn well, *KL* should appear to be a straight line, and *K* and *L* should be at the same distance from the *t*-axis. A visual examination should be adequate; don't look for minute differences. If you don't realize then that *KL* represents motion with constant speed, you need to review the subject of graphs; also see the explanation of the next question.

2. (D) Remember that, except in special cases which are clearly marked, it is customary to have increasing positive values on the axes going to the right (*X*-axis) from the point of intersection of the axes, and up (*Y*-axis). On the graph line *LM* slopes down, and therefore represents decreasing speed. The next interval, *MN*, curves up representing increasing speed, but it is not one of the given choices. The next interval which curves up is *OP*; it represents increasing speed and is choice *D*.

3. (B) The straight line *LM* shows that the speed is changing the same amount during each second. This means that the acceleration is constant. In this case the speed is decreasing, and therefore the acceleration is negative. During interval *KL* the speed and direction do not change; the acceleration is zero. For the curved portions of the graph, the speed does not change the same amount during each second; the acceleration is not constant.

4. (A) The potential energy of an object lifted through a vertical height $h = wh = mgh$.

5. (C) Displacement over time (s/t) is the velocity of an object moving with constant velocity. Mass times velocity is momentum.

6. (B) Electric charge over time, Q/t, or the rate with which electric charge moves past a point, is electric current, I. I^2R represents the power used by a resistor.

7. (A) As mentioned in explanation 5, s/t represents velocity, v. Thus s^2/t^2 represents v^2. Then ½ mv^2 gives the kinetic energy of an object with mass m moving with speed v.

8. (A) A little difficult. Multiply numerator and denominator by t; this gives Q^2Rt/t^2 which equals I^2RT, since $Q/t = I$; (see explanation 6). I^2Rt represents power multiplied by time, or energy.

9. (B) The kinetic energy of a body is given by the expression ½mv^2. Since the mass of the body is constant, its kinetic energy is proportional to v^2, the square of its speed.

10. (B) This acceleration equals v^2/r, where v is its speed and r is the radius of its circular path. Since r is constant, the acceleration is proportional to v^2, the square of its speed.

11. (A) The momentum of a body is equal to the product of its mass and its velocity. Since its mass remains constant, its momentum is proportional to its velocity.

12. (A) The acceleration of an object in equilibrium is zero. Its velocity is constant (and may be zero). Its displacement $s = vt$. In this question, $t = 1$ s, a constant. Hence, the displacement is proportional to the velocity.

13. (B) Such motion is described by the relationship, $v^2 = 2as$. Since $2a$ is constant, v^2 is proportional to s, and the displacement, s, is proportional to the square of the speed, v.

14. (E) Since the five wires are connected in parallel, they each have the same potential difference—that of the source of voltage. The current in each wire then depends on its resistance. You should remember that silver is the best conductor of electricity; copper is one of the best; nichrome is used as the heating elements in many products, including toasters and soldering irons, and is a much poorer conductor. Apply Ohm's law: $I = V/R$; since the potential difference V is the same for each wire, the current I is inversely proportional to the resistance. For a wire of given material, the resistance is proportional to the length and inversely proportional to the cross section area. Since all the wires have the same diameter, we look for the shortest wire of the best conductor. This is given by choice *E*, two feet of silver wire.

15. (E) A useful formula here for the power used by an electric device is that it is the product of its potential difference and its current ($P = VI$). In a parallel circuit each branch has the same potential difference. We already saw in question 14 that the two feet of silver wire has the greatest current; therefore it also uses the greatest power.

16. (B) Two forces which act on an object in opposite directions tend to nullify each other. The resultant of such forces is their difference, and the direction of the resultant is the direction of the larger force.

17. (D) If two forces, acting on the same point, are at right angles to each other, their resultant may be found by applying the Pythagorean theorem. $R^2 = 6^2 + 8^2$; $R = 10$ lb. You should have recognized the 3-4-5 right triangle; the two sides, 6 and 8, are twice 3 and 4 respectively, and therefore the hypotenuse, giving the value of the resultant, is twice 5.

18. (D) Potential energy = weight × height
= 10 lb × 5 ft = 50 ft-lb

19. (B) For explanations see next question.

20. (D) To keep the weights from moving, one can apply an additional 4-lb downward force to the 6-lb object. Then the tension at *A* must be 10 lb to keep the 10-lb object in equilibrium. (The tension in the other parts of the rope around the pulley is also 10 lb.) When the 4-lb force is removed, the 10-lb object starts to move down; that is, it is accelerated. Therefore, the upward pull due to the tension in the rope is less than 10 lb, the downward pull due to gravity. Similarly for the 6-lb object: the upward pull due to the tension in the rope is greater than 6 lb, the downward pull due to gravity. (But the tension is the same throughout the rope.) Therefore the tension in the rope is between 6 and 10 lb. Since the net force on each object is less than its weight, the acceleration is less than g; ($F/W = a/g$).

21. (A) For a freely falling object, the distance covered is equal to the product of one-half its acceleration and the square of the time of fall:

$$\text{distance} = \tfrac{1}{2}gt^2 = \tfrac{1}{2}(9.8\ \text{m/s}^2)(1\ \text{s})^2 = 4.9\ \text{m}$$

22. (A) Steel is denser than water. The fact that it can float is not explained by Archimedes' principle. The concept of surface tension, however, suggests that the surface of the water acts like a stretched membrane. Unless this membrane is broken, the steel wire is supported by it.

23. (C) This follows from the fact that all the angles of reflection will be equal to each other.

24. (E) Since there is no change in temperature, the heat is required only for change of state: melting.

Heat requ'd = mass × heat of fusion
= 10 gm × 77 cal/g = 770 cal

25. (E) Heat requ'd for vaporization equals mass times heat of vaporization. The heat of vaporization of lead is not furnished.

26. (B) The coefficient of expansion per F deg $= \frac{5}{9}$ (coefficient of expansion per C deg) $= 5 \times \frac{0.000025}{9} = 0.000014$.

$$\begin{aligned}\text{Change in length} &= \text{length} \times \text{coeff.} \times \text{temp. change}\\ &= 100\text{ cm} \times 0.000014 \times 10\text{ F deg}\\ &= 0.014\text{ cm.}\end{aligned}$$

27. (C) Heat lost = heat gained.

Heat lost (or gained) = mass × sp. ht × temp. change
Heat lost by block = m × 0.220 (cal/g-C°) × (100° − 30°);
Heat gained by water = 200 gm × 1 (cal/gm-C°) × (30° − 10°).
m × 0.220 × 70° = 200 × 20°
m = 260 g

Note that in problems like this one we assume that the temperature of boiling water is 100° C.

28. (A) Sound is a longitudinal wave and cannot be polarized. All the others are electromagnetic waves; these can be polarized.

29. (B) In elementary physics we usually consider illumination to be due to a point source from which light goes out uniformly in all directions. Illumination due to such a source varies inversely as the square of the distance.

30. (D) The speed of sound in air is the same for all audible frequencies.

31. (D) For motion with constant speed,

distance = speed × time.

$$t = s/v = \frac{4000\text{ ft}}{1110\text{ ft/s}} = 3.6\text{ s}$$

Noting that this is between 3 and 4 s would have been enough.

32. (D) $v = fL;\ L = \frac{v}{f} = \frac{1110\text{ ft/sec}}{222\text{ vib/sec}} = 5\text{ ft}$

33. (E) In question 31 we found that it took 3.6 sec for the sound to reach the cliff. It takes another 3.6 sec for the sound to get back to the ship.

34. (E) In the calculations it was assumed that the distance remained 4000 ft and that therefore the time required for the sound to come back is 3.6 sec. Actually, if the ship moved slowly, or moved in such a way that the distance between it and the cliff remained practically constant, the measured time would still be 7.2 s. (E) is the best choice.

35. (C) Absolute pressure = gauge pressure + atm. press. Abs.p = (135 lb/in^2) + (15 lb/in^2) = 150 lb/in^2. We end up with a gas whose vol. is 10 ft^3 and whose pressure is 150 lb/in^2 What volume would this gas occupy at atmospheric pressure? Apply Boyle's law: $p_1V_1 = p_2V_2$; (150 lb/in^2) × (10 ft^3) = (15 lb/in^2) × (V_2); V_2 = 100 ft^3. However, the tank had 10 ft^3 of air in it at the start; we had to add only 90 ft^3.

36. (C) Don't let a complex circuit scare you away from an easy question. For R_1 you know: V_1 = 20 V; R_1 = 40 ohms. $I_1 = \frac{V_1}{R_1} = \frac{20\text{ V}}{40\ \Omega} = 0.5\text{ A}$

37. (C) The electric current heats the wire ($P = I^2R$); the wire heats the liquid. The temperature of the liquid rises until it starts to boil. With further heating more and more liquid will evaporate, but the temperature of the boiling liquid remains constant.

38. (B) The resistance of tungsten (as for most metals) goes up with temperature. The tungsten is heated by the current and cooled by the liquid. When the liquid reaches its highest temperature (78° C), the wire reaches its highest temperature and therefore its highest resistance.

39. (B) The coil on the left becomes an electromagnet. Electron flow is from minus to plus outside the battery. Grasp the coil with the left hand so that the fingers point in the direction of electron flow; the outstretched thumb points in the direction of the N-pole: in this case, *W*.

40. (C) A changing current in coil *X* induces a current in the other coil. Current in coil *X* builds up slowly to its maximum value because of the large self-induced emf in coil *X*. Until this maximum value is reached, current is induced in the second coil in a direction opposing the current build-up in coil *X*. For this, the left end of the second coil should be an S-pole, because it will then tend to weaken the left magnet. Then apply the left-hand rule, as in explanation 39.

41. (D) The electric field intensity due to a point charge is proportional to the charge and inversely proportional to the square of the distance from the charge. In this question, charges *Y* and *Z* are equal. Therefore at the point midway between them the field intensity due to each charge is the same in magnitude and directed oppositely. (Recall that the direction of the field intensity is the direction of the force on an imagined positive charge.) Therefore the two field intensities cancel each other completely at the point midway between the charges.

42. (C) The force acting on a charge moving perpendicularly through a magnetic field is proportional to the speed with which the charge moves: (recall, $F = Bqv$). Therefore the force on the charge doubles if the speed doubles.

43. (D) The proton is slightly smaller in mass than the neutron. They differ by about 1 part in 2000.

44. (B) The time of one complete vibration of a simple pendulum or its period is equal to $2\pi\sqrt{L/g}$. If the length is decreased, the period is decreased. If g is decreased, by going up to the top of the Empire State Building, the period is increased. In the simple pendulum, the period is independent of the mass of the bob.

45. (D) The bubble is a gas. If we assume that the temperature is constant, we may apply Boyle's law: the volume of the gas varies inversely as the pressure. The volume of the gas at the bottom of the lake is $^1/_3$ of its volume at the top; the pressure at the bottom must be 3 times as great as at the top. The pressure at the top is approx. 1 atmosphere, produced by the atmosphere, equivalent to that produced by a column of water 34 ft high. At a depth of 68 ft in the lake, we have the pressure due to the water (equal to 2 atmospheres), plus the pressure of the atmosphere which is transmitted without loss by the water (Pascal's principle).

46. (D) The description fits water-ice; think of regelation. If enough pressure is applied to ice it will melt. In order to have it freeze under this pressure, we must lower the temperature. The freezing point is the same as the melting point.

47. (A) The test to see if an object is electrically charged is to see if it will repel an electrically charged object. An uncharged rod will attract, and be attracted by, a charged pith ball.

48. (D) The rate of heat production of a wire = I^2R. The resistance of the wire is practically constant. Therefore, the rate of heat production is proportional to the square of the current.

49. (B) This is consistent with various observations on the earth. Choices C, D, and E are completely irrelevant. Observations during a solar eclipse, as well as observations from high-flying balloons, eliminate choice A. Satellites may be used for this observation.

50. (C) Water is densest at 4° C. It expands steadily when heated above, or cooled below, this temperature. On being cooled from 7° C to 4° C water contracts; on further cooling it expands again.

51. (D) Boyle's law applies to a definite mass of gas at constant temperature. Pressure and volume, and also density (= mass/vol), change.

52. (D) Assume that the radius of the circular orbit is approximately 4000 miles (not 300). The length of orbit equals $2\pi r = 2\pi \times 4000$ mi.

$$v = s/t = \frac{2\pi \times 4000 \text{ mi}}{1.5 \text{ hr}} \doteq 16{,}000 \text{ mi/hr.}$$

53. (D) If it were not for the earth's pull, the satellite would tend to go off along a straight line tangent to its orbit—until affected by the gravitational pull of some other object.

54. (C) A perfectly black object absorbs all light. There is no visible black light, but the term is sometimes applied to ultraviolet light.

55. (C) By definition, the fundamental frequency of a vibrating object is the lowest frequency at which it can vibrate freely. A string has many such free or natural vibrations.

56. (B) The rating of the lamp is 60 watt at 110 volts. This means that the lamp uses 60 watts when used on 110 V. The lamp may be used on a lower voltage: it will not be as bright—at 60 V. it will be quite dim. It may be used on somewhat higher voltages. Higher voltage results in higher current and greater power consumption. Of course, if a much higher than rated voltage is used, the bulb will burn out quickly. However, a 10-volt rise will not shorten the lamp's life drastically; such a rise is not uncommon in the outlet voltage.

57. (E) The person is moving with constant velocity; therefore there is no unbalanced force acting on him. The earth's pull on him (his weight) is balanced by an equal upward push on him by the floor of the elevator. In accordance with Newton's third law, his push on the floor of the elevator equals the floor's push on him: his weight.

58. (C) According to Newton's second law, the acceleration is proportional to the unbalanced force. If the unbalanced force is doubled, the acceleration is doubled. The term *smooth table* implies that friction is negligible. F is then the only horizontal force and is the net force.

59. (A) The maximum force of friction equals the coefficient of friction times the weight of the object. If F is less than this, the object will remain at rest; if F is greater than this, the object will accelerate. How do we get the object to move at constant speed? Once set in motion, the object will move with constant speed if F equals the force of friction. A slightly larger force is needed to start the motion because starting friction is greater than sliding friction.

60. (E) The acceleration will be increased, but not doubled. This can be seen readily with a numerical example. Assume that F is parallel to the plane and equal to 6 lb. Opposed to F is a component of the weight of the object, and there may also be friction; assume this adds up to 2 lb. This leaves an unbalanced force of 4 lb. Doubling F increases it to 12 lb. Opposed to it will be the same 2 lb. The unbalanced force now is 10 lb, more than twice the 4 lb which produced the acceleration in the first case. The new acceleration will be more than twice the first one.

61. (B) The melting point of ice and the boiling point of water were selected to give two fixed points on the Celsius scale: 0° and 100° The other numbers on the scale between 0 and 100 are obtained by subdividing the intervening space.

62. (C) If air resistance is negligible, freely falling objects have the same acceleration. Then, after 4 sec of free fall, both will have the same speed. Their momentums ($= mv$) will then be in the same ratio as their mass; this is also true of their kinetic energy $\frac{1}{2}mv^2$).. The 10-N ball will have 5 times the momentum and 5 times the kinetic energy of the 2-N ball. Mass is a measure of an object's inertia; the greater the mass, the greater the inertia.

63. (C) May be too tricky. The conventional mirrors do not transmit light and nothing can be seen through them. If Y is a concave lens, then a virtual image of X is seen when the eye is placed on the other side of the lens; this is the conventional case. This is not obvious for the case when Y is a convex lens. Here, X is at twice the focal length from the lens. A real image is brought to a focus on the other side of the lens, 20 cm from the lens. But there is no screen there, and the light keeps traveling on, and will diverge from there as though an object were there. To the eye it will be the same as though an object were 25 cm away from it. The normal eye will see an image of X.

64. (D) In the formation of a real image with a lens, all the rays starting from one point on the object go, after refraction by the lens, to the same point of the image. All parts of the lens are usually used for this purpose. If half of the lens is covered, the other half of the lens will still function to produce each point of the image, but only half as much light will get to each point; the image will, consequently, be half as bright.

65. (D) If the angle of incidence is zero, the ray is normal to the surface. Such a ray is not deviated;

in going from air to water it would not be bent at all. The frequency does not change either, but the wave is slowed up. Hence the wavelength must change: $v = fL$. There is some reflection of the wave at the new surface. Less energy means a lower amplitude of the wave.

66. (E) All parts are conductors except the dielectric *X*. Electrons will move through the metallic conductors until the capacitor is charged.

67. (E) Light from the point source is diverging when it gets to the concave lens; the lens makes it more diverging. Do not confuse lenses with mirrors. For lenses the relationship between position of image and radius of curvature is not as simple as suggested by choice 3.

68. (A) The water near the top is very hot; the water near the ice stays cold for a long time. Why? Heat is not conducted down by the water or the glass because they are good insulators. Convection currents do not start readily because the hot water at the top, being less dense than the cold water, does not sink.

69. (B) Its kinetic energy is proportional to the square of its speed. ($E_K = \frac{1}{2}mv^2$). The square of its speed is proportional to its displacement ($v^2 = 2as$). Hence, the kinetic energy is proportional to the

70. (B) A volatile liquid vaporizes readily at ordinary room temperatures. The space above the liquid in *X* and *Y* is saturated with the vapor. Ice water dripped on *Y* causes some of the vapor in *Y* to condense; the pressure of the vapor in *Y* is reduced. The vapor pressure in *X* remains relatively high because of the higher temperature. This higher pressure pushes some of the liquid from *X* into *Y*.

71. (A) In the *photoelectric effect* the energy of the incident photon is used to pull the electron away from the metal; if the photon has more energy than is needed to just overcome the attractive force of the metal, the remaining energy gives the freed electron its kinetic energy:

$$\text{energy of photon} = \text{work function} + \text{kinetic energy}$$
$$8.0 \times 10^{-19}\ \text{joule} = 6.7 \times 10^{-19}\ \text{joule} + \text{kinetic energy}$$
$$\text{kinetic energy} = 1.3 \times 10^{-19}\ \text{joule}$$

72. (C) We are told that each photon has an energy of 8.0×10^{-19} joule and that one electron volt = 1.60×10^{-19} joule. By inspection we note that the energy of the photon is 5 times as great, or 5 electron-volts.

73. (C) The energy of a photon is equal to the product of Planck's constant and its frequency:

$$E = hf$$
$$8.0 \times 10^{-19}\ \text{joule} = (6.63 \times 10^{-34}\ \text{joule-sec})\ f$$
$$f = 1.2 \times 10^{15}\ \text{cycles per sec}$$

74. (D) The *mass number* of a nucleus, or of the atom of which it is a part, is equal to the number of protons (its atomic number) and the number of neutrons:

$$\text{mass number} = \text{atomic number} + \text{number of neutrons}$$
$$M = R + N$$
$$R = M - N$$

If you have any trouble with this, think of the mass number as approximately the total mass of the nucleus, where the unit of mass is taken as the mass of the proton. The mass of a neutron is approximately the same as the mass of a proton.

75. (A) The beta particle is an electron, which has a charge of −1. When a nucleus with a charge of +82 loses a −1 charge, it must become more positive by one more unit; that is, the charge must become +83. This charge is its atomic number. Of the given choices, only one has a subscript of 83, which is the atomic number. Actually we don't have to do anything further. But at this point, just for review, we should note that the loss of an electron does not change the mass number of the nucleus. Choice **(A)** has not only the correct subscript, 83, but also the expected unchanged superscript, 214.

PRACTICE TEST 2

Part A

DIRECTIONS: Each set of lettered choices below refers to the numbered questions immediately following it. Select the one lettered choice that best answers each question. A choice may be used once, more than once, or not at all in each set.

QUESTIONS 1-6 refer to the following laws and principles:

(A) Conservation of momentum **(B)** Conservation of kinetic energy **(C)** Boyle's law **(D)** Charles' law **(E)** None of these

KEY for questions 1-6

m = mass
v = velocity
p = pressure
V = volume
T = absolute temperature
h = vertical height
g = acceleration due to gravity

For each of the following equations select the choice which is most closely related to it.

1. $m_1v_1^2 = m_2v_2^2$
2. $m_1v_1 = m_2v_2$
3. $V_1T_1 = V_2T_2$
4. $V_1T_2 = V_2T_1$
5. $p_1V_1 = p_2V_2$
6. $mgh = mv^2$

QUESTIONS 7-13 refer to the following concepts:

(A) Work **(B)** Power **(C)** Series circuit **(D)** Parallel circuit **(E)** Illumination

KEY for questions 7-13

F = force
s = distance
p = pressure
A = area
C = intensity of light source
I = electric current
V = potential difference
R = electric resistance
v = speed

For each of the following algebraic expressions or equalities select the choice which is most closely related to it.

7. Fs
8. pAs
9. Fv
10. $V_1 = V_2$
11. $I_1 = I_2$
12. I^2R
13. C/s^2

QUESTIONS 14 and 15 refer to the following laboratory experiment. Two small, identical metal spheres are projected at the same time from the same height by two identical spring guns. Each gun provides the same push on its sphere except that one sphere is projected vertically upward while the other one is projected horizontally. Frictional losses are negligible.

(A) the same **(B)** twice as great **(C)** greater but not necessarily twice as great **(D)** one-half as great **(E)** less but not necessarily one-half as great

14. The time required for the vertically projected sphere to hit the level floor compared with that for the horizontally projected one is

15. The kinetic energy with which the vertically projected sphere hits the floor compared with that for the horizontally projected one is

Part B

DIRECTIONS: *Each of the following questions or incomplete statements is followed by five choices. Select the choice which fits best.*

16. Of the following, the smallest quantity is **(A)** 0.635 km **(B)** 0.635×10^4 cm **(C)** 6.35×10^4 m **(D)** 0.635×10^6 mm **(E)** 0.635×10^3 m

17. 55 mm is approx. equivalent to **(A)** 2.3 in. **(B)** 6 in. **(C)** 11. in. **(D)** 18 in. **(E)** 23 in.

18. Two forces of 10 lb and 7 lb, respectively, are applied simultaneously to an object. The maximum value of their resultant is, in pounds, **(A)** 7 **(B)** 10 **(C)** 17 **(D)** $17\sqrt{3}$ **(E)** 70

For 19-20. In the graphs, *v* stands for speed, *s* stands for distance, *t* for time. The acceleration of a

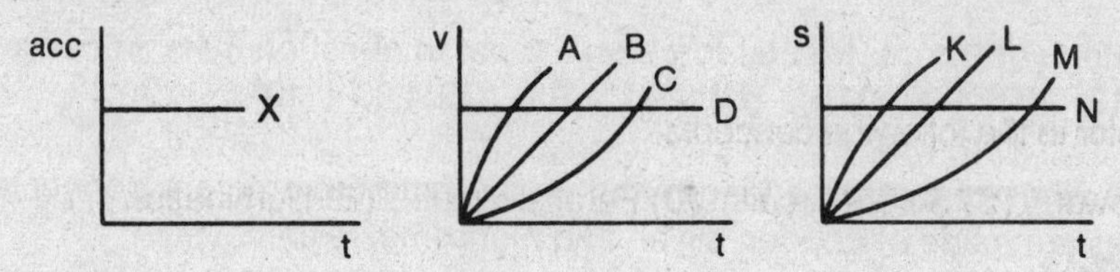

certain object starting from rest and moving with constant acceleration along a straight line is represented by graph *X*.

19. The speed of the object is represented by **(A)** graph *A* **(B)** graph *B* **(C)** graph *C* **(D)** graph *D* **(E)** none of these

20. The distance covered by the object is represented by **(A)** graph *K* **(B)** graph *L* **(C)** graph *M* **(D)** graph *N* **(E)** none of these

For 21-22. A man weighs 150 lb and exerts a force of 30N parallel to a rough inclined plane as he pushes a 90-N object 6 meters up along the plane at constant speed.

21. The force of friction is **(A)** less than 30 N **(B)** 30 N **(C)** more than 30 N but less than 90 N **(D)** 90 N **(E)** more than 90 N

22. The potential energy gained by the object is, in joules, **(A)** less than 180 **(B)** 180 **(C)** 540 **(D)** 900 **(E)** more than 900

For 23-24. An object is made of material whose specific gravity is 4. Its volume is 30 cm³. It is suspended from a string while submerged in water; it does not touch the bottom or sides of the container.

23. The tension in the string is **(A)** 3 g **(B)** 30 g **(C)** 60 g **(D)** 90 g **(E)** 120 g

24. The weight of the displaced water is **(A)** 3 g **(B)** 30 g **(C)** 60 g **(D)** 90 g **(E)** 120

For 25-26. Two small masses, *X* and *Y*, are *d* centimeters apart. The mass of *X* is 4 times as great as that of *Y*. Then *X* attracts *Y* with a force of 16 dynes.

25. *Y* attracts *X* with a force of **(A)** 1 dyne **(B)** 4 dynes **(C)** 16 dynes **(D)** 32 dynes **(E)** 64 dynes

26. If the distance between *X* and *Y* is changed to 2*d* centimeters, *X* will attract *Y* with a force of **(A)** 1 dyne **(B)** 4 dynes **(C)** 8 dynes **(D)** 16 dynes **(E)** 32 dynes

27. The period of a simple pendulum in a laboratory does not depend on **(A)** the altitude of the laboratory **(B)** the acceleration due to gravity in the laboratory **(C)** the length of the string **(D)** vibration in the laboratory **(E)** mass of the bob

28. A car is traveling on a level highway at a speed of 15 meters per second. A braking force of 3,000 newtons brings the car to a stop in 10 seconds. The mass of the car is **(A)** 1,500 kg **(B)** 2,000 kg **(C)** 2,500 kg **(D)** 3,000 kg **(E)** 45,000 kg

29. An elevator weighing 2.5×10^4 newtons is raised to a height of 10 meters. Neglecting friction, the work done is **(A)** 2.5×10^4 joules **(B)** 2.5×10^5 joules **(C)** 2.5×10^3 joules **(D)** 7.5×10^4 joules **(E)** 98 joules

30. A 10-kg rocket fragment falling toward the earth has a net downward acceleration of 5 meters per second^2. The net downward force acting on the fragment is **(A)** 5 newtons **(B)** 10 newtons **(C)** 50 newtons **(D)** 98 newtons **(E)** 320 newtons

31. A racing car is speeding around a flat, unbanked circular track whose radius is 250 meters. The car's speed is a constant 50.0 meters per second. The mass of the car is 2.00×10^3 kilograms. The centripetal force necessary to keep the car in its circular path is provided by **(A)** the engine **(B)** the brakes **(C)** friction **(D)** the steering wheel **(E)** the stability of the car

32. The magnitude of the centripetal force on the car in question 30 is, in newtons, **(A)** 1.00×10^1 **(B)** 4.00×10^2 **(C)** 4.00×10^3 **(D)** 2.00×10^4 **(E)** 4.00×10^4

For 33-34. Three wires of the same length and cross-sectional area are connected in series to a battery. The wires are made of copper, silver, and nichrome, respectively.

33. The current through the copper is **(A)** the same as through the silver and nichrome **(B)** greater than that through the silver or the nichrome **(C)** less than that through the silver or the nichrome **(D)** greater than that through the silver but less than that through the nichrome **(E)** greater than that through the nichrome but less than that through the silver

34. The potential difference across the copper is **(A)** the same as across the silver and nichrome **(B)** greater than that across the silver or the nichrome **(C)** less than that across the silver or the nichrome **(D)** greater than that across the silver but less than that across the nichrome **(E)** greater than that across the nichrome but less than that across the silver.

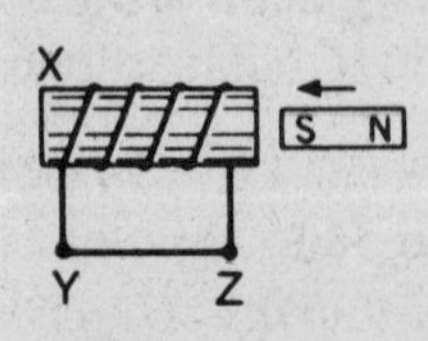

35. *X* is a coil of wire with a hollow core. The permanent magnet is pushed at constant speed from the right into the core and out again at the left. During this motion **(A)** there will be no current in wire *YZ* **(B)** electron flow in wire *YZ* will be from *Y* to *Z* **(C)** electron flow in wire *YZ* will be from *Z* to *Y* **(D)** electron flow in wire *YZ* will be from *Z* to *Y* and then from *Y* to *Z* **(E)** electron flow in wire *YZ* will be from *Y* to *Z* and then from *Z* to *Y*

36. A compass is placed to the east of a vertical conductor. When electrons go through the wire, the N-pole of the compass needle is deflected to point towards south. The direction of electron flow in the wire is **(A)** east **(B)** south **(C)** west **(D)** up **(E)** down

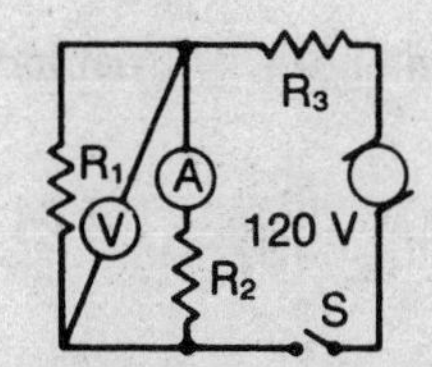

For 37-38. In the following circuit, *V* stands for a good voltmeter and *A* for a good ammeter.

37. The voltmeter should measure the potential difference across **(A)** the ammeter **(B)** R_1 **(C)** R_3 **(D)** the generator **(E)** the switch

38. The ammeter should measure the current through **(A)** the voltmeter **(B)** R_1 **(C)** R_2 **(D)** R_3 **(E)** the generator

39. One way in which a transistor differs from a semiconductor diode is that the transistor, but not the diode, **(A)** has a heated cathode **(B)** has a metal grid **(C)** can amplify signals **(D)** uses a crystal with added impurities **(E)** can be made very small

40. 200 calories of heat are added to a mixture of 5 gm of ice and 40 gm of water at 0° C. The resulting temperature will be **(A)** 0° C **(B)** 0.5° C **(C)** 3° C **(D)** 5° C **(E)** 4.5° C

41. The sound of a siren to the west of you is transmitted to your ear by air **(A)** vibrating in a north-south direction **(B)** vibrating in a west-east direction **(C)** vibrating in a vertical direction **(D)** moving continuously westward **(E)** moving continuously eastward

42. A man hears a musical sound produced by a tuning fork. The frequency of the sound may be, in vib/sec, **(A)** 4 **(B)** 400 **(C)** 40,000 **(D)** 100,000 **(E)** 400,000

43. The volume of a gas at constant pressure is directly proportional to the temperature as measured on the **(A)** Celsius scale **(B)** Fahrenheit scale **(C)** Reaumur scale **(D)** Baume scale **(E)** Kelvin scale

For 44-45. The coefficient of thermal expansion of a metal is 0.000045 per centigrade degree.

44. Its coefficient per Fahrenheit degree is **(A)** negative **(B)** 0.000025 **(C)** 0.000045 **(D)** 0.000081 **(E)** 32.00003

45. If a 100-cm rod of this material is heated from 30° C to 50° C, its change in length will be, in cm, **(A)** 0.00050 **(B)** 0.00090 **(C)** 0.00180 **(D)** 0.090 **(E)** 0.180

For 46-47. Two heating coils, *X* and *Y*, are each connected to 120 volts, dc. The resistance of *X* is twice that of *Y*.

46. The current through *X* is **(A)** twice that through *Y* **(B)** equal to that through *Y* **(C)** one-half that through *Y* **(D)** one-fourth that through *Y* **(E)** one-eighth that through *Y*

47. The rate at which *X* produces heat is **(A)** 4 times that produced by *Y* **(B)** 2 times that produced by *Y* **(C)** the same as that produced by *Y* **(D)** one-half that produced by *Y* **(E)** one-fourth that produced by *Y*

48. An atom of an element differs from an atom of one of its isotopes in the number of **(A)** neutrons in the nucleus **(B)** protons in the nucleus **(C)** valence electrons **(D)** electrons outside the nucleus **(E)** protons outside the nucleus

49. A man is 10 ft away from a plane mirror. His distance from his image is **(A)** 5 ft **(B)** 10 ft **(C)** 15 ft **(D)** 20 ft **(E)** 25 ft

50. When a person uses a convex lens as a magnifying glass, the distance that the object must be from the lens is **(A)** less than one focal length **(B)** more than one but less than two focal lengths **(C)** two focal lengths **(D)** more than two but less than four focal lengths **(E)** at least four focal lengths

51. In sound frequency determines pitch. In light frequency determines **(A)** speed **(B)** velocity **(C)** distance **(D)** the kind of polarization **(E)** color

52. The image of an erect object on the retina of the eye is **(A)** real and erect **(B)** real and inverted **(C)** virtual, erect, and enlarged **(D)** virtual, inverted, and enlarged **(E)** virtual, inverted, and reduced in size

53. The pages of this book are visible because they **(A)** absorb light **(B)** emit light **(C)** reflect light **(D)** refract light **(E)** polarize light

54. In monochromatic red light, a blue book will probably appear to be **(A)** blue **(B)** black **(C)** purple **(D)** yellow **(E)** green

55. When light emerges from water and enters air **(A)** the light will be refracted **(B)** the frequency of the light will go up **(C)** the frequency of the light will go down **(D)** the speed of the light will go down **(E)** the speed of the light will go up

For 56- 57. A series ac circuit consists of a 30-ohm resistor and a capacitor with a reactance of 40 ohms connected to an ac generator with a terminal voltage of 150 V.

56. The current through the capacitor is, in amperes, **(A)** 3.0 **(B)** 3.8 **(C)** 5.0 **(D)** 2.1 **(E)** 15

57. The potential difference across the resistor is **(A)** 60 V **(B)** 114 V **(C)** 90 V **(D)** 150 V **(E)** 250 V

58. The diagram below shows a conducting loop rotating clockwise in a uniform magnetic field.

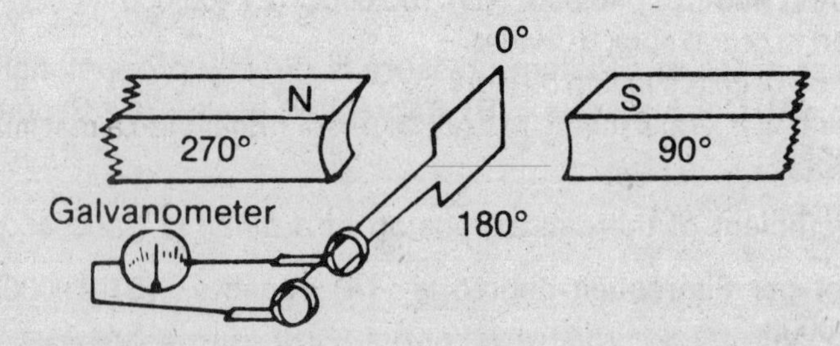

As the loop rotates, the induced voltage will be a maximum **(A)** at 0° and 90° **(B)** at 0° and 180° **(C)** at 90° and 270° **(D)** at 180° and 270° **(E)** and constant throughout the rotation

59. In question 58, the induced voltage won't be decreased by doing each of the following *except* **(A)** increasing the speed of rotation **(B)** using stronger magnets **(C)** using a rotating loop with an iron core **(D)** using a commutator **(E)** moving the north and south poles further apart

QUESTIONS 60-69 Each of these questions or incomplete statements is followed by four numbered (I–IV) statements. Select all the numbered statements which are correct and indicate these choices by the appropriate lettered answer.

60. At the red end of the visible spectrum we find

I slower waves than at the violet
II lower energy quanta than at the violet
III longitudinal waves
IV longer wavelengths than at the violet

(A) if only I, II, and III are correct
(B) if only I and III are correct
(C) if only II and IV are correct
(D) if only IV is correct
(E) if you do not select any of the above as correct. (On these practice tests, if you select E, jot down your choices so that you can compare them with the Explanations.)

61. The principle of interference is necessary to explain adequately the

I production of beats
II production of green color of leaves
III appearance of colors in thin oil films
IV appearance of a virtual image in a plane mirror rather than a real image

(A) if only I, II, and III are correct
(B) if only I and III are correct
(C) if only II and IV are correct
(D) if only IV is correct
(E) if you do not select any of the above as correct. (On these practice tests, if you select E, jot down your choices so that you can compare them with the Explanations.)

62. Of the following, the same quantity as 6.50×10^{-3} amp is represented by

I 6.50 mA
II 65.0×10^{-4} A
III 0.00650 A
IV 65.0×10^{-2}A

(A) if only I, II, and III are correct
(B) if only I and III are correct
(C) if only II and IV are correct
(D) if only IV is correct
(E) if you do not select any of the above as correct. (On these practice tests, if you select E, jot down your choices so that you can compare them with the Explanations.)

63. When a beam of light goes obliquely from a rarer to a denser medium, it does not change

I amplitude
II direction
III wavelength
IV frequency

(A) if only I, II, and III are correct
(B) if only I and III are correct
(C) if only II and IV are correct
(D) if only IV is correct
(E) if you do not select any of the above as correct. (On these practice tests, if you select E, jot down your choices so that you can compare them with the Explanations.)

64. As an object starting from rest accelerates uniformly,

I its kinetic energy is proportional to its displacement
II its speed is proportional to the square root of its displacement
III its kinetic energy is proportional to the square of its speed
IV its velocity is proportional to the square of the elapsed time

(A) if only I, II, and III are correct
(B) if only I and III are correct
(C) if only II and IV are correct
(D) if only IV is correct
(E) if you do not select any of the above as correct. (On these practice tests, if you select E, jot down your choices so that you can compare them with the Explanations.)

65. A man pulls an object up an inclined plane with a force *F* and notes that the object's acceleration is 3 ft /s^2. When he triples the force without changing its direction, the acceleration

I decreases
II increases
III remains the same
IV triples

(A) if only I, II, and III are correct
(B) if only I and III are correct
(C) if only II and IV are correct
(D) if only IV is correct
(E) if you do not select any of the above as correct. (On these practice tests, if you select E, jot down your choices so that you can compare them with the Explanations.)

66. A person standing in an elevator which goes up with constant positive acceleration exerts a push on the floor of the elevator

I equal to his weight
II whose value depends on the elevator's acceleration
III equal to less than his weight
IV equal to more than his weight

(A) if only I, II, and III are correct
(B) if only I and III are correct
(C) if only II and IV are correct
(D) if only IV is correct
(E) if you do not select any of the above as correct. (On these practice tests, if you select E, jot down your choices so that you can compare them with the Explanations.)

67. The lowest note which can be produced by a vibrating object is known as its

I fundamental frequency
II first overtone
III first harmonic
IV first basso

(A) if only I, II, and III are correct
(B) if only I and III are correct
(C) if only II and IV are correct
(D) if only IV is correct
(E) if you do not select any of the above as correct. (On these practice tests, if you select E, jot down your choices so that you can compare them with the Explanations.)

68. Assume that you have a coil of copper wire whose resistance is 100 ohms. If you want a coil of copper wire with less resistance, you may use copper wire

I which has the same length but is thicker
II which has the same thickness but is shorter
III which is thicker and shorter
IV which is thinner and longer

(A) if only I, II, and III are correct
(B) if only I and III are correct
(C) if only II and IV are correct
(D) if only IV is correct
(E) if you do not select any of the above as correct. (On these practice tests, if you select E, jot down your choices so that you can compare them with the Explanations.)

69. Small pieces of iron may be attracted by

I the N-pole of a magnet
II a negatively charged rubber rod
III the S-pole of a magnet
IV a positively charged glass rod

(A) if only I, II, and III are correct
(B) if only I and III are correct
(C) if only II and IV are correct
(D) if only IV is correct
(E) if you do not select any of the above as correct. (On these practice tests, if you select E, jot down your choices so that you can compare them with the Explanations.)

For 70-75. DIRECTIONS: *Each of the following questions or incomplete statements is followed by five choices. Select the choice which fits best.*

70. Monochromatic light with a wavelength of 6.0×10^{-7} meters is incident upon two slits that are 2.0×10^{-5} meters apart. As a result an interference pattern appears on a screen 2.0 meters away from the slits and parallel to the slits. What is the expected distance between the central maximum and the next bright line on the screen? **(A)** 1.0×10^{-2} m **(B)** 3.0×10^{-2} m **(C)** 1.7×10^{-2} m **(D)** 6.0×10^{-2}m **(E)** 1.2×10^{-1} m

71. A metal is illuminated by light above its threshold frequency. Which determines the number of electrons emitted by the metal? **(A)** color **(B)** frequency **(C)** intensity **(D)** speed **(E)** wavelength

72. Some of the energy levels of hydrogen are shown below (not to scale).

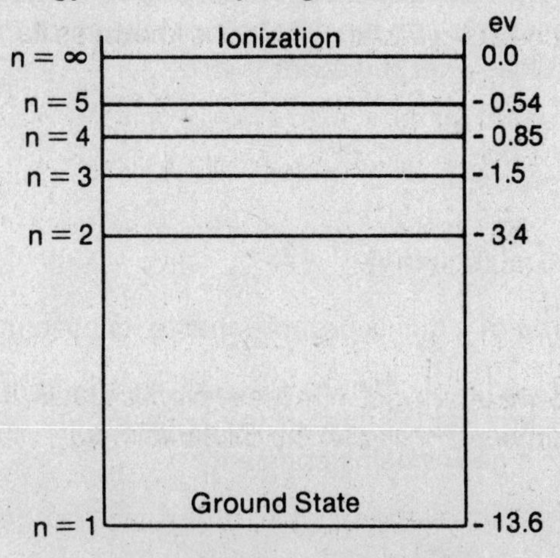

Energy Levels for Hydrogen

Which transition will result in the emission of the photon with the greatest energy? **(A)** $n = 5$ to $n = 4$ **(B)** $n = 5$ to $n = 3$ **(C)** $n = 5$ to $n = 2$ **(D)** $n = 3$ to $n = 2$ **(E)** $n = 2$ to $n = 1$

73. What is the frequency of the photon for the transition from $n = 3$ to $n = 2$, in cycles per sec? **(A)** 6.0×10^{-19} **(B)** 2.0×10^{14} **(C)** 4.6×10^{14} **(D)** 6.8×10^{14} **(E)** 6.8×10^{15}

74. If the half-life of $^{234}_{90}$Th is 24 days, the amount of a 12-gram sample remaining after 96 days is **(A)** 1 gram **(B)** 0.75 gram **(C)** 6 grams **(D)** 1.6 grams **(E)** 3 grams

75. Given the equation $^{27}_{13}Al + ^{4}_{2}He \rightarrow ^{30}_{15}P + X$; the correct symbol for X is **(A)** $_{+1}^{0}e$ **(B)** $_{-1}^{0}e$ **(C)** $^{4}_{2}He$ **(D)** $^{1}_{0}n$ **(E)** $_{+1}^{1}p$

ANSWERS FOR PRACTICE TEST 2

Answer Key for Test 2

1. B	16. B	31. C	46. C	61. B
2. A	17. A	32. D	47. D	62. A
3. E	18. C	33. A	48. A	63. D
4. D	19. B	34. D	49. D	64. A
5. C	20. C	35. E	50. A	65. E (II)
6. E	21. A	36. D	51. E	66. C
7. A	22. A	37. B	52. B	67. B
8. A	23. D	38. C	53. C	68. A
9. B	24. B	39. C	54. B	69. E (I, II, III, IV)
10. D	25. C	40. A	55. E	70. D
11. C	26. B	41. B	56. A	71. C
12. B	27. E	42. B	57. C	72. E
13. E	28. B	43. E	58. C	73. C
14. C	29. B	44. B	59. E	74. B
15. A	30. C	45. D	60. C	75. D

Explained Answers for Test 2

1. (B) The kinetic energy of an object with mass *m* equals ½ mv^2. If, as in an elastic collision, kinetic energy is conserved, the kinetic energy before collision is equal to the kinetic energy of the masses after collision: $\frac{1}{2} m_1v_1^2 = \frac{1}{2} m_2v_2^2$. If we divide both sides by ½, we get the equality of question 1: $m_1v_1^2 = m_2v_2^2$.

2. (A) The momentum of an object is the product of its mass and velocity: *mv.* In the collision of two objects or in an explosion, momentum is conserved; the momentum of the masses before the event equals the momentum of the masses after the event.

3. (E) Also see explanation 4.

4. (D) Write the relationship as a proportion:

$$\frac{V_1}{V_2} = \frac{T_1}{T_2} \quad \text{(check by cross-multiplying).}$$

This expresses Charles' law; the volume of a gas is proportional to its absolute temperature, if the pressure is constant.

5. (C) This is one way of expressing Boyle's law: If the temperature remains constant, the product of the pressure and volume of a gas remains constant.

6. (E) The formula *mgh* gives the potential energy of an object; mv^2 is *not* the kinetic energy of the object. (Kinetic energy, of course, is ½ mv^2.)

7. (A) Work done by a force is equal to the product of the force and the distance moved in the direction of the force.

8. (A) Pressure equals force per unit area; $p = F/A$; $pA = F$; $pAs = Fs$. This represents work, as stated in explanation 70. This could represent the work done in pushing the piston of a cylinder containing gas under pressure p.

9. (B) You may recognize this expression for power directly. If not, substitute for v its equivalent: s/t. Then $Fv = Fs/t$, which shows that we are talking about the rate of doing work, that is, power.

10. (D) This is always true in a parallel circuit: the voltage across one branch equals the voltage across another branch.

11. (C) This is always true in a series circuit: the current in one part of the circuit equals the current in any other part of the circuit.

12. (B) Perhaps too easy. The expression I^2R gives the electrical power used by a resistor.

13. (E) The illumination due to a point source of light is proportional to the intensity of the source and inversely proportional to the square of the distance away from the source.

14. (C) The time required for the horizontally projected sphere to hit the floor depends only on the vertical height from which it starts. The horizontal velocity does not affect the vertical velocity; the initial vertical velocity of this sphere is zero. The initial vertical velocity of the vertically projected sphere is not zero. It is gradually slowed by gravity as it rises. At some greater height its velocity will be momentarily zero, and its time of fall from that greater height is greater than the time of fall for the other sphere. In addition, of course, some time was required to reach that greater height. Not enough information is provided to calculate actual times.

15. (A) Since the guns are identical, the springs do the same work on the spheres. This results in the two spheres leaving the guns with the same kinetic energy. In addition, the spheres initially have the same gravitational potential energy because they leave the guns from the same height. As the spheres fall, this potential energy is converted to kinetic energy. In the case of the vertically projected sphere another conversion occurs first. While it rises during the first part of its motion, its initial kinetic energy is converted to gravitational potential energy. This adds to the initial potential energy.

16. (B) One way to find the smallest quantity is to first write each quantity in standard scientific notation using the same unit, such as the meter. This converts the choices to: **(A)** 6.35×10^2 m **(B)** 6.35×10 m **(C)** 6.35×10^4 m **(D)** 6.35×10^2 m **(E)** 6.35×10^2 m. Of these, 6.35×10 m is the smallest.

17. (A) An approximate calculation is satisfactory. 1 in. = 2.54 cm = 25.4 mm. Therefore, 55 mm is a little more than 2 in., and much less than 3 in.

18. (C) The resultant of two forces is greatest when they act in the same direction. Their resultant is then equal to their sum.

19. (B) For an object starting from rest and moving with constant acceleration, the speed is given by $v = at$. The graph of v vs. t for this relationship is a straight line going through the origin.

20. (C) For such an object the distance covered is given by $s = \frac{1}{2} at^2$. The graph of s vs. t is a parabola symmetrical about the s-axis.

21. (A) The 30-Nt force is required to overcome two forces: the force of friction and the component of the weight parallel to the plane. The latter is not negligible for a rough plane.

22. (A) In this case we don't know the vertical height through which the object has been lifted. However, we know that the man did 180 joules of work. This work was done for two purposes: one to overcome friction, the other to lift the object to a higher level. Since friction is not zero in this problem, only part of the 180 joules is left to raise the object. This part, of course, is the potential energy gained by the object.

23. (D) The tension in the string is the pull at either end of the string. The downward pull of the mass of the object is counteracted by the buoyant force of the water. The mass of the object = vol. x density. $M = 30\ cm^3 \times 4\ gm/cm^3 = 120$ gm.

The buoyant force of the water equals the weight of the displaced water (Archimedes' principle). The object sinks, since its specific gravity is greater than that of water, and displaces 30 cm^3 of water. This water weighs 30 gm, since 1 cm^3 of water weighs 1 gm. Therefore, the buoyant force of the water is 30 gm. The tension equals (120 − 30) gm. The preferred unit of force in the cgs system is the dyne. The weight of one gram is equal to 980 dynes.

24. (B) See explanation 23.

25. (C) In accordance with Newton's third law: If *X* exerts a force on *Y*, then *Y* exerts an equal and opposite force on *X*.

26. (B) Gravitational, magnetic, and electrostatic forces vary inversely as the square of the distance between the small objects. Since the distance is doubled, the force must become one-fourth as great.

27. (E) Recall the formula for the period of a simple pendulum: $T = 2\pi\sqrt{L/g}$. This should remind you that the period depends on the length and on the acceleration due to gravity. The greater the altitude, the less the acceleration due to gravity. Vibration in the laboratory may interfere with the point of suspension and affect the period. Changing the mass of the bob of a simple pendulum does not change its period.

28. (B) The car is decelerated by an unbalanced force of 3,000 newtons. We apply Newton's second law of motion, which can be stated as: the unbalanced force acting on an object is equal to the product of the object's mass and the resulting acceleration:

$$F = ma.$$

First we calculate the acceleration. Acceleration is defined as the ratio of the change in velocity divided by the time in which the change takes place:

$$\begin{aligned} a &= \Delta v/t \\ &= (15 \text{ m per sec})/(10 \text{ sec}) \\ &= 1.5 \text{ m/sec}^2 \end{aligned}$$

Substituting in $F = ma$,

$$\begin{aligned} 3{,}000 \text{ N} &= \text{m } (1.5 \text{ m/sec}^2) \\ \text{m} &= 2{,}000 \text{ kg} \end{aligned}$$

Another method is to use a different form of the second law: The impulse acting on an object is equal to the resulting change in momentum. The *impulse* is equal to the product of the unbalanced force and the time during which it acts. Therefore,

$$\begin{aligned} Ft &= mv \\ (3{,}000 \text{ N})\ (10 \text{ sec}) &= m\ (15 \text{ m/sec}) \\ m &= 2{,}000 \text{ kg} \end{aligned}$$

Note that the object started with a velocity of 15 m/sec and then came to rest; therefore the change in velocity is 15 m/sec.

29. (B) The work done in lifting an object is equal to the product of the object's weight and the vertical height through which it is lifted:

$$\begin{aligned} \text{Work} &= \text{weight} \times \text{height} \\ \text{work} &= (2.5 \times 10^4 \text{ N})\ (10 \text{ m}) \\ &= 2.5 \times 10^5 \text{ joules} \end{aligned}$$

30. (C) According to Newton's second law of motion, the net or unbalanced force acting on an object is equal to the product of the object's mass and the resulting acceleration:

$$\begin{aligned} F &= ma \\ &= (10 \text{ kg})\ (5 \text{ m/sec}^2) \\ &= 50 \text{ newtons} \end{aligned}$$

31. (C) When an object moves at constant speed around a circle, the centripetal acceleration is always directed towards the center of the circle. An unbalanced force acting on the object from the outside is needed to provide this acceleration. Here it is the friction between the wheels and the road that provides the centripetal force acting on the car. The engine turns the wheels to make the motion possible; the steering wheel turns the wheels in the desired direction of motion, but unless there is friction the car would merely continue sliding in a straight line.

32. (D) The centripetal force is equal to the product of the mass of the car and its centripetal acceleration:

$$F = mv^2/r$$
$$= (2 \times 10^3 \text{ kg}) (50 \text{ m/sec})^2/(250 \text{ m})$$
$$= 2 \times 10^4 \text{ N}$$

33. (A) In a series circuit the current is the same in every part.

34. (D) The potential difference across each wire is equal to the IR-product. Since the current I is the same in each of the wires, the potential difference is proportional to the resistance. You should know that, of all metals, silver has the highest conductivity (or lowest resistivity); copper and aluminum are close behind. Therefore, the resistance R of the silver wire is the least, that of the copper next, and that of the nichrome highest. The potential differences follow in the same sequence.

35. (E) Because the magnet moves with respect to the coil, an emf is induced in the coil. In accordance with Lenz's law, the direction of the induced current has to be such as to oppose the motion of the magnet. When the magnet approaches the coil, the right end of X has to become an S-pole, and the left end an N-pole. The left-hand rule indicates that the direction of electron flow is from Y to Z. When the magnet emerges and moves away to the left, the polarity of, and electron flow in, X must reverse, if the motion of the magnet is to be opposed.

36. (D) You may use your pencil as a prop. Hold it so that it is vertical. Apply the left-hand rule; the fingers point in the direction that the N-pole of the compass needle points. The thumb will point up, in the direction of the electron flow.

37. (B) The voltmeter measures the potential difference between its terminals. The voltmeter is in parallel with R_1; therefore the voltage across them is the same. Remember, in such a diagram the straight line implies zero resistance. The fact that the voltmeter is drawn along a diagonal means nothing special.

38. (C) The ammeter is in series with R_2; the current through these two is the same.

39. (C) The basic function of a diode is to change alternating current to direct current. The transistor is similar to the vacuum tube triode in that they are both capable of amplifying a signal if the common circuits are used. Neither semiconductor device uses a heated cathode. Both are made from crystals to which a small, precise amount of impurities has been added, and both can be made very small. The vacuum tube triode has a metal grid.

40. (A) All the ice has to be melted before the temperature can go above 0° C. 80 calories are needed to melt one gram of ice at 0° C; 400 calories would be required to melt all 5 gm. Since only 200 cal are supplied, half of the ice will melt. The resulting ice-water mixture will remain at 0° C.

41. (B) The energy must travel from west to east; it is transmitted by a sound wave. Sound waves are longitudinal waves: waves in which the medium vibrates back and forth in a direction parallel to the direction in which the energy travels.

42. (B) The range of audible frequencies depends on the individual, but is roughly 16-16,000 vib/sec or 16-20,000 vib/sec.

43. (E) The volume of a gas is directly proportional to the absolute temperature. The Kelvin scale and the Rankine scale are absolute scales of temperature.

44. (B) A Fahrenheit degree is 5/9 of a centigrade degree. Therefore the coefficient per Fahrenheit degree is 5/9 the coefficient per centigrade degree; this equals $^5/_9 \times 0.000045 = 0.000025$.

45. (D) Change in length = length × coefficient × temp. change = 100 cm × 0.000045/C° × (50° − 30°) C = 0.090 cm.

46. (C) Apply Ohm's law. $I = V/R$. Since the voltage V is the same for X and Y, the currents through them compare inversely as the resistance. Then X has a resistance which is twice as large; therefore it has a current half as large as Y.

47. (D) Since the voltage is the same for X and Y, it is convenient to compare their rates of heat production (or power consumption) by $P = V^2/R$. This shows that their rate of heat production varies inversely as their resistance. (You may recall that, in a parallel circuit, the device with the lesser resistance consumes the greater power.) Then X has twice as much resistance as Y; therefore it consumes ½ the power.

48. (A) This is a simple fact. Of course, in ordinary neutral atoms all the protons are in the nucleus and the number of electrons outside the nucleus equals the number of protons in the nucleus.

49. (D) The man is 10 ft in front of the mirror. In a plane mirror his image is 10 ft behind the mirror. The distance between him and his image is 20 ft.

50. (A) This is almost pure recall of a fact. You should also think of the fact that, when you use a magnifying glass, you are using a convex lens to get a virtual, enlarged, erect image. A virtual image is obtained with a convex lens only if the object is less than one focal length away from the lens.

51. (E) In light, frequency determines color. In vacuum the speed of light is independent of frequency: in vacuum all electromagnetic waves travel at approx. 186,000 mi/sec. (A) would therefore be a poor choice.

52. (B) The eye has a convex lens of fairly short focal length. In general, the objects we see are more than one focal length away from our eye; objects more than one focal length away from a convex lens always produce real, inverted images.

53. (C) Nonluminous opaque objects are seen by the light they reflect. Luminous objects are seen by the light they emit.

54. (B) Monochromatic red light contains only red light. An ideal blue object will reflect only blue light; when illuminated by monochromatic red light, there is no blue light for the object to reflect. When no light enters our eye we "see" blackness. (An actual blue book may reflect other colors, such as green, in addition to the blue. However, such colors will not be present in monochromatic red light.)

55. (E) You should know as a fact that light travels faster in air than in water or glass. In addition, the light will be refracted if it reaches the surface obliquely; it will not be refracted if it is normal to the surface.

56. (A) This is a series circuit. The current through the capacitor is the same as that through the whole circuit.

$$Z_T = \sqrt{R^2 + X_c^2} = \sqrt{30^2 + 40^2} = 50 \text{ ohms.}$$

$$I_T = V_T/Z_T.$$

(I hope you recognized the 3-4-5 right triangle.)

$$I_T = 150 \text{ V}/50\ \Omega = 3 \text{ A}.$$

57. (C) $V_R = IR = 3 \text{ A} \times 30\ \Omega = 90 \text{ V}$.

58. (C) At the instant shown in the diagram, the magnetic flux from the N-pole to the S-pole is perpendicular to the face of the loop. At this instant, the voltage induced in the loop is zero. One way to look at this is to think of the top and bottom wires of the loop. At the instant shown they are moving parallel to the magnetic flux and therefore are not cutting any lines of force; thus no voltage is induced. One-quarter of a rotation later, the top wire is at 90° and the bottom wire is at 270°. At that instant a maximum voltage is induced in them since they are moving perpendicularly across the lines of force. Another half of a rotation later, the top wire is at 270° and the bottom wire is at 90°. Again a maximum voltage is induced, but this time in the opposite direction around the loop because the wire that moved down across the field before is now moving up, and vice versa.

59. (E) The voltage induced in the coil is proportional to the flux density provided by the external magnets and the speed of rotation of the loop. The flux density decreases between the two poles when the poles are moved further apart. This decreases the induced voltage. When the loop is wound around an iron core, the flux density is increased. Using a commutator (see page 118) merely changes the current in the external circuit to direct current.

60. (C) In vacuum all electromagnetic waves travel at the same speed. In materials like glass and water, violet light travels more slowly than red light. The wavelengths of red light are greater than those of violet light. Therefore, the frequencies of red light are lower than those of violet light ($c = f\lambda$). The energy of a quantum equals Planck's constant times the frequency ($E = hf$). Hence each quantum of red light has less energy than a quantum of violet light.

61. (B) Beats are produced by the constructive and destructive interference of two different frequencies. Where the waves are in phase, they reinforce each other; where they are out of phase, they tend to annul each other. Reinforcement leads to comparative loudness; annulment to comparative quiet.

A color spectrum is produced with thin oil films and soap bubbles because light is reflected by both surfaces of the film; the beams from the two surfaces interfere with each other. At different angles of vision different colors are annulled.

Green leaves appear green in ordinary light because the leaves absorb most of the light. Green predominates in the reflected light.

62. (A) 1 mA $= 10^{-3}$ A. Therefore, 6.50 mA $= 6.50 \times 10^{-3}$ A $= 65.0 \times 10^{-4}$ A $=$ 0.00650 A. But, choice 4, 65.0×10^{-2} A $= 6.50 \times 10^{-1}$ A.

63. (D) The amplitude changes because there is always some reflection when a wave goes from one medium to another with a different index of refraction. When the energy of a given wave decreases, the amplitude of the wave decreases.

The direction changes: when a ray goes obliquely from a rarer to a denser medium, it is bent toward the normal. The frequency of the wave is the same in the two media, but the speed is lower in the denser medium. Therefore the wavelength is smaller in the denser medium ($v = f\lambda$).

64. (A) The kinetic energy of the object equals $\frac{1}{2}mv^2$ (proportional to the square of its speed). For an object starting from rest, $v^2 = 2as$. Therefore, kinetic energy $= \frac{1}{2}m \times 2as = mas$ (kinetic energy is proportional to its displacement s). Since $v^2 = 2as$, $v = \sqrt{2as}$ (speed proportional to the square root of its displacement).

Its velocity is proportional to time; $v = at$, and a is constant.

65. (E) If the net force were tripled, the acceleration would be tripled. However, the force the man exerts is not the net force. Suppose the force he exerts is 5 lb. Part of this is needed to overcome friction and another part is needed to overcome a component of the weight. Suppose these two parts add up to 1 lb. The net force is only 4 lb. Tripling the original force results in an applied force of 15 lb; the net force will be 14 lb. In this example, the net force, and therefore the acceleration, is more than tripled ($14/4 = 3.5$).

66. (C) The person's push on the floor of the elevator equals the floor's push on him. This push moves him up with the same acceleration as the elevator. The push that would be required merely to keep the man in equilibrium is equal to his weight. An additional force is required to accelerate him. According to Newton's second law, this force is proportional to the acceleration.

67. (B) First harmonic is another name for fundamental frequency. In addition to the fundamental frequency, many vibrating objects produce higher frequencies called overtones. ("First basso" is not a physics term.)

68. (A) The resistance of a wire = kL/A. If we make the length smaller, the resistance becomes less. If we make the cross-section area A larger, the resistance becomes less. Doing the opposite will increase the resistance.

69. (E) A charged rod will attract small pieces of anything, including iron. Magnetic material, which has not been magnetized, will be attracted by north and south poles.

70. (D) In the interference pattern produced by two slits, the distance between two adjacent bright lines is given by a fairly simple relationship. If x is this distance,

$$x = \lambda L/d$$

where λ = the wavelength of the light
L = the distance between the double slit and the screen
d = the distance between the two slits

Substituting, we get $x = (6.0 \times 10^{-7}\text{ m})\ (2.0\text{ m})/\ 2.0 \times 10^{-5}\text{ m})$
$= 6.0 \times 10^{-2}\text{ m}$

71. (C) The intensity of the arriving light determines how many photons arrive each second. If the frequency of the light is above the threshold frequency, the greater the number of the arriving photons, the greater the number of emitted electrons.

72. (E) This question requires only rough mental calculation. On a test it would be a big waste of time to do anything other than that. Quick inspection shows that a transition from $n = 2$ to $n = 1$ results in a release of a photon with an energy of about 10 ev. Any drop to a level other than the ground state will involve at most a release of 3.4 ev. However, to illustrate the principle, and because we need it for the next question, we'll show the calculation for the transition from $n = 3$ to $n = 2$:
$E = (E_3 - E_2) = (-1.5\text{ ev}) - (-3.4\text{ ev}) = 1.9\text{ ev}$

73. (C) The frequency of a photon is proportional to its energy: $E = hf$. Substituting we get, after changing the energy to joules,
$(1.9 \times 1.6 \times 10^{-19}\text{ joules}) = (6.6 \times 10^{-34}\text{ joules-sec})\ f$
$f = 4.6 \times 10^{14}\text{ cy/sec}$

74. (B) The *half-life* of an isotope of an element is the time required for one-half of the nuclei of a sample of the isotope to break up. In this example we start with 12 grams of thorium (Th). In 24 days only 6 grams are left. In another 24 days (a total of 48 days) one-half of the 6 grams disintegrates, and only 3 grams of thorium are left. After still another 24 days (a total of 72 days) one-half of the remaining 3 grams breaks up, leaving 1.5 grams of thorium. Finally, after another 24 days (a total of 96 days) one-half of the 1.5 grams has broken up, leaving 0.75 gram of thorium.

75. (D) When the symbol of an element is written with a superscript and a subscript, the superscript is the mass number and the subscript is the atomic number. In the case of the given isotope of aluminum, Al, the mass number is 27, and the atomic number is 13. The equation describes a nuclear change. In all nuclear changes mass number and atomic number are conserved. This means that the sum of the mass numbers on the left side of the equation must equal the sum of the mass numbers on the right. The sum of the mass numbers on the left is 27 + 4, or 31. On the right side the mass number of P is 30. Therefore the mass number of X must be 1, because 30 + 1 = 31. Since atomic number is also conserved, the algebraic sum of the subscripts on the left side must equal the algebraic sum on the right. The sum on the left is 13 + 2, or 15. The subscript for X must be zero, because 15 + 0 = 15. The neutron is the particle that has a mass number of 1 and an atomic number (or charge) of zero.

PRACTICE TEST 3

Part A

DIRECTIONS: Each set of lettered choices below refers to the numbered questions immediately following it. Select the one lettered choice that best answers each question. A choice may be used once, more than once, or not at all in each set.

QUESTIONS 1-7 refer to the following units:
(A) lumen/ft^2 **(B)** candle power **(C)** erg **(D)** lb-ft **(E)** horsepower

For each of the terms below select the choice which is most closely associated with it, either because they are both units of the same concept, such as energy, or because the choice is a unit for the given concept.

1. foot-candle
2. joule
3. torque
4. watt-second
5. kilowatt
6. illumination
7. Btu

QUESTIONS 8-13 refer to the following laws and principles:

(A) Newton's first law of motion **(B)** Newton's second law of motion **(C)** Archimedes' principle **(D)** Hooke's law **(E)** Bernoulli's principle

For each of the following situations select the choice which is most closely associated with it.

8. Circular motion with constant speed
9. Rectilinear motion with constant speed
10. A freely falling object
11. An object projected vertically up
12. The fact that a hydrometer floats higher in salt water than in freshwater.
13. The stretching of a rubber band.

QUESTIONS 14-15 refer to the formation of images using an object, a screen, and one or more of these:

I a plane mirror
II a concave lens (diverging)
III a convex lens (converging)

For the following questions consider these choices
(A) I only
(B) II only
(C) III only
(D) I and II only
(E) none of these

14. A virtual image larger than the object can be produced by using
15. A real image the same size as the object can be produced by using

Part B

DIRECTIONS: *Each of the following questions or incomplete statements is followed by five choices. Select the choice which fits best.*

16. Of the following, the one that is most like a gamma ray in its properties is **(A)** alpha ray **(B)** beta ray **(C)** X ray **(D)** electron **(E)** proton

17. The sun's energy is believed to be a result of **(A)** oxidation of carbon **(B)** oxidation of hydrogen **(C)** oxidation of helium **(D)** nuclear fission **(E)** nuclear fusion

18. The emission of one alpha particle from the nucleus of an atom produces a change of **(A)** −1 in atomic number **(B)** −1 in atomic mass **(C)** −2 in atomic number **(D)** −2 in atomic mass **(E)** +2 in atomic mass

19. Two forces of 10 lb and 7 lb, respectively, are applied simultaneously to an object. The minimum value of their resultant is, in lb, **(A)** 0 **(B)** 3 **(C)** 7 **(D)** $7\sqrt{3}$ **(E)** 10

20. Two forces act together on an object. The magnitude of their resultant is greatest when the angle between the forces is **(A)** 0° **(B)** 45° **(C)** 60° **(D)** 90° **(E)** 180°

21. Of the following, the largest quantity is **(A)** 0.047 cm **(B)** 47×10^{-4} cm **(C)** 4.7×10^{-2} cm **(D)** 0.00047×10^{2} cm **(E)** 0.000047×10^{4} cm

22. 65 mm is equivalent to approximately **(A)** 1.5 inches **(B)** 2.6 inches **(C)** 5.5 inches **(D)** 10.5 inches **(E)** 15 inches

23. The work done in holding a weight of 40 N at a height of 3 mabove the floor for 2 sec is, in joules, **(A)** 0 **(B)** 40 **(C)** 80 **(D)** 120 **(E)** 240

24. Two sound waves travel through air at the same time. One has a frequency of 200 vib/sec; the other 800 vib/sec. The speed of the first, as compared with the second, is **(A)** one-fourth **(B)** one-half **(C)** the same **(D)** twice **(E)** four times

25. Two sounds of the same frequency in air must have the same **(A)** amplitude **(B)** wavelength **(C)** intensity **(D)** overtones **(E)** loudness

26. A string 80 cm long has a fundamental frequency of 300 vib/sec. If the length of the string is changed to 40 cm without changing the tension, the fundamental frequency will be, in vib/sec, **(A)** 150 **(B)** $150\sqrt{2}$ **(C)** 300 **(D)** $300\sqrt{2}$ **(E)** 600

For 27-28. A sonar depth finder uses sound signals to determine the depth of water. Four seconds after the sound leaves the boat it returns to the boat because of reflection from the bottom. Assume the speed of the sound in water is 4800 ft/sec.

27. The speed of the sound in water, expressed in meters/sec, is **(A)** 330 **(B)** 1090 **(C)** 1460 **(D)** 14,400 **(E)** 186,000

28. The depth of the water is, in ft, **(A)** 2200 **(B)** 4400 **(C)** 4800 **(D)** 9600 **(E)** 19,200

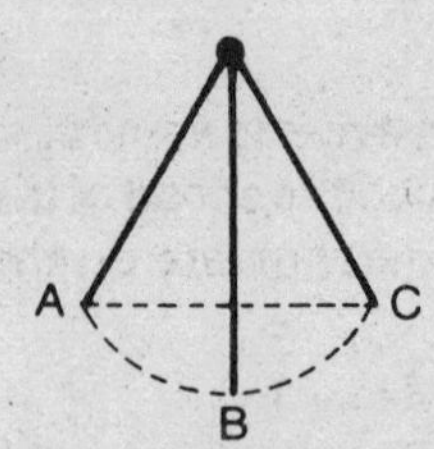

29. A pendulum, consisting of a string of length *L* and a small metal ball of mass *m* at the free end, is allowed to swing freely after being released from point *A*. The highest point it reaches on the other side is *C* and the lowest point of swing is *B*. At *B* the mass *m* has **(A)** zero potential energy and zero kinetic energy **(B)** maximum potential and maximum kinetic energy **(C)** minimum potential and maximum kinetic energy **(D)** maximum potential and zero kinetic energy **(E)** zero potential and the same kinetic energy as at *C*.

30. In question 23, if we want the period of the pendulum to be doubled, we should change the length of the string to **(A)** ¼ *L* **(B)** ½ *L* **(C)** 2*L* **(D)** 4*L* **(E)** 2*L* and the mass to 2*m*

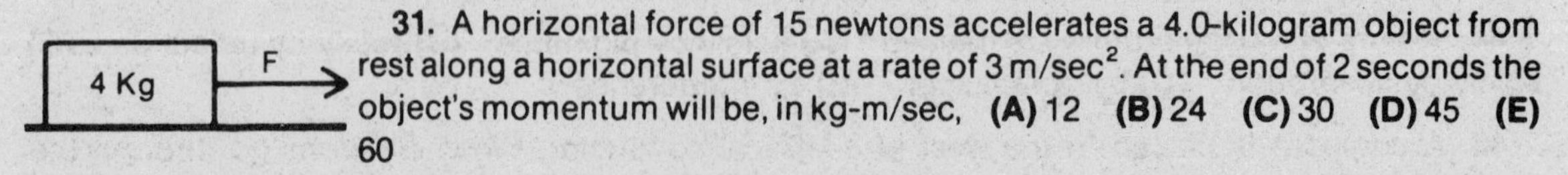

31. A horizontal force of 15 newtons accelerates a 4.0-kilogram object from rest along a horizontal surface at a rate of 3 m/sec^2. At the end of 2 seconds the object's momentum will be, in kg-m/sec, **(A)** 12 **(B)** 24 **(C)** 30 **(D)** 45 **(E)** 60

32. In question 31 what is the frictional force that is retarding the forward motion of the object, in newtons? **(A)** 3 **(B)** 4 **(C)** 12 **(D)** 40 **(E)** 60

For 33-34. A 40-N object is lifted 10 m above the ground and then released.

33. After the object has fallen 4 m and is 6 m above the ground, its kinetic energy will be, in joules **(A)** 5 **(B)** 160 **(C)** 240 **(D)** 400 **(E)** 5120

34. Just before it hits the ground, the object's kinetic energy will be, in joules **(A)** 12 **(B)** 160 **(C)** 240 **(D)** 400 **(E)** 12,800

35. Ball *X* is projected horizontally at the same time as ball *Y* is released and allowed to fall. If air resistance is negligible and both start from the same height, **(A)** *X* and *Y* will reach the level ground at the same time **(B)** *X* will reach the level ground first **(C)** *Y* will reach the ground first **(D)** we must know the mass of *X* and *Y* to tell which will hit the ground first **(E)** we must know the density to tell which will hit the ground first

36. Ball *X* is projected vertically up at the same time as ball *Y* is projected horizontally. If air resistance is negligible and both start from the same height, **(A)** *X* and *Y* will reach the level ground at the same time **(B)** *X* will reach the level ground first **(C)** *Y* will reach the level ground first **(D)** we must know the mass of *X* and *Y* to tell which will hit the ground first **(E)** we must know the density to tell which will hit the ground first

37. Object *X* is dropped at the same time as object *Y* is projected vertically upward and object *Z* is projected horizontally. Air resistance is negligible. **(A)** objects *X* and *Z* have the same downward acceleration and this is greater than *Y*'s **(B)** objects *X* and *Z* have the same downward acceleration and this is less than *Y*s **(C)** objects *Y* and *Z* have the same downward acceleration and this is greater than *X*'s **(D)** objects *Y* and *Z* have the same downward acceleration and this is less than *X*'s **(E)** objects *X*, *Y*, and *Z* have the same downward acceleration

38. A metal cube, 2 inches on edge, is suspended in air so that the top and bottom faces are horizontal. The force exerted on the bottom face by the atmosphere is **(A)** 2 lb **(B)** 4 lb **(C)** 20 lb **(D)** 40 lb **(E)** 60 lb

For 39-41. This set of graphs applies to the next three questions.

39. The graph which shows most nearly the variation of the pressure of a gas against its volume, at constant temperature, is **(A)** *A* **(B)** *B* **(C)** *C* **(D)** *D* **(E)** *E*

40. The graph which most nearly shows the variation of current through a given resistor against the voltage applied to it is **(A)** *A* **(B)** *B* **(C)** *C* **(D)** *D* **(E)** *E*

41. The graph which most nearly shows the variation of current supplied by a given dry cell as the external resistance is increased (axes: current against resistance) is **(A)** *A* **(B)** *B* **(C)** *C* **(D)** *D* **(E)** *E*

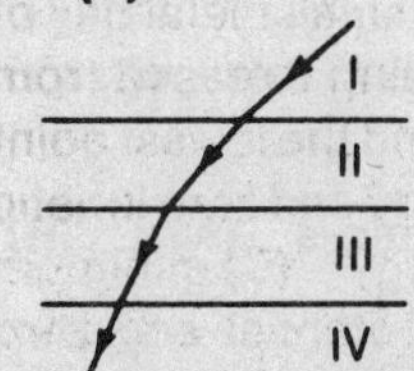

42. The diagram shows the path of a ray of light through 4 media separated from each other by horizontal surfaces. The medium in which the speed of the light is greatest is **(A)** I **(B)** II **(C)** III **(D)** IV **(E)** indeterminate on the basis of the given information

43. A bright-line spectrum may be obtained by using light from **(A)** the sun **(B)** a sodium vapor lamp **(C)** a tungsten filament bulb **(D)** a clean glowing platinum wire **(E)** a clean glowing nichrome wire

44. Colors in a soap bubble are caused **(A)** solely by pigments **(B)** solely by reflection **(C)** solely by absorption **(D)** by polarization **(E)** by interference

45. A compass is placed to the west of a vertical conductor. When electrons go through the conductor, the N-pole of the compass needle is deflected to point towards the south. The direction of electron flow in the conductor is **(A)** down **(B)** up **(C)** east **(D)** west **(E)** south

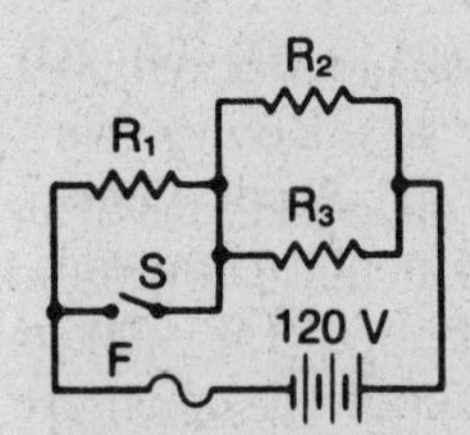

For 46-47. In the circuit shown, each of the three resistors has a resistance of 40 ohms. The fuse *F* is rated at 10 amp and has negligible resistance. The battery supplies 120 V and has negligible internal resistance.

46. When the switch *S* is open as shown, the current through R_1 is, in amperes, **(A)** 0 **(B)** 1 **(C)** 2 **(D)** 6 **(E)** 9

47. When switch *S* is closed, the current through R_2 is **(A)** 3 amperes **(B)** 6 amperes **(C)** 10 amperes **(D)** 12 amperes **(E)** zero because the fuse will blow

48. If pith ball X attracts pith ball Y but repels pith ball Z, **(A)** Y must be positively charged **(B)** Y must be negatively charged **(C)** Y may be positive or negative but not neutral **(D)** Y may be neutral or charged **(E)** Y must be neutral

49. The sp. ht. of ice is 0.5. The heat required to change 20 gm of ice at $-5°$ C to water at $0°$ C is, in calories, **(A)** 10 **(B)** 50 **(C)** 130 **(D)** 450 **(E)** 1650

For 50-53. *Base your answers on the information below.*

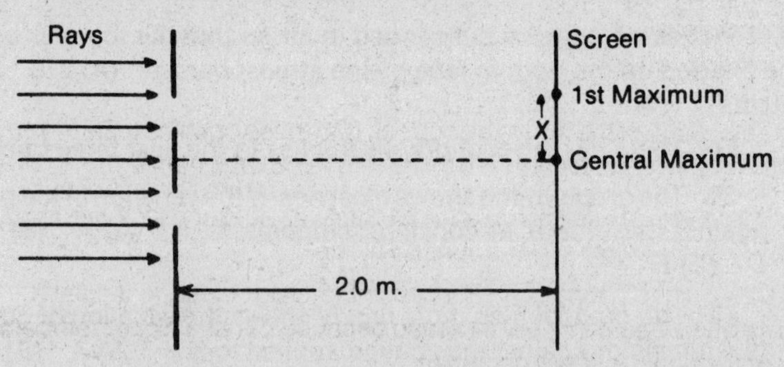

Two parallel slits 2.0×10^{-6} meter apart are illuminated by parallel rays of monochromatic light of wavelength 6.0×10^{-7} meter, as shown. The interference pattern is formed on a screen 2.0 meters from the slits.

50. The distance *X* is **(A)** 6.0×10^{-1} m **(B)** 6.0×10^{-7} m **(C)** 3.0×10^{-1} m **(D)** 3.3 m **(E)** 330 m

51. The difference in path length for the light from the two slits to the first maximum is **(A)** λ **(B)** 2λ **(C)** $\frac{\lambda}{2}$ **(D)** 0 **(E)** 3λ

52. If the wavelength of the light passing through the slits is doubled, the distance from the central maximum to the first maximum **(A)** is halved **(B)** is doubled **(C)** remains the same **(D)** is quadrupled **(E)** is tripled

53. If the screen is moved to 1.0 meter from the slits, the distance between the central maximum and the first maximum **(A)** is halved **(B)** is doubled **(C)** remains the same **(D)** is quadrupled **(E)** is tripled

54. Two hundred cm^3 of a gas are heated from $30°$ C to $60°$ C without letting the pressure change. The volume of the gas will **(A)** be halved **(B)** be decreased by about 10% **(C)** remain the same **(D)** be doubled **(E)** be increased by about 10%.

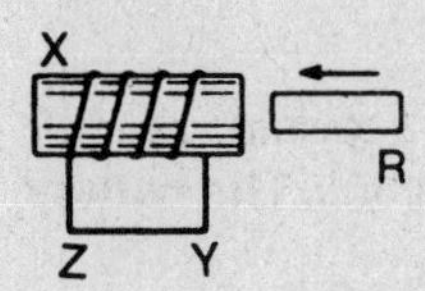

55. *X* is a coil of wire with a hollow core. Brass rod *R* is pushed at constant speed from the right into the core (and out again at the left). During this complete motion, **(A)** there will be no current in wire *YZ* **(B)** electron flow will be from *Y* to *Z* **(C)** electron flow will be from *Z* to *Y* **(D)** electron flow will be from *Z* to *Y* and then from *Y* to *Z* **(E)** electron flow will be from *Y* to *Z* and then from *Z* to *Y*.

QUESTIONS 56-60 Each of these questions or incomplete statements is followed by four numbered (I–IV) statements. Select all the numbered statements which are correct and indicate these choices by the appropriate lettered answer.

56. If yellow light emerges from a box, the light

I may contain red light
II must contain a wavelength shorter than 8000 Å. (1 Å = 10^{-8} cm)
III may be of a single wavelength
IV must contain a wavelength which by itself would appear yellow

(A) if only I, II, and III are correct
(B) if only I and III are correct
(C) if only II and IV are correct
(D) if only IV is correct
(E) if you do not select any of the above as correct. (On these practice tests, if you select E, jot down your choices so that you can compare them with the Explanations.)

57. All air is removed from the device shown after a volatile liquid, such as ether, is poured in at *Y*. Then the device is sealed and portion *X* is surrounded by ice.

I some of the liquid will appear at *X* because the vapor is condensed at *X*
II the volume of the liquid in *Y* will increase because it is at a higher temperature than *X*
III the level of the liquid in *Y* will decrease
IV no change will take place

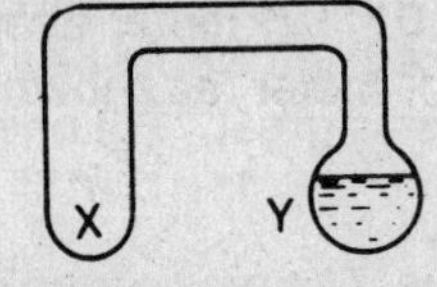

(A) if only I, II, and III are correct
(B) if only I and III are correct
(C) if only II and IV are correct
(D) if only IV is correct
(E) if you do not select any of the above as correct. (On these practice tests, if you select E, jot down your choices so that you can compare them with the Explanations.)

58. A gas is allowed to expand isothermally from a volume of 100 cm^3 to a volume of 400 cm^3. The original pressure of the gas was 4 atmospheres. The new pressure

I is 1 atmosphere
II is 16 atmospheres
III is 76 cm of mercury
IV cannot be calculated without knowing the temperature

(A) if only I, II, and III are correct
(B) if only I and III are correct
(C) if only II and IV are correct
(D) if only IV is correct
(E) if you do not select any of the above as correct. (On these practice tests, if you select E, jot down your choices so that you can compare them with the Explanations.)

59. If air is kept saturated with water vapor, it is true that its relative humidity will

I be high at high temperatures
II be high at low temperatures
III depend on the mass of water vapor in it
IV be 100%

(A) if only I, II, and III are correct
(B) if only I and III are correct
(C) if only II and IV are correct
(D) if only IV is correct
(E) if you do not select any of the above as correct. (On these practice tests, if you select E, jot down your choices so that you can compare them with the Explanations.)

60. Two halves of an iron washer are placed between two magnets as shown in the diagram. As a result

I *X* will be an S-pole and *Y* an N-pole
II *Y* will be an N-pole and *Z* an S-pole
III *Y* and *W* will be S-poles
IV *X* and *Z* will be S-poles

(A) if only I, II, and III are correct
(B) if only I and III are correct
(C) if only II and IV are correct
(D) if only IV is correct
(E) if you do not select any of the above as correct. (On these practice tests, if you select E, jot down your choices so that you can compare them with the Explanations.)

For 61-62. Select the choice which fits the question best.

61. An emf of 0.003 volt is induced in a wire when it moves at right angles to a uniform magnetic field with a speed of 4.0 m/sec. If the length of wire in the field is 15 cm, what is the flux density, in webers/m^2? **(A)** 0.003 **(B)** 0.005 **(C)** 6 **(D)** 12 **(E)** 2000

62. In question 61 if the wire were to move at an angle of 30° to the field instead of at right angles, the value of the induced emf would be **(A)** unchanged **(B)** two times as great **(C)** three times as great **(D)** four times as great **(E)** one-half as great

For 63-68. DIRECTIONS: *The next group of questions consists of a statement in the left-hand column and a reason in the right-hand column. Select*

A if both "statement" and "reason" are true and the "reason" is the cause, and the "statement" is its effect
B if both "statement" and "reason" are true statements, but the "reason" is not actually the cause of the phenomenon described by the "statement"
C if the "statement" is true, but the "reason" is a false statement
D if the "statement" is false, but the "reason" is a true statement
E if both "statement" and "reason" are false statements

STATEMENT		REASON
63. A floating object appears to weigh nothing	because	no two objects can occupy the same space at the same time.
64. When a glass rod is rubbed with silk, the glass is positively charged	because	protons go from the silk to the glass.
65. The pitch caused by a vibrating tuning fork is not affected by the motion of the observer	because	the pitch due to a vibrating object depends only on its

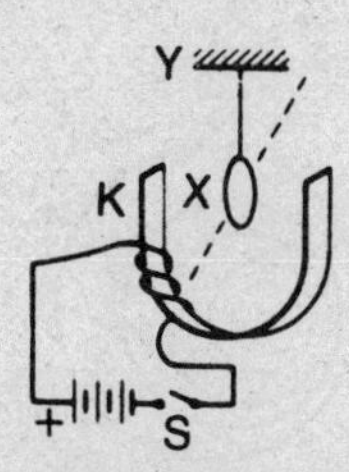

For 66-67. The pendulum pivoted at *Y* has a copper disk *X* at its end. The pendulum is allowed to swing so that the copper disk moves between the poles of an electromagnet at right angles to the imaginary line connecting the poles.

66. When switch *S* is closed, the pole at the left marked *K* will be an N-pole	because	the positive terminal of a battery always induces a magnetic N-pole near it.
67. When the switch is closed, the speed of the pendulum decreases	because	the current induced in *X* produces a magnetic field which opposes the motion.
68. The temperature at which water is densest is 4° F	because	this is one of the fixed points on the Fahrenheit scale.

For 69-75. DIRECTIONS: *Each of the following questions or incomplete statements is followed by five choices. Select the choice which fits best.*

For 69-70. In the following graph, the velocity of an object as it moves along a horizontal straight line is plotted against time.

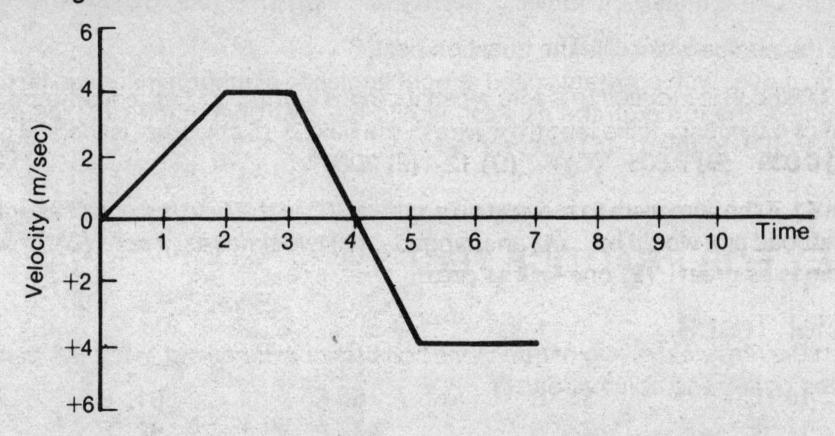

69. The distance traveled by the object during the first 2 seconds is **(A)** 2 m **(B)** 4 m **(C)** 8 m **(D)** 16 m **(E)** 24 m

70. The average speed of the object during the first 4 seconds is, in m/sec, **(A)** zero **(B)** 1.2 **(C)** 2.5 **(D)** 4 **(E)** 8

71. Which point in the wave shown in the diagram is in phase with point *A*? **(A)** *F* **(B)** *B* **(C)** *C* **(D)** *D* **(E)** *E*

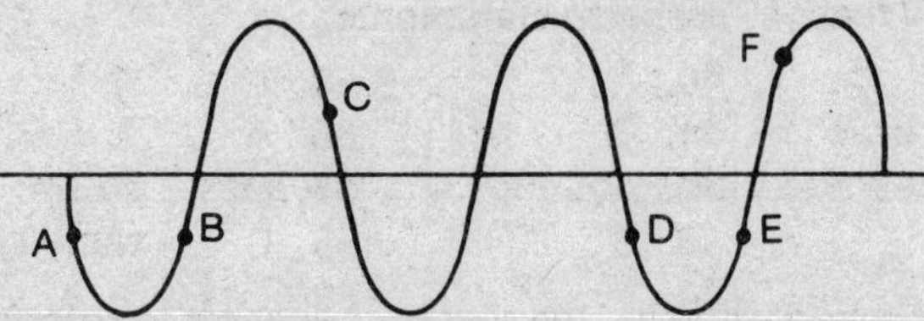

72. The diagram represents a wave pattern produced by a vibrating source moving at constant speed. Towards which point is the source moving? **(A)** 1 **(B)** 2 **(C)** 3 **(D)** 4 **(E)** perpendicularly into the page

73. If the source were to accelerate, the wavelength immediately in front of the source would **(A)** increase only **(B)** decrease only **(C)** first increase and then decrease **(D)** first decrease and then increase **(E)** remain the same

74. The graph below shows the maximum kinetic energy of the photoelectrons ejected when

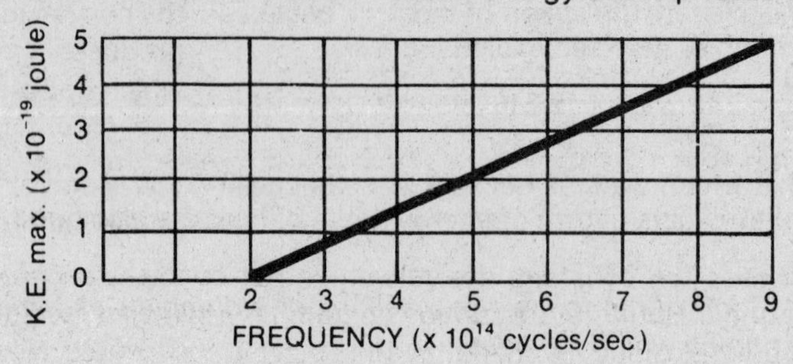

photons of different frequencies strike a metal surface. The slope of the graph is Planck's constant, h. The work function of this metal is, in joules, **(A)** $h \times 10^{14}$ **(B)** $2h \times 10^{14}$ **(C)** $3h \times 10^{14}$ **(D)** $4h \times 10^{14}$ **(E)** $8h \times 10^{14}$

75. An electron moves with constant speed at right angles to a uniform magnetic flux. If the flux density were doubled, the force on the electron would be **(A)** unaffected **(B)** reduced to one-fourth **(C)** halved **(D)** doubled **(E)** quadrupled

ANSWER KEY FOR PRACTICE TEST 3

Answer Key for Test 3

1. A	**16.** C	**31.** B	**46.** C	**61.** B
2. C	**17.** E	**32.** A	**47.** A	**62.** E
3. D	**18.** C	**33.** B	**48.** D	**63.** B
4. C	**19.** B	**34.** D	**49.** E	**64.** C
5. E	**20.** A	**35.** A	**50.** A	**65.** E
6. A	**21.** E	**36.** C	**51.** A	**66.** C
7. C	**22.** B	**37.** E	**52.** B	**67.** A
8. B	**23.** A	**38.** E	**53.** A	**68.** E
9. A	**24.** C	**39.** C	**54.** E	**69.** B
10. B	**25.** B	**40.** A	**55.** A	**70.** C
11. B	**26.** E	**41.** B	**56.** A	**71.** D
12. C	**27.** C	**42.** A	**57.** B	**72.** A
13. D	**28.** D	**43.** B	**58.** B	**73.** B
14. C	**29.** C	**44.** E	**59.** E (1, 2, 3, 4)	**74.** B
15. C	**30.** D	**45.** A	**60.** E (1, 2, 4)	**75.** D

Explained Answers for Test 3

1. (A) Know that 1 ft-cdle = 1 lumen/ft^2. You should also recognize that, if we use a different unit of length, we still have a unit of illumination; e.g., meter-candle, lumen/m^2. Students should also note that candle power is the unit for measuring the rate at which a luminous source emits liaht.

2. (C) Some units of energy: joule, erg, watt-hour, ft-lb, calorie, Btu, newton-meter.

3. (D) Torque, or moment of a force, is the product of force and distance. Therefore, a unit of torque consists of a force unit multiplied by a distance unit. Usually we don't bother much with these units of torque, but you should know that the foot-pound is used as a unit of energy, while the pound-foot is used as a unit of torque.

4. (C) 1 watt-sec = 1 joule. Also see explanation 2 .

5. (E) 1 horsepower = 746 watts = 0.746 kilowatt

6. (A) See explanation 1.

7. (C) Don't forget that heat is a form of energy; units of heat are also good units of energy.

8. (B) Although the speed is constant, the velocity is not constant because the direction is constantly changing: at one instant the object may be going north; half a revolution later it will be going south. Since the object's velocity is changing, the object is accelerated. Newton's second law is associated with accelerated motion: acceleration takes place in the direction of the net force.

9. (A) Motion with constant speed along a straight line is motion with constant velocity, motion without acceleration. This is the subject of Newton's first law: if an object moves with constant (or zero) velocity, the net force acting on the object is zero.

10. (B) A freely falling object is accelerated because of the earth's pull on it. Newton's second law is associated with accelerated motion.

11. (B) The velocity of a vertically projected object is constantly changing. Change in velocity, or acceleration, is the subject of Newton's second law.

12. (C) Archimedes' principle tells that the weight in air of a floating object is equal to the weight of the displaced liquid in which it floats. Salt water is denser than fresh water. A smaller volume of salt water than of fresh water will equal the weight of the hydrometer. When a hydrometer floats higher (as it does in salt water), it displaces a smaller volume. Bernoulli's principle deals with the flowing of a liquid or a gas. The greater the speed, the lower the pressure.

13. (D) The stretching of a rubber band illustrates the behavior of elastic substances. Hooke's law describes this: the change in length of an elastic substance is proportional to the stretching force.

14. (C) A *plane mirror* always produces a virtual image which is the same size as the object. (If we look at our image in a plane mirror as we walk backwards away from it, our image will *appear* to get smaller for the same reason that a real person will look smaller to us as the distance between us and the person increases: it is a visual effect.) A *concave lens* produces a virtual image whose size is smaller than the object and decreases as the distance between the object and lens increases.

A *convex lens* produces a virtual image larger than the object when the object's distance from the lens is less than one focal length. A magnifying glass is an application of this characteristic.

15. (C) A convex lens produces a real image the same size as the object when the object's distance from the lens is two focal lengths. (Also read the explanation for question 14.)

16. (C) Both X rays and gamma rays are usually classified as electromagnetic waves. Their wavelength is of the order of 10^{-8} cm. The others are classified as part of ordinary matter. Alpha rays are streams of helium nuclei; beta rays are streams of electrons.

17. (E) Oxidation reactions yield very little energy compared with nuclear fission or fusion. The sun's vast energy is believed to be the result of a cycle of reactions involving the fusion of hydrogen nuclei.

18. (C) Alpha particles are helium nuclei. They have a mass number of 4 and charge of +2 (they consist of 2 protons and 2 neutrons). Therefore, when an alpha particle emerges from a nucleus, the positive charge of the remaining portion has decreased by 2; that is, its atomic number, which is the charge of the nucleus, is reduced by 2.

19. (B) The resultant of two forces applied to an object at the same time depends on their direction with respect to each other. The resultant is largest when the two forces act in the same direction—at an angle of 0°. Then the resultant is equal to the sum of the two forces. As the angle between the two forces increases, the value of the resultant decreases until the angle between the forces is 180°. The resultant then has a minimum value and is equal to the difference between the two forces.

20. (A) See explanation 19

21. (E) You may be able to get the answer quickly by eliminating some of the choices after a quick inspection. Systematic comparison can be made by writing all the choices in standard form (scientific notation) or as decimal fraction. We'll use the former method. The choices then become: **(A)** 4.7×10^{-2}; **(B)** 4.7×10^{-5}; **(C)** 4.7×10^{-2}; **(D)** 4.7×10^{-2}: **(E)** 4.7×10^{-1}.

22. (B) The choices differ widely; let us estimate the answer.

$$1000 \text{ mm} = 1 \text{ meter} = 39.37 \text{ inches} \doteq 40 \text{ in}$$
$$100 \text{ mm} = 4.0 \text{ in}$$
$$50 \text{ mm} = 2.0 \text{ in}$$

Therefore 65 mm is between 2.0 and 4.0 inches (closer to 2.0). Choice B fits this.
More exactly, $65 \text{ mm} = 65 \times 4.0 \text{ in.}/100 = 2.6 \text{ in.}$

23. (A) Considering the choices, it is clear that you are asked to find the work done on the object. Although a force is exerted, the force doesn't move and the object doesn't move. Hence no work is done; work equals force times distance moved in the direction of the force.

24. (C) All sound waves in the audible range travel with the same speed through air; frequency doesn't affect speed—it does affect wavelength.

25. (B) Speed = frequency × wavelength. Since speed and frequency are the same for the two sounds, wavelength must also be the same. (Loudness and intensity depend on amplitude; two waves may have the same frequency but differ in amplitude. Two sounds of the same fundamental frequency may have the same overtones, but the overtones may also be different.)

26. (E) The frequency of a vibrating string is inversely proportional to its length. If the length is decreased, the frequency is increased. If the length is halved, the frequency is doubled.

27. (C) One way to get the answer for this question is to make the actual conversion:
$4800 \text{ ft/s} = (12 \times 4800) \text{ in/s} = \frac{12 \times 4800}{39.37} \text{ m/s}$. You can round off a little ($39.37 \doteq 40$) and get 1440 m/s as the answer.

Another way is easier, if you remember that the speed of sound in air is approx. 1100 ft/s of 330 m/s. The speed in water is given as a little more than 4 times 1100 ft/s. Hence it should be a little more than 4 times 330 m/s. More exact calculation gives 1460 m/s.

28. (D) If it takes 4 sec for the sound to go down and up again, it takes only 2 sec just to go to the bottom.

$$\text{Distance} = \text{speed} \times \text{time}.$$
$$\text{Distance} = 4800 \text{ ft/sec} \times 2 \text{ sec} = 9600 \text{ ft}.$$

29. (C) As the pendulum bob swings back and forth, at the highest point of its swing, at *A* and *C*, all of its energy is potential. At those points it is momentarily at rest. As it swings through its arc from its highest point, the potential energy is converted to kinetic. At the lowest point of the swing, at *B*, the potential energy has been converted to kinetic. Then as it swings up towards *C* the kinetic energy is converted to potential. If air friction is negligible, *C* is at the same height as *A*, and the bob has the same potential energy there as at *A*. Also, in that case, throughout the swing the sum of the potential and kinetic energy remains constant.

30. (D) The period of a simple pendulum is proportional to the square root of its length. If the length is multiplied by 4, the period is multiplied by the square root of 4, which is 2. The complete formula for the period of the pendulum is: $T = 2\pi\sqrt{L/g}$, where g is the acceleration due to gravity at that location.

31. (B) The momentum of an object is equal to the product of its mass and its velocity:

$$\begin{aligned} \text{momentum} &= \text{mass} \times \text{velocity} \\ &= (4.0\text{ kg})\ (3\text{ m/sec}^2)\ (2\text{ sec}) \\ &= 24\text{ kg-m/sec} \end{aligned}$$

We calculated the final velocity at the end of 2 seconds by multiplying the constant acceleration by the time. Note also that it did not say that the 15-N force is a net force. Therefore the quick use of the relationship that the impulse is equal to the change in momentum would not be well-advised. (Unfortunately, in some questions the absence of friction has to be assumed.) This will be more apparent from question 32,

32. (A) The horizontal force of 15 N was used to give the object acceleration and also to overcome friction. According to Newton's second law of motion, the force which gives an object acceleration is equal to the product of the object's mass and its acceleration:

$$\begin{aligned} F &= ma \\ &= (4\text{ kg})\ (3\text{ m/sec}^2) = 12\text{ N} \end{aligned}$$

The force of friction is equal to the difference between the total force and the net force: (15 N) − (12 N) = 3 N

33. (B) Potential energy = Wt × height.
At the top, 10 m above the ground, the object's potential energy = 40 N × 10 m = 400 joules.
When it is 6 m above the ground, the object's potential energy = 40 N × 6 m = 240 joules.
The difference in potential energy is the kinetic energy which the object acquired. Kinetic energy = (400 − 240) joules = 160 joules.

34. (D) Just before it hits the ground, all the object's potential energy has been converted to kinetic energy.

35. (A) This is one of the classic experiments. It shows that horizontal motion does not affect vertical motion (that is, vertical motion is affected only by a vertical force or by a force having a vertical component).

36. (C) This follows from the result of the experiment described in question 30. Ball *X* must first rise to a greater height than ball *Y;* we can think of it as then falling freely from the highest point it reaches. In the meantime ball *Y* has already started on its downward trip. Of course, mass and density have nothing to do with the time it will take to reach ground, if air resistance is negligible.

37. (E) The change in velocity of the three objects is due to gravity only; the acceleration due to gravity is the same for all objects near the surface of the earth.

38. (E) The pressure due to the atmosphere is approximately 15 lb/in.^2 The force due to the atmosphere is equal to the pressure times the area on which it acts (since $p = F/A$):
The area of the face of the cube = 2 in × 2 in = 4 in^2.
$F = pA = (15\text{ lb/in}^2)\ (4\text{ in}^2) = 60\text{ lb}$. (Of course this is balanced by the almost equal downward force of the atmosphere.)

39. (C) This refers to Boyle's law: the pressure of a gas varies inversely as its volume. The graph is a hyperbola. Choice B is not satisfactory, since graph *B* would indicate that the volume of the gas could be reduced to zero.

40. (A) This refers to Ohm's law.
$V = IR$. As the applied voltage increases, the current increases. The only graph which shows such a variation is graph *A*.

41. (B) When the external resistance is zero (a short circuit across the dry cell), the current supplied by the dry cell is large, but not infinite. (In a 6-in. dry cell this current may be about 30 amperes.) This value is represented by point *K* on graph *B*. As the external resistance becomes larger and larger, the current becomes less and less, approaching zero as the external resistance becomes that of an insulator. This cannot be the graph of a straight line, such as graph *E*.

$I = \frac{E}{r + R}$, where *E* is the emf of the cell, *r* its internal resistance, and *R* the external resistance.

42. (A) An oblique ray is bent towards the normal when it enters a medium in which the wave is slowed up. In the diagram, the ray is shown bent more and more towards the normal as it goes from medium I towards medium IV. The light travels more slowly in each of the successive media.

43. (B) A bright-line spectrum is usually obtained from a glowing gas or vapor. In the laboratory, sodium vapor is often used as the source of such a spectrum. A special sodium vapor lamp may be used; frequently ordinary salt (sodium chloride) is sprinkled into the gas flame, giving the necessary light. Glowing tungsten, platinum, or nichrome provide a continuous spectrum. The sun's spectrum contains the dark absorption lines, known as *Fraunhofer lines.*

44. (E) The rainbow colors of soap bubbles and oil films are produced by interference of light.

45. (A) Hold a pencil in a vertical position in front of you. Imagine that you are facing north. Then to the left of the pencil is west. Imagine the compass placed to the left of the pencil. According to the question, the compass needle points south. Apply the left hand rule: grasp the wire with your left hand so that your fingers, when on the left side of the wire, will point south. To do this, your thumb will have to point down; this gives the direction of electron flow.

46. (C) When switch *S* is open, it has no effect on the circuit. R_1 is then in series with the parallel combination of R_2 and R_3. The combined resistance of R_2 and R_3, two equal resistors in parallel, is $R/n = 40\ \Omega/2 = 20\ \Omega$. The total resistance of the circuit then is $R_1 + 20\ \Omega = 40\ \Omega + 20\ \Omega = 60\ \Omega$.
$I_1 = I_T = V_T/R_T = 120\ \text{V}/60\ \Omega = 2\ \text{A}$.

47. (A) The effect of closing the switch is to eliminate the effect of R_1 from the circuit. The circuit is the same as though we had only the parallel combination of R_2 and R_3 connected to the 120-volt source through a fuse. (If you wish, you can think of the zero-resistance switch in parallel with R_1; the resistance of this parallel combination is zero ohms.)
The full 120 V is then applied to R_2 (and also to R_3). $I_2 = V_2/R_2 = 120\ \text{V}/40\ \Omega = 3\ \text{A}$

48. (D) If two nonmagnetic objects repel each other, both objects must be electrified, and the charge on the two must be of the same kind; both positive or both negative. Hence, since *X* repels *Z*, *X* must be charged, but it may be positive or negative. A charged object will attract neutral objects and objects oppositely charged. Since charged object *X* attracts *Y*, *Y* may be neutral or charged.

49. (E) Heat required = heat to raise temp. of ice to 0° C + heat to melt ice at 0° C.

Heat required = mass × sp. ht. × temp. change + mass × heat of fusion
= 20 gm × 0.5 cal/gm × 5° C + 20 gm × 80 cal/gm
= 50 cal + 1600 cal
= 1650 cal

50. (A) This question refers to Young's double slit experiment. When monochromatic light passes through two small slits, a series of bright lines is seen on a distant screen. The distance *x* of the first maximum or bright line away from the central maximum can be calculated from the following relationship:

$$\frac{x}{L} = \frac{\lambda}{d}$$

where *L* is the distance from the slits to the screen
λ is the wavelength of the light
d is the distance between the slits.

Substituting, we get:

$$\frac{x}{2.0 \text{ m}} = \frac{6 \times 10^{-7} \text{ m}}{2 \times 10^{-6} \text{ m}}$$
$$x = 6 \times 10^{-1} \text{ meter.}$$

51. (A) In the interference pattern, wherever there is a bright region or maximum, the waves from the two slits must arrive in phase. This means that they must arrive a whole number of wavelengths apart, which means that the path length from the two slits to the bright regions must differ by a whole number of wavelengths. The path difference to the first maximum is one wavelength.

52. (B) From the above formula, we see that x and λ are proportional to each other. (If we increase the numerator on the left, we must also increase the one on the right to keep the two fractions equal.) Therefore, if the wavelength is increased, the distance x between the two maxima also increases.

53. (A) From the above formula, we see that x is proportional to L, the distance between the slits and the screen. (If the numerator of the fraction increases, the denominator must also increase if the value of the fraction doesn't change.) Therefore, when the screen is moved closer to the slits, the distance between the two maxima also decreases.

54. (E) At constant pressure, the volume of a gas is proportional to the absolute temperature. The absolute temperature increases from $(273+30)°$ K to $(273+60)°$ K. An increase from 303 to 333 is an increase of about 10%. The volume should also increase by about 10%.

55. (A) Brass is nonmagnetic and cannot be magnetized. Therefore its motion with respect to a conductor will not induce a current in the conductor.

56. (A) Yellow light may be light of a single wavelength, or it may be the result of a mixture of two or more wavelengths. For example, you should know that, if red and green light are mixed, yellow light is produced. The visible range of light is approximately 3900 — 7800 angstrom units.

57. (B) By definition, since the liquid is volatile, it will vaporize readily at room temperature. The space at X will be quickly saturated with the vapor, as will the space in Y. Since the space at X is kept cold by use of the ice, the saturated vapor coming to X from Y contains more vapor than it can hold at the lower temperature. Some of the vapor condenses, producing liquid at X. Since the pressure of the vapor in X becomes less than the pressure of the vapor near Y, more vapor keeps being pushed over to X, where it condenses. More and more of the liquid has to disappear from Y.

58. (B) Boyle's law applies. Since the volume is quadrupled, the pressure is reduced to one-fourth of its original value: from 4 atmospheres to 1 atmosphere. The pressure of 1 atmosphere is the same as the pressure of a column of mercury 76 cm high.

59. (E) You may be able to put up a good argument in favor of excluding choice 3 as a correct choice. This would not alter the selection of E as the correct answer. As long as the air is saturated, its relative humidity is 100%. Of course this is high, but the actual mass of water vapor in the air may be quite low: lower at low temperatures than at high. However, the relative humidity remains constant at 100% if the air is kept saturated.

60. (E) We may use the molecular theory of magnetism and the law of magnets to determine what will happen. The molecular magnets in the washer-pieces will orient themselves so that, near the poles of the permanent magnet, opposite poles will predominate: at X mostly S-poles. The opposite poles of the molecular magnets have to point in the opposite direction from X. Hence, at Y and W, N-poles will predominate. The reverse happens in the other half of the washer, which is nearer to the S-pole. Therefore Z will be an S-pole.

61. (B) An emf is induced in a wire when it cuts across magnetic flux lines. This emf is maximum when the wire cuts across at right angles to the flux, and it is then equal to the product of the magnetic flux density, the length of wire in the field, and the speed of cutting:

$$\text{emf} = BLv$$
$$0.003 \text{ volt} = B\ (0.15 \text{ m})\ (4 \text{ m/sec})$$
$$B = 0.005 \text{ weber/m}^2$$

62. (E) As stated in question 61, the emf is maximum when the angle of cutting is 90°. It is less for any other angle. Only choice (E) satisfies this. (Actually the emf is proportional to the sine of the angle. The sine of 30° is ½.)

63. (B) The reason given is a statement of the impenetrability of matter. It is true, but is not the reason for the loss in weight of objects immersed in a fluid.

64. (C) Protons do not leave a substance readily. In the rubbing process, electrons go from the glass to the silk, making the glass positive and the silk negative.

65. (E) The statement should bring to mind Doppler's principle. If the observer moves closer to, or further away from, the vibrating tuning fork, the pitch he observes will be different from the one he will hear if he remains stationary. Pitch depends on frequency, but also on the motion of the source or the observer.

66. (C) *K* will be an N-pole, as you can determine by applying the left-hand rule. However, the physical location of the battery terminals is irrelevant. We see how the coil is connected. The direction of electron flow determines the polarity of an electromagnet.

67. (A) As the copper disk cuts across the magnetic field, an emf is induced in the disk. In accordance with Lenz's law, an induced current circulates in the disk in such a direction as to oppose the motion of the pendulum. If the electromagnet is strong enough, the pendulum will stop swinging completely.

68. (E) The temperature at which water is densest is 4° C. This is approximately 39° F. This is not selected as one of the fixed points on the thermometer. We usually use the ice point and the steam point as fixed points.

69. (B) On a graph of velocity against time, the distance traveled (actually the displacement) is given by the area under the graph. Another method, which we shall use here, is to multiply the average speed by the time traveled. The graph for the first two seconds is an oblique straight line. This represents motion with constant acceleration. For such motion the average speed is the sum of the initial and final speeds divided by two. The initial speed is zero. The average speed is (final speed/2) = (4 m/sec)/2 = 2 m/sec.

$$\begin{aligned}\text{distance} &= \text{(average speed) (time)}\\ &= \text{(2 m/sec) (2 sec)}\\ &= \text{4 m}\end{aligned}$$

70. (C) We shall first find the total distance traveled during the first 4 seconds. We already know the distance traveled during the first 2 sec. During the 3rd second, the speed remains constant at 4 m/sec.

$$\begin{aligned}\text{distance} &= \text{(4 m/sec) (1 sec)}\\ &= \text{4 m}\end{aligned}$$

During the 4th second the average speed is again equal to (4 m/sec)/2, or 2 m/sec.

$$\begin{aligned}\text{distance} &= \text{(2 m/sec) (1 sec)}\\ &= \text{2 m}\end{aligned}$$

Total distance = (4 m + 4 m + 2 m) = 10 m.

$$\begin{aligned}\text{Average speed} &= \text{(total distance)/time}\\ &= \text{(10 m)/4 sec}\\ &= \text{2.5 m/sec}\end{aligned}$$

71. (D) Two points in a wave are in phase if they are going through the same part of the vibration at the same time. We can think of the diagram as

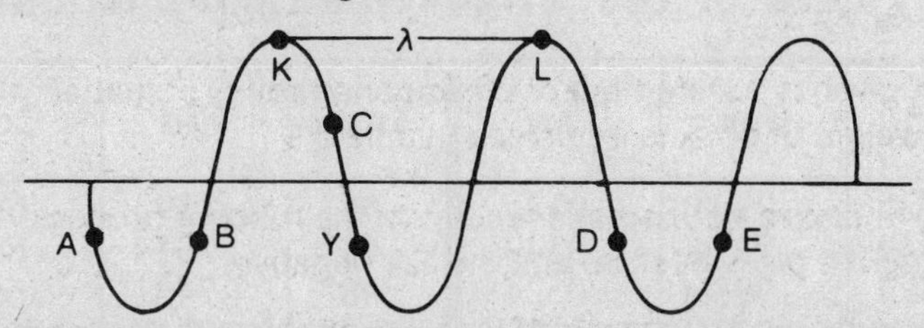

representing a periodic wave which is continuously transmitting energy to the right. *Two points in phase are then a whole number of wavelengths apart.* In the above diagram one wavelength is the distance between *K* and *L*, two adjacent peaks on the curve. This is equal to the distance from *A* to *Y*. We can think of the curve *ABKCY* as completing one cycle; then the same wave pattern is shown again by the curve *YLD*. The distance between *Y* and *D* is another wavelength. *D* is 2 wavelengths from *A*. Points *A, Y,* and *D* are going through the same part of the vibration at the same time.

72. (A) The *Doppler effect* deals with the apparent change in the frequency of a wave when there is relative motion between the source of the wave and an observer. Here we are told that the vibrating source is moving. The circles represent the spherical wave fronts produced by the source. The wavelength is the distance between two adjacent wave fronts—in this diagram, the distance between two adjacent circles. If the source were stationary, the distance between wave fronts would be the same in all directions. We notice in the diagram that the circles are much closer to each other near point 1 than anywhere else.

This shows that the source keeps moving closer to point 1 as it sends out compressions. At the same time it keeps moving away from point 2. As the wave reaches an observer at point 1, he would get a wave whose length, and therefore whose frequency, is different from what it would be if the source were stationary. Its frequency would be greater than the actual frequency of the source.

73. (B) Students are often confused about this part of the Doppler effect. In the above question the source was moving with constant velocity towards point 1. An observer at point 1 would get a wave of constant frequency and wavelength, but different from what it would be if the source were stationary. If the source now starts to move faster, it would tend to catch up with the wave front it has just sent out towards point 1. The new wave front would therefore be closer to the preceding wave front than if the source were moving with constant velocity. The wavelength therefore would decrease.

74. (B) As we lower the frequency of the radiation incident on a photoemissive material, the maximum kinetic energy of the photoelectrons becomes less and less. The energy of the photon which would just give it zero kinetic energy is the work function of this material. Its frequency would be the threshold frequency, f_0. On the graph this is the point where the graph intersects the axis, 2×10^{14} Hz. The value of the work function is equal to the product of Planck's constant and the threshold frequency:

$$\text{work function} = hf_0$$
$$= h(2 \times 10^{14}\ \text{Hz}^*)$$

*Note: One hertz (Hz) is equal to one cycle per second.

75. (D) The force on an electron moving at right angles to a uniform magnetic flux is equal to the product of the flux density, the charge on the electron, and the speed of the electron ($F = Bqv$). If the flux density is doubled, the force is doubled.

Values of the Trigonometric Functions

Angle	Sin	Cos	Tan	Angle	Sin	Cos	Tan
1°	.0175	.9998	.0175	46°	.7193	.6947	1.0355
2°	.0349	.9994	.0349	47°	.7314	.6820	1.0724
3°	.0523	.9986	.0524	48°	.7431	.6691	1.1106
4°	.0698	.9976	.0699	49°	.7547	.6561	1.1504
5°	.0872	.9962	.0875	50°	.7660	.6428	1.1918
6°	.1045	.9945	.1051	51°	.7771	.6293	1.2349
7°	.1219	.9925	.1228	52°	.7880	.6157	1.2799
8°	.1392	.9903	.1405	53°	.7986	.6018	1.3270
9°	.1564	.9877	.1584	54°	.8090	.5878	1.3764
10°	.1736	.9848	.1763	55°	.8192	.5736	1.4281
11°	.1908	.9816	.1944	56°	.8290	.5592	1.4826
12°	.2079	.9781	.2126	57°	.8387	.5446	1.5399
13°	.2250	.9744	.2309	58°	.8480	.5299	1.6003
14°	.2419	.9703	.2493	59°	.8572	.5150	1.6643
15°	.2588	.9659	.2679	60°	.8660	.5000	1.7321
16°	.2756	.9613	.2867	61°	.8746	.4848	1.8040
17°	.2924	.9563	.3057	62°	.8829	.4695	1.8807
18°	.3090	.9511	.3249	63°	.8910	.4540	1.9626
19°	.3256	.9455	.3443	64°	.8988	.4384	2.0503
20°	.3420	.9397	.3640	65°	.9063	.4226	2.1445
21°	.3584	.9336	.3839	66°	.9135	.4067	2.2460
22°	.3746	.9272	.4040	67°	.9205	.3907	2.3559
23°	.3907	.9205	.4245	68°	.9272	.3746	2.4751
24°	.4067	.9135	.4452	69°	.9336	.3584	2.6051
25°	.4226	.9063	.4663	70°	.9397	.3420	2.7475
26°	.4384	.8988	.4877	71°	.9455	.3256	2.9042
27°	.4540	.8910	.5095	72°	.9511	.3090	3.0777
28°	.4695	.8829	.5317	73°	.9563	.2924	3.2709
29°	.4848	.8746	.5543	74°	.9613	.2756	3.4874
30°	.5000	.8660	.5774	75°	.9659	.2588	3.7321
31°	.5150	.8572	.6009	76°	.9703	.2419	4.0108
32°	.5299	.8480	.6249	77°	.9744	.2250	4.3315
33°	.5446	.8387	.6494	78°	.9781	.2079	4.7046
34°	.5592	.8290	.6745	79°	.9816	.1908	5.1446
35°	.5736	.8192	.7002	80°	.9848	.1736	5.6713
36°	.5878	.8090	.7265	81°	.9877	.1564	6.3138
37°	.6018	.7986	.7536	82°	.9903	.1392	7.1154
38°	.6157	.7880	.7813	83°	.9925	.1219	8.1443
39°	.6293	.7771	.8098	84°	.9945	.1045	9.5144
40°	.6428	.7660	.8391	85°	.9962	.0872	11.4301
41°	.6561	.7547	.8693	86°	.9976	.0698	14.3007
42°	.6691	.7431	.9004	87°	.9986	.0523	19.0811
43°	.6820	.7314	.9325	88°	.9994	.0349	28.6363
44°	.6947	.7193	.9657	89°	.9998	.0175	57.2900
45°	.7071	.7071	1.0000	90°	1.0000	.0000	

Summary of Formulas

moment of a force = force × length of moment arm (p. 7)

sum of clockwise moments = sum of counter clockwise moments (p. 8)

$s = vt$ (p. 14)

$a = \frac{v_f - v_i}{t}$ (p. 14)

$v_f = at$ (p. 14)

$s = \frac{1}{2} at^2$ (p. 14)

$v_f^2 = 2as$ (p. 14)

$v_{av} = v_f/2$ (p. 14)

$\frac{F}{w} = \frac{a}{g}$ (p. 16)

$w = mg$ (p. 16)

Momentum = mass × velocity (p. 18)

Work = force × distance; $W = Fd$ (p. 23)

potential energy = wh (p. 24)

kinetic energy = $\frac{1}{2} mv^2$ (p.24)

energy produced = mc^2 (pp. 24, 145)

coefficient of friction = $\frac{\text{friction during motion}}{\text{normal force}}$ (p. 25)

work against friction = friction × distance object moves (p. 25)

$\frac{F}{w} = \frac{h}{L}$ (p. 25)

$F = w \sin \theta$ (p. 26)

power = $\frac{\text{work}}{\text{time}}$ (p. 26)

power = $\frac{\text{force} \times \text{distance}}{\text{time}}$ (p. 26)

$$\text{AMA} = \frac{\textit{resistance}}{\textit{actual effort}}; \quad \text{AMA} = \frac{F_R}{F_E} \qquad (p.\ 27)$$

$$\text{work output} = F_R S_R \qquad (p.\ 27)$$

$$\text{work input} = F_E S_E \qquad (p.\ 27)$$

$$\frac{F_R}{F_E} = \frac{S_E}{S_R} = \text{ideal MA (IMA)} \qquad (p.\ 27)$$

$$\text{efficiency} = \frac{\text{work output}}{\text{work input}} = \frac{\text{AMA}}{\text{IMA}} = \frac{\text{ideal effort}}{\text{actual effort}} \qquad (pp.\ 27,\ 28)$$

$$\frac{\text{length of plane}}{\text{height of plane}} = \text{IMA} = \frac{\text{weight of object}}{\text{ideal effort}} \qquad (p.\ 29)$$

$$\text{IMA} = \frac{\text{circumference of wheel}}{\text{circumference of axle}} = \frac{\text{diameter of wheel}}{\text{diameter of axle}} = \frac{\text{radius of wheel}}{\text{radius of axle}} \qquad (p.\ 30)$$

$$\text{IMA} = \frac{2\pi L}{\text{pitch}} \qquad (p.\ 30)$$

$$\text{density} = \frac{\text{mass}}{\text{volume}} \qquad (p.\ 34)$$

$$p = \frac{F}{A} \qquad (p.\ 34)$$

$$\text{sp. gr.} = \frac{\text{density of substance}}{\text{density of water}} = \frac{\text{weight of substance}}{\text{weight of equal volume of water}} = \frac{\text{mass of substance}}{\text{mass of equal volume of water}} \qquad (p.\ 34)$$

$$p = hd \qquad (p.\ 35)$$

$$F = hdA \qquad (p.\ 35)$$

$$\text{IMA} = \frac{F}{f} = \frac{A}{a} = \frac{(\text{diameter of large piston})^2}{(\text{diameter of small piston})^2} \qquad (p.\ 36)$$

$$\text{sp. gr.} = \frac{\text{weight in air}}{\text{apparent loss of weight in water}} \qquad (p.\ 37)$$

$$\text{sp. gr.} = \frac{\text{apparent loss in weight of solid in liquid}}{\text{apparent loss in weight of solid in water}} \qquad (p.\ 37)$$

$$F = kx \qquad (p.\ 40)$$

$$\text{change in Fahrenheit degrees} = {}^9/_5\ (\text{change in centigrade degrees}) \qquad (p.\ 44)$$

$$F = {}^9/_5\, C + 32° \quad \text{or} \quad C = {}^5/_9\, (F - 32°) \qquad (p.\ 44)$$

$$\text{change in length} = \text{original length} \times \text{coeff. of expansion} \times \text{temp. change} \qquad (p.\ 45)$$

$\frac{V_1}{V_2} = \frac{T_1}{T_2}$ or $\frac{V_1}{V_2} = \frac{C_1 + 273°}{C_2 + 273°}$ *(p. 46)*

$p_1V_1 = p_2V_2$ *(p. 46)*

$\frac{p_1V_1}{T_1} = \frac{p_2V_2}{T_2}$ *(p. 46)*

Heat lost by hot object = heat gained by cold object *(p. 50)*

Heat lost (or gained) = mass × sp. heat × temp. change *(p. 50)*

Heat required for melting = mass × H_F *(p. 50)*

Heat required for vaporization = mass × H_v *(p. 51)*

Heat gained (or lost) = *mass* × sp. ht. × temp. change

+ mass melted × heat of fusion

+ mass vaporized × heat of vaporization *(p. 51)*

$T = \frac{1}{f}$ *(p. 62)*

$v = fL$ *(p. 62)*

length = $\frac{L}{4}$ *(p. 65)*

the number of beats = the difference between the two frequencies *(p. 66)*

$L = 4l_a$ *(p. 67)*

$L = 2l_a$ *(p. 67)*

$L = 2l_s$ *(p. 68)*

decibels = 10 log $\frac{p_1}{p_2}$ *(p. 69)*

$E = hf$ *(p. 72)*

illumination = $\frac{\text{intensity of source}}{(\text{distance})^2}$ *(p. 73)*

$\frac{cd_1}{s_1^2} = \frac{cd_2}{s_2^2}$ *(p. 73)*

$f = \frac{R}{2}$ *(p. 75)*

$$\text{index of refraction} = \frac{\text{sine of the angle in the rarer medium}}{\text{sine of the angle in the denser medium}} = \frac{\text{speed of light in air (or vacuum)}}{\text{speed of light in the substance}}$$ *(pp. 79-80)*

$$\frac{1}{\text{object distance}} + \frac{1}{\text{image distance}} = \frac{1}{\text{focal length}}\ ;\ \frac{1}{p} + \frac{1}{q} = \frac{1}{f}$$ *(p. 83)*

$$\frac{\text{size of image}}{\text{size of object}} = \frac{\text{image distance}}{\text{object distance}} = \text{magnification } (m)$$ *(p. 83)*

$$\text{telescope magnification} = \frac{\text{focal length of the objective}}{\text{focal length of eyepiece}}$$ *(p. 84)*

$$\frac{\lambda}{d} = \frac{x}{L}$$ *(P. 89)*

$$F = \frac{kQ_1Q_2}{d^2}$$ *(p. 100)*

$$E = \frac{F}{Q}$$ *(p. 102)*

$$E = \frac{V}{d}$$ *(p. 104)*

$$V = \frac{\text{work}}{Q} \quad \text{or} \quad \text{volts} = \frac{\text{joules}}{\text{coulomb}}$$ *(p. 104)*

$$R = \frac{kL}{A}$$ *(p. 105)*

(Formulas for Ohm's Law and electric circuits are summarized on p. 106.)

$V_T = \text{emf} - Ir$ *(p. 106)*

$H = 0.24\ I^2Rt$ *(p. 107)*

$P = VI = I^2R = V^2/R$ *(p. 107)*

energy = power × time *(p. 107)*

$F = BIL$ *(p. 115)*

$$F \propto \frac{I_1 I_2}{d}$$ *(p. 115)*

$F = BQv$ *(p. 115)*

$I = I_{eff} = I_{rms} = 0.707\ I_p$ *(p. 126)*

$V = V_{eff} = V_{rms} = 0.707\ V_p$ *(p. 126)*

$Q = CV$ *(p. 126)*

$X_c = \frac{1}{2\pi fc}$ (p. 128)

$X_L = 2\pi fL$ (p. 128)

$Z = \sqrt{R^2 + (X_L - X_c)^2}$ (p. 128)

$I = \frac{V}{Z}$ (p. 128)

$V_L = IX_L$ (p. 128)

$V_C = IX_C$ (p. 128)

$V_R = IR$ (p. 128)

$P = I^2R$ (p. 129)

$\text{power factor} = \frac{\text{true power}}{\text{apparent power}}$ (p. 129)

$X_L = X_c$ (p. 130)

$IX_L = IX_c$ (p. 130)

$f_r = \frac{1}{2\pi\sqrt{LC}}$ (p. 130)

$\frac{\text{secondary emf}}{\text{primary emf}} = \frac{\text{number of turns on secondary}}{\text{number of turns on primary}}$ (p. 131)

power supplied by secondary = efficiency × power supplied to primary (p. 131)

$V_sI_s = V_pI_p \times \text{efficiency}$ (p. 131)

$E_k = hf - W$ (p. 143)

$W = hf_0$ (p. 143)

$p = \frac{h}{\lambda}$ (p. 146)

APPENDIX I

Physics Reference Tables

List of Physical Constants

Gravitational Constant (G)	6.67×10^{-11} newton-meters²/kg²
Acceleration of gravity *(g)* (near earth's surface)	9.81 meters/second²
Speed of light *(c)*	3.00×10^{8} meters/second
Speed of sound at STP	3.31×10^{2} meters/second
Mechanical equivalent of heat	$J = 4.19 \times 10^{3}$ joules/kilocalorie
	$\frac{1}{J} = 2.39 \times 10^{-4}$ kilocalories/joule
Mass energy relationship	1 amu $= 9.31 \times 10^{2}$ Mev
Electrostatic constant	$k = 9.00 \times 10^{9}$ newton-meters²/coul²
Charge of the electron = 1 elementary charge	1.60×10^{-19} coulomb
One coulomb	6.25×10^{18} electrons
	6.25×10^{18} elementary charges
Electron volt (ev)	1.60×10^{-19} joule
Planck's Constant *(h)*	6.63×10^{-34} joule-second
Rest mass of the electron (m_e)	9.11×10^{-31} kilogram
Rest mass of the proton (m_p)	1.67×10^{-27} kilogram
Rest mass of the neutron (m_n)	1.67×10^{-27} kilogram

Trigonometric Functions*

sine 0° = .000
sine 30° = .500
sine 45° = .707
sine 60° = .866
sine 90° = 1.000

Wavelengths of Light in a Vacuum

VIOLET	$< 4.5 \times 10^{-7}$ meters
BLUE	$4.5 - 5.0 \times 10^{-7}$ meters
GREEN	$5.0 - 5.7 \times 10^{-7}$ meters
YELLOW	$5.7 - 5.9 \times 10^{-7}$ meters
ORANGE	$5.9 - 6.1 \times 10^{-7}$ meters
RED	$> 6.1 \times 10^{-7}$ meters

*A more complete Table of Trigonometric Functions appears on p. 254

Heat Constants

	Specific Heat (average)	Melting Point	Boiling Point	Heat of Fusion	Heat of Vaporization
	—	°C	°C	kcal/*kg*	kcal/*kg*
Alcohol, ethyl	0.58 (liq)	–115	78	25	204
Aluminum	0.21 (sol)	660	2057	77	2520
Ammonia	1.13 (liq)	–78	–33	84	327
Copper	0.09 (sol)	1083	2336	49	1150
Ice	0.50 (sol)	0	—	80	—
Iron	0.11 (sol)	1535	3000	7.9	1600
Lead	0.03 (sol)	327	1620	5.9	207
Mercury	0.03 (liq)	–39	357	2.8	71
Platinum	0.03 (sol)	1774	4300	27	—
Silver	0.06 (sol)	961	1950	26	565
Steam	0.48 (gas)	—	—	—	—
Water	1.00 (liq)	—	100	—	540
Tungsten	0.04 (sol)	3370	5900	43	—
Zinc	0.09 (sol)	419	907	23	420

Absolute Indices of Refraction

($\lambda = 5.9 \times 10^{-7}$ m.)

Air	1.00	Carbon Tetrachloride	1.46	Glycerol	1.47
Alcohol	1.36	Diamond	2.42	Lucite	1.50
Benzene	1.50	Glass, Crown	1.52	Quartz, Fused	1.46
Canada Balsam	1.53	Glass, Flint	1.61	Water	1.33

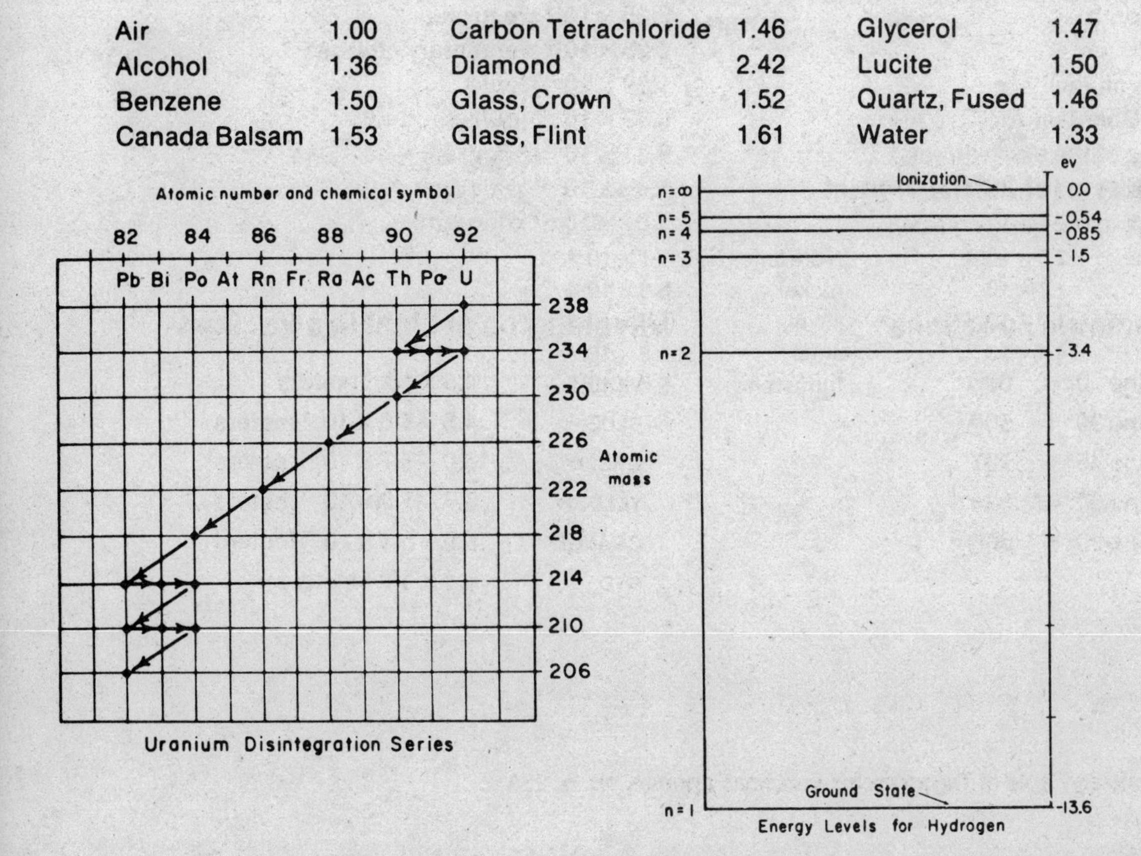

APPENDIX II

Density of Some Substances in grams/cm^3

Solids

aluminum	2.7	iron	7.8
brass	8.5	lead	11.3
copper	8.9	marble	2.6
cork	0.24	paraffin	0.9
glass	2.6	platinum	21.5
gold	19.3	silver	10.5
ice	0.92	steel	7.7
		sulfur	2.0

Liquids

alcohol	0.79	kerosene	0.8
carbon tetrachloride	1.6	mercury	13.6
gasoline	0.7	sulfuric acid	1.84
glycerine	1.3	water	1.0

Resistivity (at 20°C), ohm-m

aluminum	2.8×10^{-8}	nichrome	1.1×10^{-6}
copper	1.7×10^{-8}	nickel	6.8×10^{-8}
iron	1.0×10^{-7}	silver	1.6×10^{-8}
manganin	4.4×10^{-7}	steel	1.8×10^{-7}
		tungsten	5.6×10^{-8}

INTERNATIONAL ATOMIC MASSES*

Based on the Atomic Mass of C-12 = 12 atomic mass units (amu or u)

Name	Symbol	Atomic Number	Atomic Mass	Name	Symbol	Atomic Number	Atomic Mass
Actinium	Ac	89	[227]	Hydrogen	H	1	1.01
Aluminium	Al	13	26.98	Indium	In	49	114.82
Americium	Am	95	[243]	Iodine	I	53	126.90
Antimony	Sb	51	121.75	Iridium	Ir	77	192.2
Argon	Ar	18	39.95	Iron	Fe	26	55.85
Arsenic	As	33	74.92	Krypton	Kr	36	83.80
Astatine	At	85	[210]	Lanthanum	La	57	138.91
Barium	Ba	56	137.34	Lawrencium	Lw	103	[257]
Berkelium	Bk	97	[247]	Lead	Pb	82	207.19
Beryllium	Be	4	9.01	Lithium	Li	3	6.94
Bismuth	Bi	83	208.98	Lutetium	Lu	71	174.97
Boron	B	5	10.81	Magnesium	Mg	12	24.31
Bromine	Br	35	79.90	Manganese	Mn	25	54.94
Cadmium	Cd	48	112.40	Mendelevium	Md	101	[256]
Caesium	Cs	55	132.91	Mercury	Hg	80	200.59
Calcium	Ca	20	40.08	Molybdenum	Mo	42	95.94
Californium	Cf	98	[251]	Neodymium	Nd	60	144.24
Carbon	C	6	12.01	Neon	Ne	10	20.18
Cerium	Ce	58	140.12	Neptunium	Np	93	237.05
Chlorine	Cl	17	35.45	Nickel	Ni	28	58.71
Chromium	Cr	24	52.00	Niobium	Nb	41	92.90
Cobalt	Co	27	58.93	Nitrogen	N	7	14.01
Copper	Cu	29	63.55	Nobelium	No	102	[259]
Curium	Cm	96	[247]	Osmium	Os	76	190.2
Dysprosium	Dy	66	162.50	Oxygen	O	8	16.00
Einsteinium	Es	99	[254]	Palladium	Pd	46	106.4
Erbium	Er	68	167.26	Phosphorus	P	15	30.97
Europium	Eu	63	151.96	Platinum	Pt	78	195.09
Fermium	Fm	100	[257]	Plutonium	Pu	94	[244]
Fluorine	F	9	19.00	Polonium	Po	84	[210]
Francium	Fr	87	[223]	Potassium	K	19	39.10
Gadolinium	Gd	64	157.25	Praseodymium	Pr	59	140.90
Gallium	Ga	31	69.72	Promethium	Pm	61	[145]
Germanium	Ge	32	72.59	Protactinium	Pa	91	231.04
Gold	Au	79	196.97	Radium	Ra	88	[226]
Hafnium	Hf	72	178.49	Radon	Rn	86	[222]
Hahnium	Ha	105	......	Rhenium	Re	75	186.2
Helium	He	2	4.00	Rhodium	Rh	45	102.91
Holmium	Ho	67	164.93	Rubidium	Rb	37	85.47

*A value given in brackets is the mass number of the most stable known isotope.

Name	Symbol	Atomic Number	Atomic Mass	Name	Symbol	Atomic Number	Atomic Mass
Ruthenium	Ru	44	101.07	Thallium	Tl	81	204.37
Rutherfordium	Rf**	104	....	Thorium	Th	90	232.03
Samarium	Sm	62	150.35	Thulium	Tm	69	168.93
Scandium	Sc	21	44.95	Tin	Sn	50	118.69
Selenium	Se	34	78.96	Titanium	Ti	22	47.90
Silicon	Si	14	28.09	Tungsten	W	74	183.85
Silver	Ag	47	107.87	Uranium	U	92	238.03
Sodium	Na	11	22.99	Vanadium	V	23	50.94
Strontium	Sr	38	87.62	Xenon	Xe	54	131.30
Sulfur	S	16	32.06	Ytterbium	Yb	70	173.04
Tantalum	Ta	73	180.95	Yttrium	Y	39	88.91
Technetium	Tc	43	[99]	Zinc	Zn	30	65.37
Tellurium	Te	52	127.60	Zirconium	Zr	40	91.22
Terbium	Tb	65	158.92				

**Another suggested name is Khurchatovium.

1. These are also known as atomic weights.
2. These are based on assigning the carbon-12 isotope an atomic mass of 12u.
3. Most of the listed atomic masses are for naturally occurring mixtures of the isotopes.

INDEX

INDEX

INDEX